Environmental Reflections on the Anthropocene

Incorporating the intellectual history of disciplines from across the humanities, including environmental anthropology, philosophy, ethics, literature, history, science and technology studies, this volume provides a select orientation to the experience of nature from the ancient world to the Anthropocene.

Taking its momentum from the emerging environmental humanities, this collection integrates Western, Indigenous, postcolonial, feminist and eco-spiritual perspectives that address pressing environmental concerns and reimagine the place of humans within the natural world. Across thirteen chapters, the contributors discuss the blending of environmental concerns with political and moral questions and encourage collaborative methods across disciplines to address dialectical tensions between culture and nature. They draw on a wide range of critical perspectives, provide a historical framework and speak to global environmental pressures from multiple standpoints. The global approach adopted throughout highlights the various realities of the growing ecological crisis experienced across the world.

Written to appeal to a broad range of readers across the environmental humanities, this edited book will be particularly useful to academics, scholars and researchers in philosophy, anthropology, literature, history and critical theory.

Gabriel R. Ricci is Professor of Humanities at Elizabethtown College, USA, and teaches environmental ethics, political philosophy and ancient philosophy. He has published on phenomenology and time consciousness, and politics, technology and ethics. Recent publications with Routledge include *Natural Communions* (2019) and *The Persistence of Critical Theory* (2017).

Routledge Environmental Anthropology

Environmental Anthropology explores historic and present human-environment interactions, highlighting the link between human-caused environmental problems such as climate change, species extinction, and pollution, with the complex cultural, political, and economic systems that have created them. This series aims to contribute to the growing subfield by providing a comprehensive survey of contemporary topics in environmental anthropology, including food procurement, ethnobiology, spiritual ecology, resilience, nonhuman rights, architectural anthropology, industrialism, and education.

The *Routledge Environmental Anthropology* series welcomes submissions that combine strong academic theory with practical applications, and as such is relevant to a global readership of students, researchers, policy-makers, practitioners, and activists. Please contact Grace Harrison (*Grace.Harrison@tandf.co.uk*).

Ecological Epistemologies and Spiritualities in Brazilian Ecovillages
In the Labyrinth of an Environmental Anthropology
Luz Gonçalves Brito

Building the Critical Anthropology of Climate Change
Towards a Socio-Ecological Revolution
Hans A. Baer and Merrill Singer

Environmental Reflections on the Anthropocene
Nature Transformed
Edited by Gabriel R. Ricci

For more information about this series, please visit: https://www.routledge.com/Routledge-Environmental-Anthropology/book-series/REA

Environmental Reflections on the Anthropocene

Nature Transformed

Edited by Gabriel R. Ricci

Routledge
Taylor & Francis Group
LONDON AND NEW YORK

earthscan
from Routledge

First published 2025
by Routledge
4 Park Square, Milton Park, Abingdon, Oxon OX14 4RN

and by Routledge
605 Third Avenue, New York, NY 10158

Routledge is an imprint of the Taylor & Francis Group, an informa business

British Library Cataloguing-in-Publication Data
A catalogue record for this book is available from the British Library

ISBN: 978-1-032-47230-0 (hbk)
ISBN: 978-1-032-47241-6 (pbk)
ISBN: 978-1-003-38520-2 (ebk)

DOI: 10.4324/9781003385202

Typeset in Times New Roman
by SPi Technologies India Pvt Ltd (Straive)

Contents

Figures

Contributors

Manuel Arias-Maldonado is Professor of Political Science at the University of Málaga, Spain. He has worked extensively in the field of environmental political theory and, in recent years, devoted his research to the concept of the Anthropocene and its political and normative implications. He is the author of *Real Green. Sustainability after the End of Nature* (Routledge 2016) and the co-editor of *Rethinking the Environment for the Anthropocene* (Routledge 2019) together with Zev Trachtenberg. He has been a visiting researcher in Berkeley, New York University and the Rachel Carson Center in Munich.

J.M. Bernstein is University Distinguished Professor of Philosophy at the New School for Social Research. His writings include *The Fate of Art: Aesthetic Alienation from Kant to Derrida and Adorno* (1992); *Recovering Ethical Life: Jürgen Habermas and the Future of Critical Theory* (Routledge 1995); *Adorno: Disenchantment and Ethics* (2001); *Against Voluptuous Bodies: Late Modernism and the Meaning of Painting* (2006). His most recent book is *Torture and Dignity: An Essay on Moral Injury* (2015). He is working on a manuscript with the tentative title: *Of Ecocide and Human Rights: Political Morality in the 21st Century.*

Jean-Daniel Collomb is Professor of American Studies at Université Grenoble Alpes, in France. His research is focused on environmental politics and policies in the United States. He is the author of *John Muir, écologie et parcs nationaux* (Presses universitaires de Bordeaux 2013) and *Une histoire de la radicalité environnementale aux États-Unis* (Presses universitaires de Bordeaux 2018). He has also written several articles about the opposition between the American Right and the US environmental movement from the early 1980s to the present time.

Jennifer Evans is a senior Queer Dharug scholar with dual connections to Dharug and palawa country. They are appointed to the College of Health and Medicine in the role of Aboriginal Health Leadership at the University of Tasmania. In 2022 they won the Tasmanian Premier's Literature Award for their academic work regarding Queer Indigenous voices. Jennifer is a multidisciplinary academic whose research and advocacy blends technology, country and queerness to create safe places for Indigenous methodological work and the BlaQ scholarly body. Through their work, Jennifer passionately gives voice to Queer narratives about the agency of country, cultural practices, identity, belonging and sovereignty.

Mary D. Garrard, Professor Emerita of Art History at American University, Washington, D C, is best known for her groundbreaking scholarship on Artemisia Gentileschi, which launched modern studies of the now-famous artist. Her extensive writings on Italian Renaissance and Early Modern art include *Brunelleschi's Egg: Gender, Art, and Nature in Renaissance Italy* (2010), in which she analyzed the changing relationship of art and nature in a critical historical period, when the entities were cast by art theorists as gendered competitors in an escalating rhetoric. With her colleague Norma Broude, Garrard also created and edited three books on feminism and art history that have become basic texts in art history and women's studies courses.

Phoebe Godfrey is Professor-in-Residence in Sociology at University of Connecticut. She teaches Social Theory, Society and Climate Change, Sustainable Societies, Sociology of Food, Sociology of Education and more. She is co-editor of *Systemic Crises of Global Climate Change: Intersections of Race, Class and Gender and Emergent Possibilities for Global Sustainability Intersections of Race, Class and Gender* (Routledge 2016) and of *Global [Im]-Possibilities: Exploring the Paradoxes of Just Sustainabilities* (2021). She is the author of *Understanding Just Sustainabilities: A Case Study of a Share-Use Kitchen in Connecticut* (Routledge 2021). She lives in Willimantic, CT, with her wife Tina and three cats.

Patricia Likos Ricci is Professor of the History of Art at Elizabethtown College. Her research focuses on Renaissance revivals and interdisciplinary studies of the relationship between art and science. She has published *The Inquiring Eye: Illustration and the Production of Knowledge* (2019); *Edwin Austin Abbey's The Passage of the Hours: Astronomy as History* (2015); *Joseph Priestley House* (2012); *Natural Laws: The Literary and Artistic Roots of American Environmentalism* (2006); *Lux ex tenebris: Etienne-Louis Boullée's Cenotaph for Sir Isaac Newton* (2005); and *The State as a Work of Art: Design, Technology and Social Reform 1876–1917* (2002).

David Martínez (Gila River Indian Community) is a Professor of American Indian Studies and Transborder Studies at Arizona State University. He is the author of *Dakota Philosopher: Charles Eastman and American Indian Thought and Life of the Indigenous Mind: Vine Deloria Jr and the Birth of the Red Power Movement* (2009). He has also published papers on O'odham history and culture in *Journal of the Southwest* and *The Arizona Journal of History*. Recently, he became founder and director of the Institute for Transborder Indigenous Nations.

Christopher Schliephake (Dr. phil. habil.) is Assistant Professor in Ancient History at the University of Augsburg. He specializes in ancient environmental history, the environmental humanities and premodernity, as well as studies in ancient divination and Plutarch. His recent publications include the co-edited volume, with Evi Zemanek, *Anticipatory Environmental (Hi) Stories* (2023) and *Conversing with Chaos in Greco-Roman Antiquity: Writing and Reading Environmental Disorder in Ancient Texts*, with Esther Eidinow (2024).

Caroline Smith is Senior Lecturer in education at the University of Tasmania, specializing in science and sustainability education. As a regular science writer for *EarthSong Journal* she has provided perspectives on ecology, spirituality and education for a wide Australian audience. She has been active in the permaculture community for some thirty years as a teacher and designer, and co-edited *Permaculture Pioneers, Stories from the New Frontier* (2022). Currently, she is working with a University of Queensland–based agriculture project developing Plant Health Clinic training for the Pacific Islands that aims to help farmers reduce their dependence on pesticides.

Kirsty Stewart is an independent researcher based in Edinburgh who received her DPhil in Byzantine studies from the University of Oxford in 2016 with a thesis entitled "Nature and Narratives: Landscapes, Plants and Animals in Palaiologan Vernacular Literature". Her research takes an ecocritical approach to literature of the later Byzantine period, with a particular focus on humor, religion and concepts of female beauty in relation to the natural world. She is currently working on the production and rhetoric of cosmetics in Byzantium.

Nick Towle began his professional journey working as a doctor in emergency medicine before pursuing education and research opportunities at University of Tasmania (UTAS). His UTAS work spans medical education, health advocacy, planetary health and climate change and health. He is currently a senior lecturer in Systems Thinking at UTAS. In parallel, Nick has been teaching and practicing Permaculture design since 2009. The focus for much of his work is around translating theoretical concepts into practical community action, such as co-founding the RESEED Centre (Penguin, Tasmania) as a hub for sustainability education and community resilience building.

Carol Wayne White is Professor of Philosophy of Religion at Bucknell University, specializing in Poststructuralist Philosophies, Process Philosophy, Religious Naturalism, and Critical Theory and Religion. Her books include *Poststructuralism, Feminism, and Religion: Triangulating Positions* (Humanities Press 2002); *The Legacy of Anne Conway* (1631-70): *Reverberations from a Mystical Naturalism* (SUNY Press 2009); and *Black Lives and Sacred Humanity: Toward an African American Religious Naturalism* (Fordham University Press 2016). Her work in philosophy and critical religious thought has also appeared in *Zygon: The Journal of*

Religion and Science, The American Journal of Theology and Philosophy, Philosophia Africana, and Religion & Public Life. White has received a Science and Religion Grant from The John Templeton Foundation and a NEH Fellowship. Her new book project explores the insights of religious naturalism expressed in contemporary North American nature poets and writers.

Jacqline Wolf Tice is an Adjunct for the University of Oklahoma College of Law Indigenous Peoples Law program. She researches issues at the nexus of tribal sovereignty, environment and culture. Settled in the Lenapehoking, she is active in stream monitoring programs in the Delaware River watershed as a Master Watershed Steward for Penn State Extension. She has served on the board and as Equity Chair of the Pennsylvania Chapter of the Sierra Club. She is the Tribal liaison for the steering committee of the proposed Delaware River National Park and Lenape Preserve, and regularly writes on Indigenous issues for the nationally recognized Stroud Water Research Center.

Brian Yothers is Professor and Chair of the Department of English at Saint Louis University. He is the author of *Why Antislavery Poetry Matters Now* (2023) *and Sacred Uncertainty: Religious Difference and the Shape of Melville's Career* (2015), among other books. He is the editor of *Leviathan: A Journal of Melville Studies*.

Acknowledgments

I owe a debt of gratitude to Grace Harrison at Routledge for shepherding my original proposal, gathering encouraging and valuable reviews and setting a course to produce a collection of essays on the experience of nature in the Anthropocene. Over the past two years Matthew Shobbrook has helped shape the manuscript and prepare it for production. Together their efforts have transformed a fledgling idea into a cohesive literary undertaking. Of course, what ideas I began with would not have been realized without the company of contributors who, from multiple perspectives, produced a collective orientation to the ways that nature has been interpreted and conceptualized over time. The personal advantages from my exchanges and communications with contributors during the development of this project now hopefully will accrue to the readers of this volume. I am especially grateful to my wife and colleague, Patricia, who first inspired the idea of producing a book that is just not a subject I have taught for many years but something that is close to home.

Introduction

Gabriel R. Ricci

In 2005, Bill McKibben added a new introduction to *The End of Nature*, first written in 1988. In it he reiterated that nature as an independent force had been displaced by implacable forces that are manmade. Rachel Carson's clarion call of the compromising forces of radiation and DDT in *Silent Spring* would have another iteration. McKibben's warning was not restricted to quashing birdsong in spring, but a more insidious condition produced by geophysical forces that threaten the globe and have skewed our perception and understanding of nature. Carson gave us the image of a spring without birdsong, but McKibben illustrates our alienation from nature with an image of being disturbed at the sight of a discarded Coke can while viewing a sunset beside a pond. He contrasts the shock of this encounter with coming across a pile of deer droppings. The picture throws into relief the difference between humans who casually dispose of their refuse and the deer who was simply doing what comes naturally. What is at issue is that we humans have the capability of self-restraint but we have obviously chosen to ignore it (McKibben, xx). The deer may not have the capacity to live according to the idea of a law, but it and the environment are well served by the laws in nature.

What McKibben referred to as the end of nature now resounds in the expression Anthropocene, a geological designation referring to the cultural and artifactual displacement of nature: There are the scientific efforts to identify geophysical markers embedded on the earth, there are the historical calculations identifying colonialization, with its elimination of vast populations in the Americas; and there is the political and physical evidence from the cold war and the proliferation of radionuclides. Whatever standard is used to delineate the time period, it is synonymous with human activity that was known to Arrhenius, the nineteenth-century chemist who first noted how increased levels of carbon dioxide raised the surface temperature of the Earth, and it was emphatically noted by George Perkins Marsh who detailed the way in which humans had altered the physical conditions of the globe. The evidence Marsh mustered in *Man and Nature* in 1864 meant that humans had not just modified the natural world; they had been actively determining it. And then there is the more contemporary marker set by Paul Crutzen, the atmospheric chemist and Nobel laureate, who sounded the alarm about ozone depletion. Whatever

DOI: 10.4324/9781003385202-1

standard we rely upon the International Union of Geological Sciences, after some deliberation and discussion of competing locations, has selected, at the writing of this essay, a site that will serve as a representation for the reality of the Anthropocene. Crawford Lake in Ottawa, Canada, has won the dubious designation perhaps because of the geological record at the bottom of this deep Canadian lake which reveals a laminated history of human activity mixed with evidence of natural evolutionary processes. Though this effort will not be a definitive calculation of when the Anthropocene began, the lake provides a composite picture of the varied periods of hyper-human pollution.

Finding a definitive starting point for having crossed this threshold seems elusive and might be decided by one's worldview, but a safe and more striking answer might be that the Anthropocene is coeval with the human imagination, it is not just a modern industrial and/or political problem. The evidence, then, need not be gathered by scientific experiment and calculation; it might require an examination of another sort of documentation, the written word. Fortunately, I recently stumbled upon a review essay in the NYRB by Robert McFarlane on the latest translation of *Gilgamesh* (Yale Press, 2021, translated by Sophus Helle) that put the notion in my head, the idea of perhaps the original written documentation of humanity's exploitation and destruction of nature, an archetypal ecocide nested in an epic story of a "heroic" figure originating in the second millennium BCE; scratched in cuneiform on clay tablets: *Gilgamesh*.

As foundational texts the *Bhagavad Gita* and the *Iliad* share a story of the motivations and misgivings about war and combat with *Gilgamesh*. The *Gita* adds a metaphysical overlay identifying the ontology of the self and the *Iliad* is distinguished by the anticipation of the polis in Scroll 18 which describes the elaborate design of Achilles' shield upon which a public legal scene depicts a competition for a straight judgment in a blood libel case. The war in *Gilgamesh*, while aimed at uprooting a powerful protective spirit, Humbaba, is really a war against nature. To overcome their chosen foe Gilgamesh and his companion, Enkidu, will eradicate an entire forest; what amounts to an ecocide with the expressed goal of garnering future fame and renown. In the balance is human narcissism and the despoilation of a vast Cedar Forest. Perhaps as a trophy Enkidu and Gilgamesh select a champion cedar that touched the sky and was grand enough to make a door 120 feet high, 40 feet wide and 2 feet deep, to grace a temple dedicated to the Sumerian god Enlil. The spoils from the war would also be gathered to fashion a raft to transport the head of Humbaba, the erstwhile guardian of the Cedar Forest. Gilgamesh's assault on Humbaba was so powerful that an initial blow divided Mt. Heron and created a valley, after which he and Enkidu produced a wasteland with their outsized axes. They were intent on crushing Humbaba's mind (Gilgamesh, Tablet V, ln 267), the spirit that provided the fortitude and psychic connection to guard such a commanding sacred space. To murder Humbaba the entire forest needed to be eradicated. Any surviving trees would be the germ of a future forest. This explains why Enkidu called to crush the mind of Humbaba. It was the idea (in Humbaba's mind) of a grand Cedar Forest that Gilgamesh and Enkidu wanted to destroy. Ironically, Enkidu, who appears the most militant, was once a Wild Man who grazed with gazelles and had no human contact, nor country. The sight of him cavorting with the beasts was enough to frighten humans, and his physical stature kept people at bay. Enkidu was once the embodiment of the spirit of nature but a conspiracy to initiate him into the ways of humans, and the charms of a woman in particular; followed by physical combat with Gilgamesh, inspired a zealotry for political engagement. His antagonism toward Humbaba and the Cedar Forest was a denial of his origins in the ways of the wild. His socialization alienated his previous companions, sabotaged his physical abilities, and he unwittingly exchanged his natural animal instincts for reason and the ability to think (Gilgamesh, Tablet 1). Their campaign against Humbaba first reconfigured the landscape and then annihilated the sacred Cedar Forest.

If reliance on foundational narratives seems too fanciful there has been growing support for an early anthropogenic hypothesis that according to William Ruddiman, an early proponent of this theory (2003), began with agriculture and its progressive deforestation. The evidence gathered from archaeologists and paleoecologists have introduced a human element, alongside the findings of glacial geology and geophysical accounts of orbital variations, to explain large-scale changes in early landscapes and global warming. The wider global picture is presented in Peter Frankopan's *The Earth Transformed* (Knopf 2023), a 700-page tome which comes with 200 pages of notes on its own webpage. Evidence for the early anthropogenic transformation of the Earth through land use is amply identified in these notes and challenges the predominant model of natural Holocene warming from the previous century.

However the status of the Anthropocene is resolved, Manuel Arias-Maldonado in the opening chapter, "Nature in the Anthropocene," opens a window onto a contemporary conversation that helps sort out the different ways in which nature is interpreted under the new light of the Anthropocene. As to be expected, there is some wrangling on the ontological status of the Anthropocene generated by considerations of a factual basis and the conceptualization of how humans have disrupted the function of the planet in which they live. Maldonado argues that unless one adopts a strong constructivist position, nature is indeed out there. Moreover, the Anthropocene is not interpreted unanimously as saying this or that; on the contrary, there are several competing views about the subject. With competing designations like the Capitalocene, the Chtulucene or even the Narcisscene Maldonado identifies three approaches: (1) that nature has simply come to an end, (2) a hybridization thesis in which humans and nature are so entangled that it no longer makes sense to distinguish between them and (3) a pluralistic conception since nature and socio-natural relations in the current epoch are just too complex and multifaceted to be exhausted by a single explanation. Maldonado leans to the latter, provided that it remains connected to the empirical foundation that the natural sciences provide and maintains that any experience of nature is rooted in evolutionary history in which human beings, in the geological scheme of things, are recent newcomers.

For his part, McKibben presents incontrovertible evidence that human activity has challenged the idea that nature is a fundamental constant with abiding foundational elements which moderate through an evolutionary dynamic, and persists through time. His subsequent webpage 360.org monitors the ever-increasing CO_2 levels that threaten the Earth; since its introduction projected alarming thresholds have been surpassed. The overturning of nature as a fundamental constant obviously unsettles the conventional understanding of nature, whether humans are considered part of or separate from nature. What McKibben's *The End of Nature* does is to entertain that ideas, like living flora and fauna, can become extinct. What can it mean to say that nature, as an idea, is extinct? Reality has not been wiped out, the living world is not defunct and we, as humans, are still here. Simply put, the domination of humanity with its destabilizing footprint has led to the consideration that nature has been so adulterated by anthropogenic forces that it has been displaced by cultural superimposition. In equally sinister terms nature is what humans say it is; it is a linguistic artifice fashioned to suit human needs and desires. This is played out in various social constructionist interpretations of nature which rely on the medieval realism/nominalism debate to suggest that cultural forces have superseded natural ones to justify saying that nature is a linguistic designation, having merely a nominal status assigned by humans. If nature is what humans say it is, its ontological status is now spontaneously fluid, and, to say the least, environmental ethics is likewise destabilized and basically up for grabs; displaced by expedient technical solutions, or perhaps even moot.

The early and persistent conservation ambition, in the West, to legislate the constancy of nature by setting aside wilderness areas acknowledged that nature was fragile. While early conservationists' efforts intended to protect the natural world, it may also be construed as an early attempt to control and define nature. Imagining that the natural order could be preserved intact, say in the form of national parks, is to create a cultural artifice presenting as relief from the drudgery of civilization. To imagine an idyllic setting undisturbed by human intervention is to suffer from a golden-age syndrome that the environmental historian William Cronin identified with Thoreau's nostalgia for vast virgin forests that had been disturbed by the cultural and financial advancement of agricultural industriousness.

Thoreau's 1851 lecture *Walking* announces his desire to speak on behalf of nature. In his role as advocate, he prescribes how to walk, after the manner of a pilgrim to the Holy Land, sauntering and deliberately taking in the land, and in this case preserving at once the Wild in the land and in one's spirit; deliberately suspending concern from the civil world. Thoreau's instruction that "In wildness is the preservation of the world" refers as much to the wilderness as it does to the dual nature of humanity, part wild and part civil. Thoreau was not advocating a physical regimen to stay in shape, for him walking is a condition and context for stimulating thought that enlivens the wild spirit in being human. Our legs must involuntarily transport us, Thoreau advises, but to be spiritually present we should be thinking of the woods we have entered.

Bracketing the ballast of the civil world and its preoccupations is a prerequisite for recovering the wild in us. Landscapes that have been subdued by human "improvements" must be avoided if purely human affairs are to be forgotten. Thoreau is keen on the nutritional and spiritual value of tonics from the forest and recommends a touch of arbor vitae and spruce in our tea. In an allusion to Romulus and Remus these wild ingredients are just the elements upon which the political community should be founded. In the end, if we wish to consider the redemption of humanity, Thoreau favored the salutary benefits of woods and swamps over the virtue of men.

Following Cronon's work which introduced that wilderness too is an idea, a human fabrication captured by the fantasy of a golden-age standard in which humans lived in harmony with nature, the delineation between natural and human has been obscured. Moreover, wilderness as the centerpiece of conservation efforts in the United States has been challenged on various fronts. Ramachandra Guha and others, representing the Global South, have argued that American environmentalism has been so mesmerized by this fantastic standard, that the real causes of environmental degradation, consumerism and the militarization of society, have been ignored. Political ecology, particularly Ecofeminism, has laid the blame at the feet of patriarchy, whose economic suppression of women derives from the domination of nature, both representing companion reproductive forces that are subjugated by the constructive elements of the growth-economy mandate associated with the patriarchy.

J Bernstein's "On the Severance of Production from Reproduction: Simone de Beauvoir and Ecofeminist Critical Theory" invokes a political ecology embodied in the domination of women, patriarchy and the domination of nature by a threatening instrumental reason, following the critique of Enlightenment reason in Horkheimer's and Adorno's *The Dialectic of Enlightenment*. Here, natural science, technology and free market economies are entangled such that the domination of one cannot be separated fully from the domination of the other; we are subjugated by a culture industry that meets us at every turn.

Bernstein's governing thesis is that the experience of delegitimated nature and women as the other of man have been ideologically secured through the comprehension of women and nature as primarily sources of biological reproduction. Consequently, the liberation of women cannot be complete without the liberation of nature (and vice versa): The fate of feminism and the fate of the environment have become entwined.

In the context of the history of art and science Mary Garrard's "Nature, Art and Gender in Renaissance Italy: A Counter-Narrative" makes related claims. Garrard's narrative traces the parallel tracks of the scientific revolution and the development of Renaissance art as catalysts for the transition of nature from creative agent to a passive product of creation. For male artists, especially, the consequence became a competition between their creative powers and nature's, but the art-nature dynamic was informed by nature's established female identity which mimicked the conventional inferior status of the female. The celebration of the genius of the Renaissance male artist, like God conceived as a

Divine Artificer, Garrard finds displayed in the epic achievement of the design for the dome of Florence Cathedral, by Brunelleschi, who, like Daedalus, turned his attention to the arts to technologically overcome the laws of nature. Tragically for Icarus, gravity was only momentarily defied; Brunelleschi's victory over nature still keeps watch over Florence. Garrard emphasizes that art's victory over nature preceded science's role in mastering nature, a struggle which pitted nature and art as gendered competitors.

Under the influence of patriarchy Garrard demonstrates how the idea of nature moved from a view of the earth as a living, intelligent organism imbued with spirit to the scientific conception of the world as a machine. The shift produced two standards for the human–nature relationship: One of detachment and one of connectedness. In the first instance the Earth must be conquered and subdued and in the second we are co-extensive with nature. The model of man's conquest of nature dominated the early history of science, and it informed the privileging of Florentine art in twentieth-century art history. But now, in growing realization that the exploitation of nature has brought us to the brink of ecological disaster, climate change scientists have parted ways with Earth's conquerors, and, sometimes in collaboration with artists, they seek a new model that puts nature and culture, nature and science, in partnerships of mutual self-interest.

As a proponent of and practitioner of being in the wilderness, Gary Snyder is a counterpoint to all attacks on the credibility of experiencing the wild and being committed to the reality of nature. His commitment is reflected in the detailed glossary, in *The Practice of the Wild*, replete with etymological insights and the variety of ways that the terms 'wild,' 'wilderness' and 'nature' have been catalogued in literature and through his personal experience. This might sound typical for a poet inspired by a mythical intimacy with nature, but Snyder eloquently expresses that, no matter how urbanized or industrialized we become through the management and conduct of the human world, the wild lies dormant in our physical space and our mammalian character. Snyder knows full well that the wilderness in North America has practically been extirpated, but it is the wilderness, as with Thoreau, that can be spontaneously excited through perception and consciousness he is concerned with. So, from the bowels of an oil tanker, an emblem of the industrialized world, he is able to bear witness to the "wild system" in the crickets he observed in the corners of his confined workplace in the mid-Pacific (Snyder, 16). We need not visit a pristine, undeveloped landscape to experience the wilderness, we are immediately in the wild when we cultivate bacteria-laden soil, consume the bacillus in yogurt or awaken the natural yeast for a sour dough "mother" by simply mixing flour and water and waiting. If Enkidu is the exemplar for a human in the wild, we must at least acknowledge how we are routinely nourished by the "ghost wilderness that hovers around the planet," according to Snyder, no matter the scale, they are living proof of Thoreau's dictum that in wilderness is the preservation of the world.

Thoreau's pronouncement is captured in Brian Yother's essay "Universal Application: The Natural World as Metaphor and Phenomenon in Melville,

Thoreau, and Dickinson" in which he shows how these authors make strategic use of nature as a unifying metaphor but still preserve it as the ground for autonomous ecological phenomena and the source of concrete attention to the natural world. Still, they kept faith with the Emersonian impulse to understand nature through a web of correspondences with the human mind, as professed in his poem "Each and All": "All are needed by each one; Nothing is fair or good alone"; unable to resist "the perfect whole." In keeping with Hester Blum's bracing reminder in her work on Oceanic Studies that "the Sea is not a metaphor," this chapter traces the ways in which Dickinson, Thoreau and Melville affirm the metaphoric power of nature and resist allowing natural phenomena to be subordinated fully to human consciousness. Similarly Patricia Likos Ricci's "The Raging Torrent: Myth, Metaphor and Technology" illustrates not only the independence of Nature but its moral force on the human world. Long before the waterfall was celebrated as a site of the Sublime, a natural wonder or a source of hydroelectric energy, it was a symbol of moral order in the natural environment. Through the various stages in the history of the topos of the raging torrent in literature, painting and technology, Likos Ricci shows how the moral metaphor of the waterfall was transmitted through the classical tradition and remained intact from Antiquity to the twentieth century. However they were depicted, by artists seeking sublime natural settings or as material for technological development, the public in the end, she argues, resisted the pressure to sacrifice them to the gods of Progress; resulting in their physical preservation from commercial exploitation in national parks.

Christopher Schliephake's "The Color Purple in the Ancient World and Its Social-Ecological Dimensions" challenges the modern binary conceptualization of nature/culture derived from ancient sources. Going beyond these sources, associated with the pastoral tradition and the idea of a pristine wilderness, has disclosed entanglements that exist between the nonhuman environment and human social systems. Tracing the history of the material production of the color purple, associated with social luxury and political power, makes it possible, Schliephake argues, to uncover the environmental dimension of a key ingredient of ancient life by illustrating the way that the extraction of resources, the cultural perception of matter, and the particular ecologies of the Mediterranean interacted.

Further, Kirsty Stewart in "The Byzantine Experience of the Natural World" demonstrates how a worldview could promote a practical, pious and pleasurable understanding of nature, perceiving Creation as both a physical and a spiritual benefit to humankind. Her comprehensive view of the medieval Greek Orthodox world looks at Byzantine texts and art to discuss the role of nature in day-to-day life and the correspondence between reality and spiritual ideology. In contrast to Lynn White's analysis of the negative impact of Christianity's worldview on the natural world, there is not the clear sense of humans exploiting and ruling nature in the Byzantine world. Stewart finds, particularly in commentaries and homilies of the hexameral literature, an expression that knowledge of God and knowledge of nature are reciprocally intertwined.

Aldo Leopold, whose conservation efforts singled out a middle ground between wilderness and urbanization, adopted a dual evolutionary and ecological perspective to express humanity's role and place in nature. His mental image of the Land Pyramid in *A Sand County Almanac* represents the intricate chains and connections that make humans a mere member alongside other mutually interdependent citizens of the biotic community. Carol Wayne White's "An Eco-Spirituality of Wonder: The Anthropocene Paradox and New Ways of Seeing, Knowing, and Acting" aligns with Leopold's elaboration of the biotic community in which humans are embedded alongside nonhumans in a matrix of tangled exchanges and interactions that determine ecological equilibriums and perturbations. Wayne White first acknowledges the triumphalist view of humans outside of nature, then recommends, as a solution, an eco-spirituality grounded in the tenets of religious naturalism. This representation reframes humans as natural processes in relationship with other forms of nature, in the manner that Leopold's appeal for humans to overcome the status of conqueror is mitigated in the biotic community. Like Leopold, Wayne White relies on humanity's embeddedness in materiality and its constitutive relationality to activate aesthetical-ethical responses to current forms of ecological degradation and environmental injustices. Leopold tantalizes his readers with a final prescription in *The Land Ethic*: A thing is right when it tends to preserve the integrity, stability and beauty of the biotic community which Wayne White captures in the promise of an eco-spirituality of wonder in which humans enact evolutionary capacities as relational organisms that can love and engage in multilayered processes of changing behaviors, values and relationships that promote the betterment of nature.

To consult Leopold's almanac in Part I of *A Sand County Almanac* is to discover a naturalist's guide to personal encounters of the wild throughout the year. The month of February, in particular, is a lesson in how to resuscitate connections with the land that have been obscured by our reliance on cultural artifacts. Living on a farm, he recommends, reminds us that breakfast does not just come from the grocery, or heat from the furnace. Planting a garden can avoid the first vulnerability and laying up a chord of wood will correct the second. He devotes the bulk of his account of February to felling a tree, and laying up wood, as he recounts the legal and environmental history in Wisconsin archived in the rings of the tree. Three tools, the axe, the wedge and the saw, he concludes, are necessary for good oak and good history (Leopold, 19).

In the civilized world, cultural foundations have shifted from the wild and wildlife, but Leopold is still able to press out vestiges of this primordial cultural base from his vantage point in mid-twentieth-century and mid-western America. Our wild roots, as he calls them, are retained by cultural values that excite connections with wild things. The challenges of being a scout in those days would have reconnected youth to a national drama of pioneering and living off the land. Leopold casts this insight as a case of ontogeny recapitulating phylogeny. The lessons of living on or near a farm keeps us informed on the value of the food chain "and the fundamental organization of the biota."

Leopold adds a third value, borne of his own habits and experience, on the ethical guidelines of "sportsmanship" and wildlife management. Leopold illustrates that there is no real dichotomy between culture and the wild. He is certain that wilderness has provided the original ground for culture and that primordial culture was predicated upon wildlife and life in the wild which remain dormant and ready for reactivation when these values are prompted.

When *A Sand County Almanac* appeared the study of ecology was first making its way into university curricula in the United States, thanks to the painstaking efforts of Eugene Odum at the University of Georgia. Over time he was able to convince his colleagues in biology that ecosystems were worth studying because they were more than the sum of their parts and thus deserved intellectual attention on their own merits, one of them being the context in which value systemically emerges in the first place, thus dissolving the fact/value dichotomy.

For Leopold a holistic approach to the land required an understanding of ecology. He bemoaned the fact that ecology was yet to become part of education, but it may well be, he surmised, that the study of ecology might be uncovered in tangential disciplines like geography, botany, agronomy, history or economics. (Leopold, 262) With the latter two disciplines attention to a human narrative informed by political and financial arrangements would receive ecological attention; hence, the systematic way relationships are established and maintained, as in the case of the biotic community, becomes the focus of study. Citing the discipline of geography invokes the embryonic political ecologies in the work of Peter Kropotkin and Alexander von Humboldt, both of whom challenged social hierarchical structures and promoted indigenous peoples' knowledge and interaction with the land (Robbins, 25–30). Like Leopold, the contemporary geographer Carl Sauer devoted attention to the nexus between social processes and the physical realm (Robbins, 36). Both were impressed by soil conservation efforts in the thirties and the need to do more than instruct farmers on techniques to mitigate soil erosion. The monocropping of wheat had as much to do with events like the Dust Bowl as the weather conditions beyond human control. Planting wheat, which displaced the drought-tolerant native Buffalo Grass, repeatedly tortured the soil. Not seeing the forest through the trees, Sauer would argue, derived from "the suicidal qualities of our current commercial economy" (Robbins, 37). Leopold's normative treatment of ecology likewise begins with understanding landscape configurations in terms of economically-driven human intervention and the need to transcend commercial expediency.

In his own way, Leopold attached an incipient social ecology to his naturalism. Promoting a sustainable ethos, Leopold knew that any expedient use of the land, comprehensively defined as soils, waters, plants and animals, must overcome mere economic determinism. First, so much of what is integral to the environment does not have obvious monetary value; and second, there are complementary aesthetic, spiritual and personal penchants on which land use turns.

What social ecology is implied in his positing of a biotic community, to which all life evolutionally converges, was not the conventional political or social ecology that first advises an analysis of the hierarchical organization of political and institutional relationships in order to understand parallel ecological degradation and the domination of nature. What correspondence and/or orientation may exist between human arrangements and nature is bypassed with Leopold's conception of the biotic community which resolves the human–nature dichotomy by acknowledging the biotic union between humans and the natural order. What convincingly links Leopold's biotic worldview to political ecology is the call for humans to move from their perch as conquerors to plain members of the biotic community, with the additional plea to respect the concomitant communalism. Like Murray Bookchin, a more overt social ecologist, Aldo Leopold critiqued the hollow pieties of conservationism and environmentalism guided by superficial technological solutions; accordingly both challenged the human entitlement to dominate nature. In 1980, Bookchin, favoring a direct political ecological assessment of society, viewed the nonhierarchical challenge of feminism as a source to articulate and overcome the domineering tendencies of our culture (Bookchin, 75–83); while in 1949, Leopold's concept of community emphatically resolves the nonhierarchical demands of political ecology by abolishing the conqueror role of *Homo sapiens*. The conqueror is always certain what makes the community function and what is of value or not; turns out humans, even scientists, Leopold adds, find biotic mechanisms too complex to fully understand. Leopold demonstrates this fusion of the natural order and the human world with an ecological interpretation of history, that is, no history of human settlement and organization is complete without a simultaneous narrative of plant succession; exchanges between the land and people. Running on parallel tracks, human initiative and the constitution of the land determine that humans are team members. (Leopold, 241) Using several key moments of settlement and pioneering in the United States to make his point Leopold anticipated history will be correctly recorded and taught "once the concept of the land as community really penetrates our intellectual life" (Leopold, 243). But, as Leopold passionately argued, in order to release the full evolutionary potential of an ecological ethic, land use must not be regarded as a mere commercial problem, perhaps the greatest challenge to our conventional mindset.

A novel outlook has been mapped out by the eco-philosopher Glenn Albrecht, whose *Earth Emotions, New Words for a New World* (Cornell, 2019) anticipates a new geological epoch to follow the Anthropocene. In it he proposes a new vocabulary that describes the malaise brought on by the Anthropocene, and further prescribes concepts that will foster the impending epoch he calls the Symbiocene. The period promises to dispel the malignant forms of development introduced in the Anthropocene by calling for a conviction that Leopold advanced when he expressed the philosophical inclination to describe ethics in ecological terms. Whether we discuss ethics in terms of restrictions on our behavior in the struggle for life or normative distinctions of acceptable and unacceptable conduct, Leopold saw the same dynamic; interdependent entities creating ways of cooperating. The political and economic processes we have devised have only displaced a primordial competitive landscape with more collaborative arrangements. These are only examples of symbiotic connections that mark ecological evolution. Albrecht's Symbiocene fulfills Leopold's assessment that ethics might very well be a "community instinct in-the-making" (Leopold, 239).

Albrecht's book is first a catalogue of emotions and psychological reactions in the face of the loss of a sense of home and the overturning of a sense of place due to lived environmental disruption. His initial laboratory was coal country in Australia where he vividly witnessed the phenomenon of homesickness while still at home, solastalgia, a neologism combining solace, desolation and nostalgia, to refer to a tumultuous psychic response to negative environmental change and the loss of home. Albrecht asks us to think of Munch's fiery *Scream*, a response to the volcanic eruption of Krakatoa in 1883, as a depiction of this intense emotion. (Albrecht, 45) Since 2003 his psychoterratic vocabulary has extended to include the anticipated epoch, Symbiocene, that will displace the Anthropocene and its upending of the natural order, along with its negative emotional responses. He imagines a new world in which emotions like euiterra, a sense of oneness with the Earth, and ghedeist, a spiritual feeling of interconnectedness, will supplant global dread and meteoranxiety, fear of extreme weather events. Introducing a novel vocabulary opens up a new world that abandons well-intentioned conservation inclinations and even sustainability as a fundamental goal, since each subscribed to a dualism that set humans against nature in the first place. He even calls to relinquish the term environment which logically entails the control and domination that marks the Anthropocene. Instead he relies on the term symbiocent to overcome the Anthropocene rift and imply an integrated experience of total immersion of symbiotic interactions that, I think and I imagine Albrecht would agree, was presaged in Leopold's land pyramid, the mental image of ecological complexity and layered trophic scales that in 1949 he substituted for the antiquated and inadequate idea of the balance of nature.

Albrecht can imagine a period when humans will be so integrated into life supporting systems without a measurable eco-destructive footprint, given his intimacy with a primordial mythic, life-affirming cosmology of indigenous populations. In Elyne Mitchell's *Soil and Civilization* (Angus & Robertson, 1946) Albrecht found a companion piece to Leopold's *A Sand County Almanac*, both of which establish symbiosis as a lens to envision an ecological ethos. Leopold's professional world was delineated by wild life and forest management and, like Mitchell, soil remediation. Mitchell's vantage point is uniquely framed by the cosmogonic knowledge of aboriginal Dreaming and foundational narratives that link the spiritual and physical world to cosmic origins that shape nature and a moral code; stories that are critical to maintaining cultural heritage and the world itself.

Jen Evans's "The Sovereign Body of Country" examines the cosmogony of Aboriginal culture in the phenomenon of Country, a sentient living entity within itself. While we might refer to it as a region beyond urban settings, Country is a place of relationality, where Indigenous People have a deep sense of belonging; home and place are one with no incommensurable differences. The Dreaming Law and creation stories are alive in Country and invite relationships of mutual obligation, where humans and nonhumans enjoy multifaceted deep time relations. Country has agency and power, able to hold relations and shape and provide for and reawaken culture. All these phenomena are distinct from western concepts of wilderness, nature, bush, preservation and conservation. Yet Country is rarely discussed in terms of sovereignty, packed with its own sense of rights. The spirit of Country informs the 1984 film *Where the Green Ants Dream* by Werner Herzog, loosely based on a real legal case about Indigenous land rights, in which a cultural clash turns on a living mythology which incorporates a cosmogonic time in which ants dream the cosmos into existence. In the film, a western industrial invasion of a sacred space does not just threaten a living environment but the creative forces that sustain the world and have the force of law.

Albrecht teases out a proto-Symbiocene from the ancient Aboriginal culture that occupied Australia for eighty thousand years. He is certain that it can be reproduced by sumbiocentric humans in a union of sumbiocracies (Albrecht, 126). If Albrecht's language seems futuristic, he has found evidence that this symbiocent spirit can crop up regionally in the prevailing Anthropocene. For example, in a study of the Margaret River region of South West Australia he uncovered evidence of solophilia, the opposite feeling of despairing solastalgia, a positive sense of place and the political will to fight for it. In this case the result was fulfilling a latent "cultural terroir," as he calls it, that included organic viticulture thanks to a spirited opposition to government-imposed coal fields (Albrecht, 122). While Albrecht's novel language may take some time getting used to, we can rely on extrapolating from the established ecocentric vision of Aldo Leopold's biotic community to chart the way.

A developing trend in the United States captures some of the elements of Albrecht's enthusiastic bio-regionalism and symbiocent energy. A recent agreement between the US National Park Service and the Cherokee Nation, to set aside traditional lands in the Buffalo National River Park for Cherokee Nation Medicine Keepers to gather traditional medicinal, culinary and craft plants for relocation and repopulation in western Oklahoma, is a glimmer of symbiocent sentiment (Adam Kemp, PBS NewsHour, July 13, 2023). Restoring these traditional plants relies on ancient indigenous knowledge that was disrupted in the colonial violence that uprooted and displaced the Cherokee culture. Cherokee knowledge exemplifies the Symbiocene because it is integrated into original bio-regional cultural sources whose usufruct does not appreciably alter the ecosystem. Similar arrangements between the National Park Service and Native tribes to co-manage park lands are in line with the Symbiocene.

David Martinez's "When Coyote Stole Rabbit's Heart, O'odham Himdag, Environmental Sovereignty, and the End of the American Empire" provides an indigenous account of an original union between the natural world and an original people that also begins with the knowledge that an autochthonous culture was uprooted and the land divided and distributed using destructive, alien methods that ignore the unique fealty and kinship which established primal relations between humans and nonhumans. As an O'odham thinker, Martinez's story is an argument that environmental ethics begins with the consideration that the land is sacred. The same pattern of disruption as evidenced in Australia is playing out today in the United States, where government land use ignores traditional mores and laws rooted in indigenous cultures. For example, where some might simply see the indulgence of political ambitions in the current cry to build a wall on the southern border, locals who have lived in the region for hundreds of years see the destruction of sacred spaces in the Sonoran Desert. The Tohono O'odham Nation have had movement on their own land inhibited by the construction. Sacred springs and endangered species and archeological sites dating back 16,000 years like Quitobaquito have been threatened by construction activity, and water has been redirected to make concrete. Public protests, from the point of view of the O'odham, are justified by what is essentially a First Law entailed in the gift of the land by the creator. The story that Martinez tells explains the impetus for such political responses and vindicates the ambition to restore balance at a hallowed place less than fifty meters from the US-Mexico border.

Albrecht's vision is ambitious: First the Symbiocene requires a cessation of the routine human impact on the planet and then maintaining this position by integrating everything humans do into the life support systems and leaving no lasting trace. This status will be perpetuated by the form of government Albrecht calls sumbiocracy whose constitution originates in the awareness of the interconnectedness of all life and a heightened sense of empathy. Yet another neologism designates this consciousness, ghedeist, which incorporates Hegel's universal Geist or Spirit. Albrecht makes his point about the inherent spirituality and psychology of this form of consciousness by referring to an early fragment by Hegel from 1798, subsequently titled "Love," in which he anticipates his historical dialectic with the assertion that nothing independently possesses the cause of its own being, one only exists in relation to the other. (Albrecht, 146–147) This condition of love is even more poetically expressed by Rudolf Steiner, according to Albrecht, with the example of bees and honey production that represents the spirit of shared life among coexisting organisms in what Albrecht identifies as the biocomunen (Albrecht, 144–146). Using the expression environment, then, will give way to the biocomunen because it phases out the sense of boundaries, the difference between inside and outside, between the subject and object, between organism and ecosystem, that the word environment implies. Steiner referred to the existence of bees as "love life" because they transform the symbiotically derived life of plants into honey. Casting the process in biological terms, Albrecht distills what could be a complex scientific analysis of a collective creation into the unadorned description that "Honey is a holobiont's product of life; there is more to honey than the bees" (Albrecht, 145).

Albrecht's analysis is eclectic: Clearly the holobiont character and interconnectedness of life expressed in the activity of bees is a biological metaphor and it is supplemented in a reference to a popular cinematic phenomenon, *Avatar*, which contains the same proto-Symbiocene features Albrecht cites in Aboriginal culture. At the center of the symbiotic empathy in the Na'vi culture, the analysis goes, is the tree of life which becomes a target in the lust for commercial development. The plot is emblematic of environmental engagement. This is the archetypal sacred tree, like the Norse Yggdrasil, that grounds cultural narratives. They are the axis mundi that connect the heavens and the earth and, like the Yggdrasil, are supported by root systems that extend to nourishing life sources. The tree of souls in *Avatar* is the vital connection to the all-encompassing life spirit, what Albrecht dubs the ghedeist. If these organic metaphors seem removed from Hegel's historical dialectic, we need only recall how, in the Phenomenology, the unity of life and the historical dialectic are expressed in the temporal manner a bud disappears only to become a blossom and the blossom fades only to produce the fruit. Though we must acknowledge a process of negation, it would be inaccurate not to say we also witness the unity of life in such a progression. Each contradictory phase is equally necessary and equally produces the whole (of life). Albrecht's early work in philosophy which explored the principle of organicism, how society resembled the structure of an organism, first linked Hegel and pre-Socratic thought to environmental philosophy which culminated in *Earth Emotions*. From the outset of his philosophical journey, Hegel meeting Gaia, as he put it, was not a chance encounter (Albrecht, 20).

The ecological worldview in Albrecht's intuition of the world of the Symbiocene fans out to include Aldo Leopold's biocentric community, indigenous spirituality, and the organicist dialectic of Hegel, Rudolf Steiner and others. If we are troubled by McKibben's assessment that nature might be at an end, Albrecht's striking vision adds the obligation to reconsider who we are; what do we mean when we think of ourselves. This commitment parallels Thich Nhat Hanh's advice in the meditation *The Sun My Heart*, in which he wields his own innovative language to describe existence and ecological conscience from the point of view of Buddhism. The revelations about the multiple interpenetrations of time, space and being in this reflection requires us to realize that our bodies, like the holobiont character of the bees, are not restricted to what lies within. While invoking Walt Whitman's Song of Myself, Thich Nhat Hanh resolves that there can be no fast delineation between the inside and the outside, and so caring for the environment means at once taking care of the self (Armstrong & Botzler, 255–258). In this light Godfrey's and Tice Wolf's "Re-Centering the Ancient-Enduring Indigenous Lens" presents a spiritual exercise and dialogue using Suzuki's related claim that there is no environment out there, we are constructed of the four sacred elements of earth, air wind and fire. The conversation invokes the spirit of Turtle Island, the ancient collective civilizations that functioned with codes of conduct, governance structures, cultural mores, political dynamics, trade and natural resource practices guided by a relational ethic with the Earth and all living things. This relational ethic has been displaced by colonial constructs which have decoupled human communities from ancient embedded spiritual and relational interdependence with the living environment. This chapter thus illustrates how the homogeneity of Western ideological patterns of civilization has uncoupled and uncivilized ancient nature-relative systems, while piecing together, in a modern context, a facsimile under colonial terms and values. It ends by calling for more attention to the distinct differences 'culture' makes in terms of how the nature–human connections are conceptualized and manifested, in order to reimagine more, and advocate for culturally inclusive pathways into the future that re-center the ancient-enduring Indigenous lens and the deep philosophical sense that our bodies belong to the earth.

Thich Nhat Hanh impresses on us how to rethink our being to make sense of this ontological interdependency. If Albrecht's new language presents problems, Thich Nhat Hanh's use of the term 'intrabeing,' a kind of interfacing and interweaving, to describe our kind of existence obliges us to carry the world with us at all times. Walt Whitman's lines "I believe a blade of grass is no less than the journey-work of the stars"; "I am large, I contain multitudes," from Song of Myself, he advises, does not require a philosophical effort, but a meditation from deep inside the soul. The burden of being an environmentalist, Thich Nhat Hanh proposes, and Albrecht agrees, is not as simple as adapting to a new nomenclature. Thich Nhat Hanh was an activist, renowned for his anti-war engagement and The Plum Village community which promotes the art of living in harmony with one another and with the Earth. This environmentalism is consistent with the practice of mindfulness, the energy produced when the mind unites with the body and clarifies what is going on in the present moment, within us and around us. This is the source of activism expressed in the sense of being in the meditation *The Sun My Heart*.

The final two chapters embrace the activist spirit expressed in the obligation to reconsider who we are, and what do we mean when we think of ourselves and being fully present for ourselves and whatever we are doing. Thich Nhat Hanh's prescription for mindfulness means that the way we walk on the Earth has real influence; to avoid being sleepwalkers, he admonishes, means that we should walk with mindful steps. Like Thoreau, to prepare to walk and saunter into the world, we must settle our affairs, and detach from relatives and friends. Following either prescription, having no particular self or particular place, means that we will be equally at home everywhere.

Smith and Towle recommend the practice of permaculture and Collomb applies a burgeoning technology in an ironic twist to mitigate technologically-generated pollution with a technological solution. These are not isolated solutions. Among the many ways to activate ecological consciousness there is the ever-growing appeal to create green corridors, like the Y2Y corridor connecting Yellowstone Park to the Yukon to benefit keystone species. These efforts are part of the wider movement of rewilding as in the extraordinary efforts of the Brazilian couple, Sebastião and Lélia Salgado, who over 20 years have reforested a barren landscape; returning wildlife and reviving a once depleted ecosystem, the community efforts in Scotland with the Carrifran Wildwood, and the rewilding project at the Knepp estate in West Sussex where a renovated, unenclosed 3,500-acre dairy farm now attracts endangered species and invites beavers to perform their engineering marvels while still accommodating sustainable farming and a supply of local produce. The Knepp Wildland, in particular, is living proof of the wood-pasture hypothesis of Frans Vera, the Dutch conservationist, who theorized that large herbivores once maintained open space alongside closed-canopy forest areas at the end of the Pleistocene.

Smith and Towle's chapter "Permaculture as a System for Designing Sustainable Human Settlements. Ahead of Its Time or Impossible Dream?" provides the background, principles and place of permaculture as part of the journey to a resilient, flourishing future, and contemplates its relevance in today's climate-challenged world. This promising practice originates within the broader countercultural landscape and environmental awareness of the early 1970s, particularly through the efforts of Bill Mollison and David Holmgren whose creative synthesis of ethics, systems thinking and design principles is based on an understanding of the processes of natural ecosystems, traditional small scale mixed agriculture, low impact technology, and redistributive social justice into an adaptive, interconnected dynamic design system for creating and implementing self-sustaining human settlements. Practitioners become active participants in the creation and evolution of the system, inhabiting and learning from it in dynamic and reflexive praxis and seek to lay the foundation for a new way of being on earth.

The success of permaculture can be seen in the Transition Movement, local community efforts to encourage self-sufficiency through sustainable food production and to ultimately mitigate the effects of a peak oil scenario and the after-shocks of economic insecurity and climate degradation (see Transition Network, https://transitionnetwork.org/about-the-movement/what-is-transition/principles-2/).

No one effort is sufficient to move toward sustainability and Collomb's "A Paradox of the Anthropocene: The Radicalization of Techno-Scientific Modernity and the Future of Solar Geoengineering" suggests that we ought not rule out high-tech solutions, like solar geoengineering which might provide another effort to relieve global warming.

Collomb reconsiders the viability of the modern technological dualism in which humans are regarded as separate from and controlling of nature, the very same worldview that produced our ecological crisis in the first place. Using the case of solar engineering, panned by critics as exaggerated hubris and proponents as a stop-gap measure that could temporarily compensate for the lack of enthusiasm for emissions reduction, Collomb investigates how these emerging technologies might persist alongside alternative conceptions of nature embracing transitions to sustainability, emissions reduction and a new economic order. Geoengineering is available to us as the result of the techno-scientific vision of nature which is, at the same time, the root of the problem;

so at first blush it does not recommend itself. There have been prominent critics, like James Lovelock, who have questioned our ability to keep the Earth in homeostasis. But there have been proponents like Paul Crutzen, who alerted us to the damage to the Ozone layer caused by the modern techno-ethos. The technique of cool-roofing is a benign measure that is readily available but the regional deployment of aerosol methods, for example, is fraught with political problems including the weaponization of the science. Above all, the techno-ethos of such environmental manipulation is the product of Western culture and it would be, quite rightly, read as yet another intrusion into undeveloped nations. And there would be no guarantee that the use of the science would foster a radical reorientation of our ways of inhabiting and interacting with the natural world. Collomb even raises the specter of corporate culture to double down in their quest to dominate and control nature. But there is also the likelihood, absent any other alternatives, of an alliance of the ingenuity of a renovated nonthreatening dualism, which acknowledges dependence on natural cycles and the stark reality of physical limits to economic development, with the autonomy of the evolutionary and ecological forces in a movement of mutual preservation.

Bibliography

Albrecht, Glenn A. *Earth Emotions, New Words for a New World.* New York: Cornell University Press, 2019.

Bookchin, Murray. *Toward an Ecological Society*. Montreal: Black Rose Books, 1980.

Frankopan, Peter. *The Earth Transformed, An Untold Story*. New York: Alfred A. Knopf, 2023.

Gilgamesh. Translated by Sophus Halle. New Haven: Yale University Press, 2021.

Leopold, Aldo. *A Sand County Almanac*. New York: Ballantine Books, 1966.

Marsh, George, Perkins. *Man And Nature*. Weyerhaeuser Environmental Classics, edited by David Lowenthal; foreword by William Cronon, Seattle: University of Washington Press, 2003.

McKibben, Bill. *The End of Nature*. New York: Random House, 1989.

Robbins, Paul. *Political Ecology*, 2nd edition. West Sussex, UK: John Wiley and Sons Ltd, 2012.

Ruddiman, William F. *Plows, Plagues & Petroleum, How Humans Took Control of Climate*. Princeton and Oxford: Princeton University Press, 2005.

Thich Nhat Hanh, "The Sun My Heart," *in Environmental Ethics, Divergence and Convergence*, edited by Susan J. Armstrong and Richard G. Botzler. New York: McGraw Hill, 2004, 255–258

Thoreau, Henry David. "Walking," in *Environmental Ethics, Divergence and Convergence*, edited by Susan J. Armstrong and Richard G. Botzler, New York: McGraw Hill, 2004, 121–128.

1 Nature in the Anthropocene

Manuel Arias-Maldonado

Introduction

On the face of it, to talk about 'nature in the Anthropocene' seems preposterous, as the phrase gathers two categories that do not *belong* together: whereas nature is a material reality, the Anthropocene is an idea or a concept. In particular, the latter is meant to designate the geological or historical period in which the accumulated impact of human activity on natural systems leads to the disruption of the Earth state and/or leaves a mark on the fossil record.[1] Alternatively, the Anthropocene may just refer to a state of socionatural relations—a particular moment in them—that features such pervasive impact. Be that as it may, the Anthropocene is less a fact than a particular conceptualisation of how humans have disrupted the functioning of the planet in which they live. On its part, nature is made up of causal structures, material entities and living beings. Needless to say, it is also a concept—that which encompasses those structures, entities, and beings. There is a difference, though: we might conclude that the separation between the Holocene and the Anthropocene is meaningless, and perhaps the notion of Anthropocene fails in capturing public imagination, so that we end up considering it an anecdote in intellectual history. And yet the same cannot be said of nature. Unless a strong constructivist position is adopted in which nothing exterior to human beings exists or can ever be known, nature is *out there* while the transition from the Holocene to the Anthropocene is a scientific convention that is not yet even acknowledged by all scientists.

At the same time, it makes no sense to argue that nature and the Anthropocene are at odds with each other. The Anthropocene, after all, is about nature—even if the concept usually refers to planet Earth, ecological systems, or the fossil record. But it is nature, understood as that which has not been created by human beings, that is at stake. Because it is nature, according to empirical data, that has changed. This means that nature is *changeable* in the first place—it changes on its own and it changes as a result of humans' (themselves a part of nature) interactions with it. The range and type of such interactions is vast, including both deliberate interventions *on* nature and changes *in* nature that can be deemed as side effects (both anticipated and unexpected) of social activity. Still, the

DOI: 10.4324/9781003385202-2

Anthropocene is concerned with anthropogenic change, not with the kind of natural change that happens without human intervention or influence. Once the Anthropocene is properly understood as a concept that has a strong empirical foundation and hence is not just a 'discourse' or an 'ideology', but rather an explanation of what has happened to that aspect of physical reality that we call 'nature', then the pairing of nature and the Anthropocene—the question of what happens to nature *in* the Anthropocene— makes perfect sense.

Then again, this chapter is not concerned with the description of how nature has changed in the Anthropocene—or to be more accurate, it does not intend to offer an account of the empirically observable disruptions and transformations that, either deliberately or indirectly, human beings and social activity have induced in natural systems, entities, landscapes, or beings. It is the amount and quality of such changes that justify the claim that the Holocene has ended, and the Anthropocene has begun. But detecting, measuring, and quantifying such changes is a task for natural scientists—the rest of us depend on their scientific methods and results. In that respect, the Anthropocene is a concept that incorporates a narrative. It says that the anthropogenic disruption of natural systems is significant enough to merit a new historical or geological periodization. Up to this century, though, such story had not been produced and thus the disruption of natural systems lacked a unifying narrative. As the conversation about the Anthropocene makes clear, in fact, this narrative has been contested by a few commentators for whom the reference to the *anthropos* should be suppressed, as it puts the blame on a universal human subject that does not exist. Instead, a few alternative formulations have been offered—from the Capitalocene to the Chtulucene or even the Narcisscene.[2] With the exception of the latter, which openly questions the human capability to disrupt the Earth, the others do not deny the validity of the descriptions that natural scientists provide, but rather dispute the *meaning* that is to be assigned to them—questions are raised about who has caused the Anthropocene or by which means, when did it start, or what should it be done about it.

In the same vein, I will deal in this chapter with the visions of nature that emerge from the conversation about the Anthropocene. This means that I will explore the different ways in which the core thesis of the Anthropocene concept—that human beings have become a geophysical force since their accumulated impact on natural systems has disrupted the state of the Earth—has been assimilated as far as nature and socionatural relations are concerned. To put it simply, how are nature and socionatural relations interpreted under the new light of the Anthropocene? What does the latter *say* about nature and the human relation to nature? Admittedly, the Anthropocene does not 'say' anything—we are the ones who talk. It follows that the Anthropocene is not interpreted unanimously as saying this or that, on the contrary there is several competing views about the subject. Still, they all react to the fact that the human impact on the natural world has hugely increased in the last three centuries, up to a point where even the climate system has changed because of social activity. That is what they have in common, even though this starting point can lead to very different conclusions.

The chapter is organised as follows. The second section presents the view that nature might have simply ended, insofar as none of it remains free of human influence and hence it makes no sense to talk about 'nature' anymore. The third section discusses the claim that it makes no sense to distinguish between humans and nature anymore, either because both human and non-human entities can be seen as part of the same neo-materialist ontology or because humans and nature are now so entangled with each other that the latter should be conceptualized as a hybrid. The third section takes note of how the geological dimension of the Anthropocene calls up a vast natural history in which human beings are a trivial anecdote, a viewpoint suggesting that there is a sense in which deep nature can never end. Finally, the last section argues that nature and socionatural relations in the Anthropocene are just too complex and multifaceted to be exhausted by a single explanation, so that a pluralistic conception of them is both realistic and advisable—provided that the latter remains firmly attached to the empirical foundation that the natural sciences provide.

The Anthropocene as the End of Nature

Insofar as the Anthropocene designates a state of socionatural relations in which anthropogenic influence on natural systems has become so pervasive as to be found everywhere, it lends new credit to the idea that nature has 'ended'. Actually, this is not a new idea—the Anthropocene would just seem to confirm its plausibility. But this proposition has to be properly understood, since taking it at face value would render it meaningless in view of the kind of seemingly 'natural' events that keep on affecting both individuals (diseases) and societies (floods or droughts). There is obviously a sense in which nature cannot possibly ever end, as much as there is another one in which it can be said to have already ended.

Dieter Biernbacher's distinction between nature in a *genetic* sense and nature in a *qualitative* sense is helpful in this context.[3] The former refers to nature as an entity that exists without human intervention, an essence that Kate Soper identifies with 'deep' nature: the casual powers and physical structures that operate in the world and with which human beings themselves must operate when trying to deliberately intervene in natural processes or beings.[4] On the contrary, there is nature in a qualitative sense, i.e., the external appearance of natural forms. This is Soper's 'shallow' nature, the nature of our everyday experience—material resources, animals, our bodies, landscapes, ecological processes, natural systems. Deep nature is *ontological*, whereas shallow nature is *historical* and *phenomenological*. The former provides the conditions for human interventions in the natural world, so that no matter how 'deep' they go, any transformation presupposes constancy of causal powers and physical structures.[5] Understood as a deep structure of causation that underpins existence, therefore, nature cannot end and the Anthropocene—the material changes it registers—does not change that.

And yet there is a sense in which to claim that nature *can* end makes perfect sense, and the same goes for the additional argument that the Anthropocene confirms the validity of such claim. It is not an ontological claim, but rather a historical one. It does not concern nature in isolation, as it existed for instance before the human species emerged and thrived on the planet, but socionatural *relations* and thus the effect of such relations *on* nature. It follows that the end of nature happens when natural processes and systems cease to be independent from human beings *and* they become influenced by human activity to a high degree. It is indifferent whether such influence is visible or not, and the same goes for the intentionality of it—the final result is what counts. In this regard, nature *was* autonomous from human beings and this separation provided the material basis for establishing an ontological and cultural opposition between humanity/society/culture and nature. The human liberation from necessity required in turn a liberation from nature's constraints. But the autonomy of nature cannot be taken for granted anymore—the history of socionatural relations has unfolded in a manner that actually supports the opposite claim.

As a matter of fact, the notion that nature has ended predates the formulation of the Anthropocene hypothesis. Environmental philosopher and activist Bill McKibben spoke of a 'post-natural world', while reflexive modernisation theorists Anthony Giddens and Ulrich Beck endorsed the idea that we now live in a 'created environment' where the antithesis between nature and society has become meaningless, and others lamented the 'end of the wild' to convey the loss of a formerly independent, untamed natural world.[6] The validity of this claim can be empirically tested, and that is why the Anthropocene can reinforce the proposition that nature has ended—the intensification of anthropogenic influence on the natural environment is at its core. Whether this historical transformation ends up changing how nature is understood at the ontological or metaphysical level remains to be seen. Clive Hamilton seems to be thinking ontologically when he argues that the Anthropocene represents a rupture that forces us to forget what we knew about nature, since the latter is no longer what it used to.[7] This seems to mean that nature has not disappeared but morphed—it is no longer autonomous from human beings.

On his part, biologist Earl Ellis has argued that a 'natural nature' does not exist anymore: human-dominated ecosystems now cover more of Earth's surface than 'wild' ecosystems, so that biomes (the unit of ecological analyses) should now be regarded as 'anthropogenic biomes'.[8] Natural biomes were never the norm, he cautions—meaningful anthropogenic influence on soils and landscapes are almost as old as the human species itself. Then again, the Anthropocene represents a further increase in the spread and intensity of such influence. Hence Ellis' conclusion: 'From a philosophical point of view, nature is now human nature; there is no more wild nature to be found, just ecosystems in different states of human interaction, differing in wildness and humanness'.[9] What 'the end of nature' means is thus clear—a nature that is not independent from society, a nature that can meaningfully change as a result of socionatural interactions. This entails the disruption of natural systems (climate change), an accelerated rate of species extinction and the resulting loss of biodiversity, the displacement of species to habitats different than their owns, an amazing increase in the sheer number of those animals we breed to feed ourselves, the presence of human debris all over the planet (traces of microplastics have been found in deep waters), as well as the massive removal and transportation of raw materials such as sand or concrete, the paving of vast amounts of terrain or the construction of all kinds of transport infrastructures.[10] Small wonder that there is a feeling of nostalgia for the wild nature of yore.

It should also be noted that deliberate attempts on the part of human beings to exert an influence on processes that belong to 'deep' nature are stretching the boundaries of the possible regarding the latter. Let us think of CRISPR, a technique that makes it possible to edit strands of DNA in individual cells, the genome of which can thus be manipulated. Medicine and agriculture can profit from this development, for instance, by producing crops resistant to disease or drought. Moreover, both genetic engineering and synthetic biology have been suggested as means for achieving environmental goals via 'de-extinction', which aims to restore lost species and ecosystem functions, or 'precision conservation', which improves the chance of survival of wild species through gene tweaking.[11] The relevant point here is that such techniques make possible the reengineering of primary teleological processes, so that they serve human ends.[12] The distinction between deep and shallow nature somehow dissolves at this point, and yet it would be going too far to claim that human beings are creating causal powers and structures *ex novo*—they are rather manipulating them in unprecedented ways. Resorting to Biernbacher's distinction, nature in a genetic sense is tweaked and touched upon, but neither created nor replaced, by human beings. Insofar as the Anthropocene refers to the anthropogenic impact on natural systems, this kind of intervention should not be overlooked, and they actually reinforce the case that an 'independent' nature does not exist anymore.

Then again, it remains true that nature *cannot* possibly end. The causal powers and deep structures upon which human life and social activity ultimately depend do not disappear just because they are increasingly touched upon or even disrupted by anthropogenic impacts and deliberate manipulations. At the same time, the enhanced powers of human beings do not allow them to 'control' nature, as the side effects of climate change come to show. Humans attempt to control their interactions with nature, rather than 'nature' itself, but they can only partially succeed. Valerie Plumwood speaks of those 'elements of independence' that attest to the irreducibility of nature.[13] The COVID-19 pandemic has been the last reminder of nature's autonomy, making it clear that socionatural relations just cannot be sanitized or cleansed altogether, as frictions and accidents cannot possibly be prevented.[14] It might then be concluded that the Anthropocene strengthens the pre-existing argument that nature as we knew it does not exist anymore and hence can be said to have 'ended' in a meaningful sense. Nature is no longer autonomous from society, except for the causal powers and material structures that underpin the physical reality that we have access to. The 'human epoch' is that in which nature cannot be defined anymore as that which is separate from society, especially since human manipulative powers are increasingly able to touch upon 'deep' nature. As climate change demonstrates, though, an anthropogenically influenced nature is not controlled by human beings and can thus threaten the stability or well-being of human societies.

Socionatural Hybridisation and Its Discontents

There is another vision of nature that emerges from, or is reinforced by, the rise of the Anthropocene: that of a socionatural hybrid that results from a long history of mutual entanglements between human beings and the nonhuman world. Nature has not 'ended', but rather morphed into a socionatural assemblage in which it becomes impossible to tell when society ends and nature beings or vice versa. In turn, this can be a novelty, the final outcome of a long process of hybridisation that takes place in history after the *homo faber* started to thrive. Or it can be interpreted as the intellectual collapse of the separation between society and nature, which was never 'real' but has been revealed as purely imaginary in the face of climate change and other phenomena of the Anthropocene. In the former case, there is a material process that changes the ontology—meaning how we see ontology. In the latter, the 'truth' about ontology is recognised thanks to that same material process.

Hybridisation is reflected in the social sciences, where concepts such as second nature, social nature, or hybrid nature have been used as an alternative to the all too simple 'nature'.[15] And it is understandable that the Anthropocene has been taken by many as a chance—even an injunction—to rethink Western intellectual categories that foster a separation between humans and nature.[16] If social and natural systems are 'coupled' in the Anthropocene as a result of the penetration of the former into the latter,[17] then Hamilton is right when he claims that 'social, cultural, and political orders are woven and co-evolve with techno-natural orders of specific matters and energy flow at a global level'.[18] The Anthropocene thus abound in hybrid phenomena, although the term 'hybrid' can be used in different ways when applied to the former.

On the one hand, there is the global warming caused by the liberation of CO_2 into the atmosphere throughout the industrial age. Human activity thus disrupts a planetary system, albeit unintentionally, the result being a socionatural phenomenon—sort of a hybrid—rather than a purely 'natural' one. The same applies to the 'anthropogenic biomes' referred to above, in which natural biological units have been penetrated by human influence. It is debatable whether this is a case of actual hybridisation, because nothing *new* has been produced as a result of the intermingling of the social and the natural. But other phenomena of the Anthropocene *do* produce hybrids. Take so-called 'plastiglomerates', a new kind of rock found in Hawaiian beaches that result from plastic melted in fires lit by humans who were camping or fishing and which are cobbled together from plastic, volcanic rock, beach sand, seashells, and corals.[19] On the other hand, human activity boost diversity in novel ways, for instance creating new anthropogenic habitats that contain species previously rare or absent or putting into contact formerly separated species, the overall picture being that of a process in which 'speciation by hybridization' takes place.[20] The notion that current 'nature' has hybrid qualities has implications for ecosystem management, as the existence of a 'baseline' to be kept or to be adopted as a goal is increasingly contested.[21]

The assumption that the separation between society and nature has collapsed in the Anthropocene—or else that the separation was a fiction from the beginning and what happens in the Anthropocene contributes to make it clearer—adopts a more radical expression in neo-materialist philosophy. New materialists claim that there is no such thing as a separation between society and nature, but an 'ecology of things' in which both humans and non-humans, but also human objects and inorganic matter, share the ability to transform the world.[22] Because the world is made up of dynamic assemblages of multiple entities of all kinds, which act and interact by their own volition or without any volition at all, all sharing the same ontology. The impact of Bruno Latour's notion of a 'flat ontology', as well as his distinction between human actors and non-human 'actants', can be felt.[23] As it happens, neo-materialists believe that even the smallest portion of matter can be related to a larger whole in which the reproduction of life through countless metamorphoses is more important than the barriers between beings.[24] The claim that nature constitutes a separated realm is rejected by neo-materialists as an anachronism. The COVID-19 pandemic would have also confirmed this worldview: society and nature are entangled in multiple and unexpected ways, the former being subjected to the possibility of suffering a severe disruption just because a virus has jumped from one species to the other in a remote location in the heart of China.

The neo-materialist conception of nature is, however, compatible with the view that nature has ended. On the face of it, there was never any 'nature' that stood separated from 'society' and thus the former cannot 'end' at all. But neo-materialist thinking allows for a distinction between *nature* and *matter*, the latter lying *below* nature or at least being contained *within* it. They are not identical—the natural world is a phenomenon of matter, an appearance under which all entities and beings share the same physical foundation. As long as the universe continues to exist, matter cannot 'end', but nature can—nature as perceived by human beings and nature as a historical entity subjected to the influence of social activity. Ontologically, it is matter that underlies everything, but a distinction between nature and society can be maintained. It is a distinction that evolves throughout history, as humans increase their powers and exert an ever-growing influence on natural systems, the end point of which so far is that what we call the 'Anthropocene'. In this regard, as we will discuss later, neo-materialist philosophy fails to account for the uneven distribution of agency among human actors and non-human actants: the former has proven to be much more influential than most of the latter.

In this vein, it is worthwhile to offer an account of ecomodernism's view of nature, since it seems to reveal some of the inevitable contradictions involved in the claim that nature has ended as a result of socionatural hybridisation. Ecomodernism's stance is significant because ecomodernism itself is a doctrine or a philosophy—it is unclear yet what it is or can become—that is attuned to the current state of socionatural relations and fully acknowledges the realities of the Anthropocene. Ecomodernists present us with an articulated view of socionatural relations, while also offering a number of prescriptions for securing their sustainability in the near future. According to the *Ecomodernist Manifesto*, the term 'nature' refers to landscapes, seascapes, biomes and ecosystems, some of which belong to what is often called 'wild nature'.[25] But nature has been regularly disrupted by human influence over millennia, up to the point at which nature and society are now hybridised and nature itself can be said to have 'ended'. In order to preserve nature, human societies must 'decouple', which means that they have to stop having an environmental impact or at least significantly diminish it despite continuing to grow. Modern technologies are the means through which this can be achieved, both using resources more effectively and leaving natural areas relatively untouched. At the same time, a conservation politics and a wilderness movement should exist that pushes in that same direction. Interestingly, they assume that humans could prosper in a planet with less biodiversity and wild nature, but 'this is not a world we want nor, if humans embrace decoupling processes, need to accept'.

The most radical proposal to this effect was made by the late Edward O. Wilson, who suggested that no less than half of the planet's surface should be liberated from human presence, so that biodiversity can recover, and a vast number of species may flourish within it.[26] On its part, the practice of 'rewilding' attempts to mitigate the negative impacts of humanly driven environmental change by restoring lost species or habitats.[27] In both cases, either by act or omission, human beings actively *intervene* in order to minimise the anthropogenic impacts on the natural environment or to help natural systems to regain their stability or habitats to recover their past integrity. What interests me here is to point out how inescapable the distinction between nature and society is, even when the end of their separation—be it on an ontological or a historical level—has been announced. As Boersma has argued, ecomodernism combines the ontological claim that nature has ended as a result of past human interventions with a normative call to intervene in that very nature.[28] And the same goes for hybridity: the claim that nature and society are now deeply entangled coexists with the plan to protect nature from society.

However, classical environmentalists have reacted against the notion of socionatural hybridisation, as they did against the claim that nature had ended when it first appeared—a reaction that is nonetheless compatible with the sad acknowledgement of the fact that wild nature has been tamed and diminished by human beings. Yet the conception of nature that is behind these claims does not represent a novelty. It is rather the traditional view of classical environmentalism, according to which the natural world constitutes a reality that is separated from human beings, even though human beings are a part of nature themselves. Nature is not created by humans and has a *telos*—it unfolds in its own disparate ways and the fact that it can be transformed or destroyed by human beings does not make it identical with them. In fact, the vocabulary of the Anthropocene—the talk of transformation of hybridisation—may serve to hide the fact that the natural world is colonized and destroyed by human beings.[29] As the dualistic separation between nature and society is discredited, the resulting lack of an ontologically thick natural reality and its replacement by an intrinsically disordered world allows a new urge towards the mastery of nature to gain ground.[30]

Interestingly, environmentalist thinkers used to see dualism as a cultural fabrication that facilitated human domination of nature by creating a distance between human beings and the natural world, thus alienating the former from the latter.[31] From this viewpoint, nature does not 'end'—it is killed by mechanistic philosophies and modern technologies.[32] The belief in the superiority of ontological monism over dualism, for instance on the part of neo-materialist thinkers, ends up being detrimental to the cause of environmentalism, so much so that a dualistic position must be recovered if there is to be any hope for establishing a harmonious socionatural relation in the future.[33] Moreover, the human inability to prevent the dangerous side effects of nature's colonisation—such as climate change and its widespread consequences—come to show precisely that society and nature are not identical: the latter cannot be appropriated by, nor be subsumed in, human power.[34] Thus the 'return of nature' on the scene of human history, in the form of environmental catastrophes or biological dysfunctionalities.[35]

Deep Time and Inhuman Nature in the Anthropocene

If the focus is kept on the ongoing disruption of the Earth system and on the anthropogenic transformation of habitats and landscapes, a key dimension of the Anthropocene may be neglected. I am referring to its geological aspect, which is essential to understand the cultural resonance of the concept, and which is directly related to the emergence of yet another view of nature and socionatural relations—one that, while not lacking historical precedents, has gained unprecedented strength in the context of the Anthropocene debate.

To say that the Anthropocene has a geological aspect is not the same as arguing that the validity of the Anthropocene hypothesis depends on its official recognition by the International Commission on Stratigraphy. This latter claim can fail for reasons that pertain to the inner workings of stratigraphy as a scientific discipline—the geological periodisation is not tampered with so easily. But the fact remains that the material realities of the Anthropocene are connected to the deep time of Earth's history, thus making it clear that there are temporalities—those of the Earth itself, of biological life, of the climate system—that are much vaster than the human species, as they existed before the latter emerged and which will continue to exist after we have gone extinct. At the same time, though, the whole Anthropocene idea rests upon the claim that humanity *has* significantly changed 'shallow nature' and disrupted the state of the Earth. Humans have become, albeit unintentionally, a geophysical power.

By emitting greenhouse gases, humans did not know that they were delaying the next glaciation and changing the planet's climate for hundreds of thousands of years, thus having an impact that can be compared to that of fluctuations in Earth's orbit.[36] In that sense, humanity becomes 'colossal'.[37] At the same time, though, humans are forced to deal with phenomena whose sheer scale underlines their insignificance: climate change, waste production, biodiversity loss. Moreover, the Anthropocene establishes a link between the current moment and the long history of the Earth, thus contextualising the life of our species in a new manner. The planet seems to have been formed 4,543 billion years ago and the so-called Cambrian explosion that boosted life took place around 542 million years ago, but humanity appears much later and much recently—around 11:48 PM in the last day of December if that whole history could be compressed in the span of a human year, according to Carl Sagan's calculations. Needless to say, the powerful human agency that has 'created' the Anthropocene seems much weaker if deep time is considered. It turns out that humans are not just biological beings, but also inhabitants of a planet in which they depend on a complex set of organic and inorganic entities and processes.[38] Furthermore, many telluric phenomena can severely threaten the well-being or the life conditions of human societies—from asteroids to earthquakes, hurricanes, floods, droughts, viruses. In neo-materialist jargon, the human actor can be helpless when confronted with these nonhuman actants. Following Stengers, we are witnessing the 'intrusion of Gaia'.[39]

The dark side of the Anthropocene, that becomes apparent as soon as its geological aspect is considered, does not only relate to the deep history of the planet. It also concerns the latter's deep future, the speculations about which are key to understand how a different conception of nature and socionatural relations can emerge from the reflection upon the geological aspect of the Anthropocene. It should be reminded that not even global warming will last forever—a day will come in which the glaciers will advance, and the world will shiver.[40] But a new ice age is just an intermediate step in the march towards CO_2 starvation and the disappearance of the biosphere.[41] Later on, tectonic violence will erupt, and the continents will be torn apart and reunited and torn apart again.[42] There will be, as there was once, a 'world without us'.[43] In fact, our extinction will be unremarkable: countless other animal species were wiped out by some dramatic event in the course of past mass extinctions. Their occurrence shows how fragile the Earth can be, since every time that it was pushed outside a set of surface conditions, it became nearly sterilized.[44] Hence Sagoff's warning that we should not believe too much in the Anthropocene, which he renames 'Narcisscene', lest we experience the delusion that we can *really* disrupt the planet.[45] If the Earth has the capacity to shift from one state to another, this capacity is independent of human existence.[46] In turn, this does not prevent the possibility that *this time* it is the *anthropos* that has altered the Earth state and hence the transition from the Holocene to the Anthropocene. A deep time perspective provides, however, a view of the Earth that underscores the variability of its conditions and hence the fragility of the planet as a stable home for the human species. Telluric and cosmic events are manifestations of an 'inhuman nature' to which humans are exposed, so that we are 'inherently liable to being thrown off course by the eventualities of our planet'.[47] Hence the notion of a 'defiant Earth' that challenges our sense of stability by exposing the planet as a potentially dangerous—even deadly—habitat.[48] The Anthropocene is thus as a reminder of the planet's essential otherness, and hence of the dangerous aspect of a natural world that is oblivious to human interests. Admittedly, this is not the shallow nature that appears to us in everyday life, but a potentially devastating nature that manifests itself from time to time and threatens to destabilise human societies. But the Anthropocene also reminds us that there has been, and there will be, times in which the *ordinary* state of the Earth is itself inimical to human beings. The vastness of the natural world becomes apparent, as does its lack of morality. At the same time, a geological reading of the Anthropocene makes it clear that human beings are a passing form of nature, a transient inhabitant of a planet that may have been changed by us *this time*, but which has changed without us and will do so again in sometime in the future.Rather than an articulated conception of nature, geology provides a view of nature *and* socionatural relations that revolves around its 'inhumanity', i.e., a lack of attunement with human beings and their well-being that is revealed in deep time, in extreme telluric phenomena, and in the certainty that humanity will perish long before the Earth itself ends. What normative injunctions derive from this view, if any, remains to be discussed. However, this 'inhuman nature' is similar to neomaterialism's 'matter' in that both exhibit a remarkable autonomy from human beings—it seems to operate as if they, us, did not exist. It follows that from a geological perspective nature cannot 'end', nor can it be described as a socionatural 'hybrid'. As the next and last section will suggest, however, these two perspectives are not incompatible.

Conclusion: Towards a Pluralistic Conception of Nature for the Anthropocene

The concept of the Anthropocene attempts to make sense of a material reality that can be empirically observed. It suggests that human beings have had such an impact on their natural environment that the state of the Earth is shifting—abandoning the relatively stable Holocene—and that such anthropogenic impact will be discernible in the planet's fossil record. The global environmental change that human beings have brought about can be measured: loss of biodiversity, species invasion, landscape transformation, mining activity, waste production, infrastructure building, changes in patterns of biomass distribution, and so on.

The Anthropocene hypothesis is thus grounded on three different, albeit interrelated, material phenomena: the anthropogenical disruption of planetary systems; the manifold and intense transformation of the natural world at all levels; and the geological scale of such changes, as they leave their imprint on the fossil record. They are all subject to discussion—that is how science works. Yet the claim that a global environmental change caused by human activity is going on seems warranted. And the Anthropocene is a concept that aptly encapsulates such change. What this chapter has in turn attempted to elucidate is whether, and how, this notion affects the way in which nature is conceived of.

To such end, I have outlined a number of responses that can be summarised in three positions: (i) the Anthropocene confirms the end of nature as an entity that is autonomous from human beings; (ii) the Anthropocene is the culmination of a long process of hybridisation that collapses the traditional separation between society and nature; and (iii) the Anthropocene reveals the inhumanity of nature, as it establishes a connection between the current disruption of natural systems and the deep time of the planet, that in which human beings are meaningless and extreme telluric events abound.

None of them must be tied to a single interpretation, as they are all grounded on different aspects of the material reality of socionatural relations. As I have pointed out above, in fact, some critics of the Anthropocene claim that the latter is but another name for the accelerated destruction of the natural world in modern times, thus suggesting that nature should be seen as an autonomous entity in possession of its own *telos*. It can be colonised or destroyed by human beings, but it cannot be neither mastered nor controlled by them—it is non-identical to us, and this radical otherness will manifest itself sooner or later. Such otherness is implicit in the geological side of the Anthropocene, which is however compatible with 'the narrative of interdependence' between society and nature.[49] Blurring the boundaries between the human and the natural, a point on which posthuman thinkers insist, questions the framework of reference for humanity's conception of itself.[50] But even if nature is said to have 'ended', its disappearance is not complete: it only means that nature has ceased to operate independently from human beings, as it is now (mostly) part of a socionatural entanglement in which it is often hard to say when one begins and the other ends.

Likewise, neo-materialist thinkers do not even recognise the validity of the distinction between humans and nature, but in the end they are forced to talk of humans and nature. For one thing, there is no way to make normative claims if the exceptionality of human beings as actors that think more profoundly about their situation than other species is not acknowledged.[51] For another, the dualism nature-society does not disappear just because the socionatural entanglement is empirically observed. Albeit philosophically weakened, this distinction is methodologically and strategically required for making sense of socionatural relations themselves. Hence Tynan's claim that we should see it as an aporia that cannot ever be resolved.[52]

The rise of the Anthropocene thus suggests the need to avoid conceptual reductionism as far as nature is concerned, because even the most bombastic announcements—the end of nature, for once—are actually nuanced enough to allow for a degree of compatibility with apparently rival conceptions. As much as the narrative of the Anthropocene itself is dynamic and involves a plurality of meanings and tensions,[53] the same applies to the conception of nature that goes with it. Ideally, they should be grounded on the contributions of natural science: while the Anthropocene narrative can be deemed as 'ideology',[54] the data on which the narrative is based must be confronted. Rather than sanctifying the 'objective truths' of science, the judgements made by science are to be appraised in a reasoned way, thus contributing to a more mature understanding of Anthropocene science.[55]

In this regard, how natural scientists frame their object can be a matter of contention. The view of the Earth as a system that is in turn made up of different subsystems—such as the climate—encourages a view of nature as a holistic and integrated entity that has been rejected, for instance by chemist James Lovelock and sociologist Bruno Latour, on the grounds that the Earth is rather an anti-system, i.e., a fragile bundle of agents and processes that are always on the verge of losing their equilibrium.[56] Whether this controversy can be easily solved, is a different story, but seeing nature as a system is very different from seeing it as chaotic, especially if we are looking at ways to ensure that socionatural relations are sustainable into the future and the Earth remains unhabitable for our species.

In sum, how does the Anthropocene impact our conceptions of nature? What is the novelty that it provides? How would a synthesis of the positions described above look like?

Socionatural relations have featured from the start a tendency towards human colonisation and hence also towards socionatural hybridisation, a reflection of the fact that the human species adapt to its environment in a transformative, aggressive manner. Since the industrial revolution, the human impact on natural processes, habitats, forms, and beings have become much stronger. Nature cannot be understood any longer as a reality that operates independently from human beings and social activity, even though extreme telluric phenomena take place without us causing them and the geological aspect of the Anthropocene is a reminder of the insignificance of the human species in the deep time of the planet. However, the recognition of this latter fact is compatible with the assertion that the species has become a geophysical actor in the last few centuries. Then again, humans are not the only relevant actors: things other than humans have impact on the unfolding of social relations and thus also in the relations between society and nature—a distinction that must be kept for analytical and normative reasons. However, the primacy of human agency should be acknowledged, since that is the core of the Anthropocene narrative: we have contributed to a significant disruption of natural systems, while transforming landscapes and pushing countless species towards extinction. The power to drive global environmental change also means that humans have a heightened responsibility in keeping the planet inhabitable and preserving wild nature.

Notes

1 See Steffen, Will et al. "The Anthropocene. From Global Change to Planetary Stewardship". *Ambio* 40, no. 7 (2011): 739–761. doi: 10.1007/s13280-011-0185-x; Zalasiewicz, Jan, "Science: Old and New Patterns of the Anthropocene", in *Altered Earth. Getting the Anthropocene Right* (Cambridge: Cambridge University Press, 2022).
2 Moore, Jason, "The Capitalocene Part I: On the Nature and Origins of Our Ecological Crisis", *Journal of Peasant Studies*, 44, no. 3 (2014): 594–630. doi: 10.1080/03066150.2016.1235036; Haraway, Donna, *Staying with the Trouble. Making Kin in the Chthulucene* (Durham and London: Duke University Press, 2016); Sagoff, Mark, "Welcome to the Narcisscene", *The Breakthrough Journal*, 9 (2018).
3 Birnbacher, Dieter, *Natürlichkeit* (Berlin: Walter de Gruyter, 2006).
4 Soper, Kate, *What is Nature?* (Oxford: Blackwell, 1995).
5 Benton, Ted, "Ecology, Socialism and the Mastery of Nature: A Reply to Reiner Grundmann". *New Left Review*, 194 (1992): 55–74, p. 66.
6 McKibben, Bill, *The End of Nature* (New York: Anchor Books, 1990); Giddens, Anthony, *The Consequences of Modernity* (Cambridge: Polity, 1991); Beck, Ulrich, *Risk Society. Towards a New Modernity* (London: Sage, 1992); Meyer, Stephen, *The End of the Wild* (Cambridge: MIT Press, 2006).
7 Hamilton, Clive, *Defiant Earth. The Fate of Humans in the Anthropocene* (Cambridge: Polity, 2017).
8 Ellis, Erle, "Anthropogenic transformation of the terrestrial biosphere". *Philosophical Transactions of the Royal Society A*, 369 (2011): 1010–1035.
9 Ellis, Erle, "Anthropogenic transformation of the terrestrial biosphere". *Philosophical Transactions of the Royal Society A*, 369 (2011): 1027.
10 See Zalasiewicz, *Science*.
11 See Adams, William, "Geographies of conservation I: De-extinction and precision conservation". *Progress in Human Geography* 41, no. 4 (2016): 534–545. doi: 10.1177/0309132516646641.
12 Trachtenberg, Zev, "CRISPR as Niche Construction: An Aristotelian View", *Inhabiting the Anthropocene*, no. 19, October (2016). Available at: https://inhabitingtheanthropocene.com/2016/10/19/crispr-as-niche-construction-an-aristotelian-view/.
13 Plumwood, Valerie, "The Concept of a cultural landscape: nature, culture and agency in the land", *Ethics & The Environment*, 11, no. 2 (2016): 115–150, p. 135. doi: 10.2979/ete.2006.11.2.115.
14 See Arias-Maldonado, Manuel, "What's in a Pandemic? Covid-19 and the Anthropocene". *Environmental Values*, 32, no. 1 (2023): 45–63. doi: 10.3197/096327122X16452897197793.
15 Pollini, Jacques, "Bruno Latour and the Ontological Dissolution of Nature in the Social Sciences: A Critical Review", *Environmental Values*, 22, no. 1 (2013): 25–42, p. 30. doi: 10.3197/096327113X13528328798 1.
16 Castree, Noel, "Geography and the Anthropocene II: Current Contributions". *Geography Compass*, 8, no. 7 (2014): 450–463. doi: 10.1111/gec3.12140.
17 Liu, Jianguo et al., "Complexity of Coupled Human and Natural Systems", *Science*, 317, no. 5844 (2007): 1513–1516.
18 Hamilton, Clive, Bonneuil, Christophe, Gemenne, Françoise, "Thinking the Anthropocene", in *The Anthropocene and the Global Environmental Crisis. Rethinking Modernity in a New Epoch* (Abingdon: Routledge, 2015), 4.
19 See Corcoran, Patricia, Moore, Charles, Jazvac, Kelly, "An anthropogenic marker horizon in the future rock record". *Geological Society of America Today*, 24, no. 6 (2014): 4–8. doi: 10.1130/GSAT-G198A.1.

20 See Thomas, Chris, "The Anthropocene could raise biological diversity", *Nature*, 502, no. 7469 (2013): 7. doi: 10.1038/502007a.
21 See Marris, Emma, *Rambunctious Garden* (New York: Bloomsbury, 2013).
22 Bennett, Jane, *Vibrant matter: A political ecology of things* (Durham: Duke University Press, 2010).
23 See Latour, Bruno, *We have never been modern* (Cambridge: Harvard University Press, 1993).
24 Coccia, Emmanuel, *Metamorphoses* (Cambridge: Polity, 2021).
25 Asafu-Adjaye, John, et al., *An Ecomodernist Manifesto* (2015). Available at: http://www.ecomodernism.org/.
26 Wilson, Edward. *Half-Earth. Our Planet's Fight for Life* (New York: Liveright, 2016).
27 See Lewis, Simon and Maslin, Mark, *The Human Planet. How We Created the Anthropocene* (London: Penguin, 2018).
28 Boersma, Keje, "The Anthropocene as the End of Nature? Why Recognizing Interventionism is Key in Coming to Terms with the Anthropocene". *Environmental Ethics*, 44, no. 3 (2022): 195–219. doi: 10.5840/enviroethics202292643.
29 See Fremaux, Anne & Barry, John, "The 'Good Anthropocene' and Green Political Theory: Rethinking Environmentalism, Resisting Ecomodernism", in *Anthropocene Encounters. New Directions in Green Political Thinking* (Cambridge: Cambridge University Press, 2019).
30 See Pellizzoni, Luigi, *Ontological Politics in a Disposable World. The New Mastery of Nature* (Surrey: Ashgate, 2015).
31 Plumwood, Valerie, *Feminism and the Mastery of Nature* (London: Routledge, 1993).
32 See Vetlesen, Arne, *The Denial of Nature: Environmental Philosophy in the Era of Global Capitalism* (London: Routledge, 2015).
33 See Saito, Kohei, *Marx in the Anthropocene: Towards the Idea of Degrowth Communism* (Cambridge: Cambridge University Press, 2022).
34 See Fremaux, Anne, "The return of nature in the Capitalocene: a critique of the ecomodernist version of the 'Good Anthropocene'", in *Rethinking the Environment for the Anthropocene. Political Theory and Socionatural Relations in the New Geological Epoch* (London: Routledge, 2019), 20.
35 See Fremaux, Anne, "The return of nature in the Capitalocene: a critique of the ecomodernist version of the 'Good Anthropocene'", in *Rethinking the Environment for the Anthropocene. Political Theory and Socionatural Relations in the New Geological Epoch* (London: Routledge, 2019), 29.
36 Chakrabarty Dipesh, "Anthropocene Time", *History & Teory* 57, no. 1 (2018): 5–32, 7. doi: 10.1111/hith.12044.
37 Raffnsoe, Sverre. *Philosophy of the Anthropocene: The Human Turn* (Palgrave Macmillan, Nueva York, 2016).
38 See Clark, Nigel, *Inhuman nature* (London: Sage, 2011); Yusoff, Kathryn, "Anthropocene: Geographies of Geopower", in *Handbook of Geographies of Power* (Cheltenham: Elgar, 2018).
39 Stengers, Isabelle, "Autonomy and the Intrusion of Gaia", *South Atlantic Quarterly* 116, no. 2 (2017): 381–400. doi: 10.1215/00382876-3829467.
40 Brannen, Peter, *The Ends of the World* (New York: HarperCollins, 2017).
41 See Scherbaum, Frank, Bounama, Christine, Von Bloh, Werner, "Causes and Timing of Future Biosphere Extinctions", *Biogeosciences*, 3 (2006): 85–92. doi: 10.5194/bg-3-85-2006.
42 Ord, Toby, *The Precipice: Existential Risk and the Future of Humanity* (London: Bloomsbury, 2020).
43 Weisman, Alan, *The World Without Us* (London: Picador, 2007).
44 Zalasiewicz, Jan & Waters, Colin, "The Anthropocene", in *Oxford Research Encyclopedia of Environmental Science*, 2015, 16. DOI: 10.1093/acrefore/9780199389414.013.7.

45 Sagoff, "Welcome to the Narcisscene".
46 Clark Nigel & Szerszynski, Bronislaw, *Planetary Social Thought. The Anthropocene Challenge to the Social Sciences* (Cambridge and Medford: Polity, 2021), 20.
47 Clark, *Inhuman nature*, xiv.
48 Hamilton, *Defiant Earth.*
49 Dürbeck, Gabriele, "Narrative des Anthropozän—Systematisierung eines interdisziplinären Diskurses". *Kulturwissenschaftliche Zeitschrift* 2, no. 1 (2018): 1–20.
50 Müller, Oliver, *Selbst, Welt und Technik. Eine anthropologische, geistesgeschichtliche und ethische Untersuchung* (Berlin and New York, NY: De Gruyter, 2014).
51 Connolly, William, *The Fragility of Things: Self-Organizing Processes, Neoliberal Fantasies, and Democratic Activism* (Durham: Duke University Press., 2013).
52 Tynan, Aidan, "Aporias of Ecological Thought". *Theory, Culture & Society*, 39, no. 5 (2021): 3–21. doi: 10.1177/02632764211039.
53 Lidskog, Rolf & Waterton, Claire, "The Anthropocene: A Narrative in the Making", in *Environment and Society: Concepts and Challenges* (Basingstoke: Palgrave, 2018), 31.
54 Baskin, Jeffrey, "Paradigm Dressed as Epoch: The Ideology of the Anthropocene". *Environmental Values*, 24 (2015): 9–29. doi: 10.3197/096327115X14183182353746.
55 Castree, Noel, "The 'Anthropocene' in Global Change Science: Expertise, the Earth, and the future of Humanity", in *Anthropocene Encounters. New Directions in Green Political Thinking* (Cambridge: Cambridge University Press, 2019), 29.
56 See Höfele, Philipp, Müller, Oliver, Hühn, Lore, "Introduction: The role of nature in the Anthropocene – Defining and reacting to a new geological epoch". *The Anthropocene Review*, 9, no. 2 (2022): 129–138. doi: 10.1177/20530196221110.

2 On the Severance of Production from Reproduction

Simone de Beauvoir and Ecofeminist Critical Theory

J.M. Bernstein

> Violence and calculated competition pervade the global economy, a system in which the living habitat is sacrificed for a few; and so called 'minorities' of gender, race, and class, are unfairly used. However, sex/gender is the hardest of these discriminations to resolve, because it has the longest history.
>
> (Ariel Salleh)

Ecofeminists departed from classical Marxist theory in proposing that the primary contradiction of capitalism is its severance of economic production for the sake of profit and increasing wealth from the material conditions, natural and social, for reproduction. Capitalist wealth creation occurs through the blind destruction of the ecological integrity of living nature, and by means of the often-violent exploitation of those responsible for providing the means of natural and social reproduction – colonized bodies, racialized bodies, female-gendered bodies, especially as those bodies are located in the Global South. When Ariel Salleh states that sex/gender domination is the most difficult to resolve, it is not the simple fact of its being the oldest of the forms of domination that she is considering. As she states in the sentences directly following that claim, "Women's subjection is supported by an old belief that they are 'closer to nature' than men. No coincidence that in German, the word for mother, *Mutter*, is linguistically tied to words for mud, sludge, and swamp."[1] Salleh's extended thesis is that the formation of domination and dualism that is most acute and irresolvable is sex/gender domination because it justifies this domination by determining that women are 'closer to nature' than men because they have a primary and non-avoidable responsibility for bearing the burden of reproducing species life in society; simultaneously, as a corollary of the work of reproducing species life, women have been assigned the role of securing the "meta-industrial labor" of provisioning "intergenerational sustainability," that is, securing the conditions of social reproduction by doing the care work necessary for it.[2]

Global warming and the series of environmental crises that have already arisen from it – including unprecedented heat waves, forest fires, glacial melting, sea level rising, flooding, and increased severe storm activity to ecological

DOI: 10.4324/9781003385202-3

breakdown and mass species extinction – entails that capitalist production is destroying its own material conditions of possibility together with the material conditions for social reproduction generally. Hence, *the primary contradiction of capitalism* is now the severance of the conditions for wealth creation through industrial-financial production from the conditions of ecological, species and social reproduction. Because that severance secures the conditions for destruction and violent exploitation of the ecological integrity of living nature as well as the expropriation and exploitation of the legion of female and feminized racialized, colonized bodies carrying out the labor reproduction, what Stefania Barca aptly terms the combined natural and human "forces of reproduction,"[3] then late capitalism can be portrayed as patriarchal domination becoming a self-standing economic formation. In patriarchal capitalism the exploitation and destruction of the forces of reproduction displace the exploitation of wage labor as the primary contradiction. Free market capitalism always had practices of the exploitation of living nature as well as patriarchal domination as a condition of possibility; living nature and domesticated or feminized reproductive labor provided virtually free background resources that permitted the focal work of industrial production for profit to develop. In late-twenty-first-century capitalism, what was the background condition for industrial production has become the *critical* foreground: the crises of climate change and the collapse of Holocene nature into the Anthropocene make the orienting structure of capitalism appear as a structure of patriarchal domination, the patriarchal power that produced the outsized exploitative expansion of the capitalist economy has become the primary profile, the central contradiction of capitalism as living nature quakes with climate turmoil, disasters and species extinction, while the suffering of those responsible for reproductive labor increases while that suffering becomes increasingly unnecessary and unjustifiable. In broad terms, this is the defining structural thesis of ecofeminism; call it "the patriarchal economy thesis," or simply "patriarchal capitalism."[4] It follows from the patriarchal capitalism thesis that we cannot resolve the problem of the relation of humanity to living nature, that is, the very idea of developing a post-capitalist form of life, without solving the problem of sex/gender domination, where a comprehension of what is necessary for the overcoming of sex/gender domination may provide an essential clue to the overcoming of liberal capitalism by limning what it might *mean* to forge a post-liberal-capitalist form of life; and conversely, only in a corresponding post-liberal capitalist form of life can gender domination be definitively overcome.

This argument assumes that the fully articulated sex/gender domination we consider patriarchal domination is, understood aright, a precise anticipation of late capitalism in its radical severance of production from reproduction – wealth-creating production continuing now solely and visibly through the evisceration and destruction of the forces of reproduction. Although not yet fully theorized, this is the constitutive theorem of self-conscious ecofeminism as it first emerged in 1974 in Françoise D'Eaubonne's *Feminism and Death*. She argues that ecofeminism is a "new humanism" premised on the thesis that the threat of planetary death has been caused by "male power that brought about first agricultural overexploitation and then lethal industrial expansion."[5] The plausibility of ecofeminism's patriarchal economy thesis turns on two subtheses: first, that the correct analysis of sex/gender domination comes to have the constitutive structure of a severance between male production for profit and female-gendered life reproduction;[6] and second, that in fact the essential contradiction of present-day late capitalism is that the conditions of capitalist wealth-creating production undermine and destroy its human and natural conditions of reproduction.[7] In what I want to argue is an emphatic anticipation of ecofeminism, the first thesis is worked out in fierce and compelling detail in Simone de Beauvoir's *The Second Sex*. Beauvoir's entire theorization of patriarchal domination and the depositing of woman as the absolute Other turns on the idea of a severance of male production-work-creation from female reproduction-labor-repetition, what Beauvoir terms the categorical division between transcendence and immanence.[8] What gives Beauvoir's account of sex/gender domination its acute power is its structuring of the domination of women by men through Hegel's master-slave theory of domination. Beauvoir argues that sex/gender domination is the most acute and intransigent structuration of all master-slave relations *because* it concerns the necessity for reconciling self-*determining humanity with humanity as irrevocably but a living part of living nature*.[9] This way of conceiving of sex/gender domination explains why the patriarchal economy thesis is anticipated in the pre-history of capitalism, and why late capitalism is the most fully articulated form of patriarchal domination. Overcoming sex/gender domination and overcoming the separation and domination of nature by capitalist production are component parts of the same project.

Defending Beauvoir's thesis that sex/gender domination is the exemplary form of social domination because the exemplary form of the master-slave relation involves demonstrating that Hegel's master-slave argument is only fulfilled in Hegel's own austere terms by sex-gender domination. My argument will thus have the following structure: First, an account of Hegel's analysis of the master-slave relation as having a dual structure that concerns both some humans dominating others and that domination being a mechanism for the domination of the human over and against nature-life. Second, an account of Beauvoir's three part development of the master-slave analytic: a genealogy of patriarchal domination out of a prior more-or-less spontaneous and broadly egalitarian sexual division of social labor that replaces Hegel's master-consumption versus slave-production analysis; then, a theorization of that structure of domination through the categorical division between immanence and transcendence; and finally, an existential phenomenological depiction of the lives of modern women as, precisely, women bearing the burden of species reproduction versus male value creative individualism. Beauvoir's account of women's lives depicts the experience of sex/gender domination in an industrially advanced society that nonetheless positions women as responsible for species life and social reproduction.

If the argument here is valid it should ground, first, the thesis that Beauvoir's critical theory provides the conception of patriarchal domination necessary for ecofeminism;[10] second, that Beauvoir's critical feminism is incomplete without ecofeminism; and third, the first two theses together provide a necessary first step in defending the ecofeminist thesis of patriarchal capitalism, which is to say, political economy must now give way to political ecology.

Hegel on Self-Consciousness: The Self as Other, Nature as Other

Beauvoir's deployment of the logic of master-slave as the structure for understanding gender relations and domination draws upon its philosophical adoption and development by Hegel.[11] The purpose of Hegel's account of master and slave in *The Phenomenology of Spirit* is to demonstrate that human self-consciousness, our self-awareness of ourselves as self-determining beings living in the midst of living nature; hence, our most intimate and inward sense of ourselves as somehow self-sufficient beings is, in fact, mediated through our bodily mediated relation to living nature and each one's relation to other self-consciousnesses: to be is to be alive; to be is to be recognized. We are essentially living social beings or, in Hegel's austere sense, spiritual beings. "*A self-consciousness*," Hegel argues, "is *for a self-consciousness*. Only thereby is there in fact self-consciousness, for it is only therein that the unity of itself in its otherness comes to be for it ... [Spirit is:] The *I* that is *we* and the *we* that is *I*."[12] On Hegel's plangent reading, the human "I," at least in the West, has been driven to repudiate its double dependency-structure: I repudiates its dependence on other subjects, other I-sayers, and hence on society in general: *I* opposes *you* and *we*; *and* I, as desire or mind or will or reason or soul or spirit, opposes its bodily constitution as a living part of living and dying nature. First, then, the controlling conceptual work of Hegel's argument concerns the desire of self-consciousness (to this very day) to be free, self-determining, and authoritative, possessing full self-sufficiency, as opposed to being, initially, an anonymous presence, thing among things; but later, as opposed to being unfree, determined by the will of another, and hence being non-self-sufficient, being radically and irrevocably dependent. It is the experience of absolute dependence that generates the overwhelming desire to be a self-sufficient master rather than a slave, where being a slave is the intransigent model for what it is for a human being to be a dependent being. Second, however, a continuing coordinated theme in the master-slave narrative is the idea and ideal of human independence as involving humanity freeing itself from dependence on living nature, on life: self-consciousness is freedom. While being dependent on others initially appears as the core meaning of non-self-sufficiency, in fact, almost everywhere in Hegel's argument, dependence on others is coordinated if not utterly entangled with dependency on life and living nature – *the double-dependency structure of the human* (my term) whereby each self is mediated through its relations to other selves and mediated through its bodily placement in living nature.

Both the form of consciousness Hegel calls "self-certainty," the primitive pre-formation of self-consciousness, and the drive of self-consciousness to demonstrate its self-sufficiency concern independence from life. In the case of self-certainty, its attempt to demonstrate that its desire is authoritative for the meaning of what is as, broadly, things to be used or consumed by self-consciousness that would provide satisfaction, comes to ruin because the experience of satisfaction turns into its opposite, into the experience of the self-sufficiency of the object of desire: both the desire and its satisfaction depend on the object to be consumed, the delicious apple; and that satisfaction is only ever temporary, the satisfaction in fact "re-engenders the object as well as the desire";[13] satisfaction bequeaths dissatisfaction and need for the object. Human desire cannot find full or authoritative or final satisfaction of its standing in relation to living nature.[14] This lesson that self-consciousness is dependent on living nature, on its bodily living, is one it learns, forgets, denies, represses, and must relearn throughout the *Phenomenology*. Life will rankle and ultimately ruin self-consciousness's claim to self-sufficiency.

Self-certain self-consciousness avoids this lesson, taking itself simply to not have to look in the wrong direction for true satisfaction. It comes to believe that it can attain its satisfaction of being authoritative only in another self-consciousness, that is, through that object that has the power to negate itself in the face of individual self-consciousness's claim to sovereign authority. But here too, *sotto voce*, the relation to the other runs through the relation to life. In this stretch of narrative, self-consciousness begins as surrounded by others, as being "*self-sufficient* shapes absorbed with the *being* of *life*," that is, as being a desiring animal amongst other like desiring animals, each pursuing their sovereign self-preservation. Hence, in order to demonstrate itself as a true self-sufficiency, an authoritative being for itself, it must exhibit itself "as the pure abstraction … the pure negation of its objective mode … in showing that it is fettered to no determinate existence, that it is not at all bound to the universal singularity of existence, that it is not shackled to life."[15] In this now familiar narrative, two contending self-consciousnesses must undergo a "life and death struggle [in order that] that each *proves its worth* to itself, and that both *prove their worth* to each other."[16] Each self-consciousness must attempt to demonstrate to itself and to its other that life is not the highest good; that higher than life is its capacity *to set a value on life*, to have a self-worth that is higher than life. Nonetheless, this means for achieving validation of intrinsic self-worth falters: there can be no fully satisfying resolution to the life and death struggle because in order for one to prove their worth to another, the other must remain alive to provide recognition of worth beyond life. Hence, the primary and unsurpassable lesson "self-consciousness learns" through the life and death struggle for recognition of worth is "that life is as essential to it as is pure self-consciousness."[17] Because Hegel argues that rational human life is free and self-determining, this passage is routinely overlooked: life is *as essential* to self-consciousness as its being a pure self-consciousness. Both the future master and the future slave must recognize the necessity of life for self-consciousness:

the master lets the slave live to provide him with recognition as master, and the slave surrenders in order to live to fight another day.

In a fundamental way, the continuation of the master-slave narrative is equally consequential for comprehending how both the mediation of self-consciousness through the other and the mediation of self-consciousness through its relation to life are equal components of the structure of human constitutive dependence. At the level of self-consciousness itself, being a master depends on being recognized as having the status and standing of being a master by the slave, while each slave recognizes himself as a slave because he acknowledges the authority of the master's will. Each self-consciousness is mediated through the other, albeit unequally and hence dissatisfyingly. At the level of life, the slave's being allowed to live is granted by the master; this makes vivid a level of universal dependence: each of us is allowed to live because, finally, the other grants us the standing or right to live. There is no self-consciousness without life, and no life without the other(s) granting life: life is finally the gift of the other. That dependence of individual life on the other is suppressed until the conclusion of *The Phenomenology*. The ongoing misrecognition of the necessity of life and other is the point of Hegel's genealogical narrative of the emergence of modern sovereign self-consciousness. Recall, the master entered into the life and death struggle because unequivocal satisfaction of the claim to the sovereign authority of self-consciousness could not be found in relation to the natural world. The master lets the slave live in order that the slave provide a mediation between itself and living nature: the slave alone must now struggle with and subordinate living nature to human desire. The slave works on living nature in response to the will and desire of the master, and the master's will and desire becomes thereby the will and the desire of the slave. The slave's role in society is to mediate between living nature and the autonomous will of the master. The slave, we might say, is responsible for the self-preservation and reproduction of the master's social world, while the master is only responsible for sustaining the spiritual worth of his society. Once one highlights the continuing role of nature throughout the master-slave narrative, then the master-slave relation becomes, *sotto voce* or at least anticipatorily, a sexual division of labor.[18]

Man is to Woman as Master Is to Slave as Self-Consciousness Is to Nature

> Woman was the first human being that tasted bondage. Woman was a slave before slavery existed.
>
> (Friedrich Engels, The Origin of the Family, Private Property and the State)

Beauvoir's reconstruction of the master-slave relation as constitutive of the sexual division of social labor draws out and elaborates on what is already there in Hegel's argument, while giving that structure the social and historical specificity that Hegel's allegorical presentation patently lacks.[19] For Beauvoir, despite the fact that *a priori*, neither ontologically nor biologically does the perpetuation of the species "entail sexual differentiation" (24),[20] as a matter of fact, from the outset, evolutionarily and historically, the sexual division of labor has cast the male in the role of individual and the female as responsible for species existence, maintenance, and reproduction. Hence, even in the chapter on "Biological Data," Beauvoir presents mammal life as consigning the female to bearing the burden of species life even when she is otherwise the biological equal of the male: "the female abdicates it [individuality] for the benefit of the species that demands this abdication" (36).[21] After noting that from insects to higher animals, the male's destiny is to separate himself and be confirmed in himself (36), she states that "Hegel is right to see the subjective element in the male while the female remains enclosed in the species" (37). Directly following her canvasing of the biological givens of female existence, to which we shall return, Beauvoir states that they "do not suffice to constitute the basis for a sexual hierarchy; they do not explain why woman is the Other; they do not condemn her forever to this subjugated role" (44). In broad terms, Beauvoir insists the human life is a formation of spirit, humanity is not a given [species], but makes itself what it is; man "is a historical idea" (45). From this it follows that woman "is not a fixed reality but a becoming; she has to be compared with man in her becoming; that is, her *possibilities* have to be defined" (45). Or, again, in a recognizable Hegelian statement,

> society is not a species: the species realizes itself as existence in a society; it transcends itself toward the world and the future; its customs cannot be deduced from biology; individuals are never left to their nature; they obey this second nature, that is, customs in which the desires and fears that express their ontological attitude are reflected.
>
> (47)[22]

Neither biology nor custom as such consigns women to the status of being Other, of undergoing patriarchal domination from virtually the beginning of humanity's historical existence. Beauvoir turns to Hegel's analytic of the master-slave relation to explain how woman's role in species reproduction *became* a forever subjugated social destiny.

Beauvoir reworks the idea of the man-master subordinating the woman-slave in four distinct locales: (i) as originally being set in place during humanity's nomadic, hunter-gather history (71–75); (ii) as taking hold and becoming emphatically hierarchical during the agricultural revolution, until "the stage of writing its mythology and laws, [when] patriarchy is definitively established" (88); (iii) at the beginning of the chapter on "Myths" (159–163); and (iv) in the construction of female subjectivity at the beginning of the chapter on "Childhood." Throughout Beauvoir's account, the master-slave structure operates as fully joining the double-dependency structure that Hegel forges and represses, where woman is the *other self* to man while simultaneously being the bearer of *species life* in the midst of customary historical life. It is because woman is *simultaneously* Other-self and Other-nature, she becomes the paradigm of Otherness, the absolute Other.[23] This, I will argue, is why Beauvoir claims that certain "passages where Hegel's dialectic describes the relationship of master to slave would apply far better to the relationship of man to woman" (74). Because woman embodies and socially exemplifies the double-dependency structure of the human that explodes into the full-blown severance of production from reproduction under capital – contra Beauvoir's thesis that "the production-reproduction balance always finds a way of stabilizing itself" (72) – there can be no resolution to the crisis of patriarchal Anthropocene capitalism except through the recognition of woman-nature.

Beauvoir's theory assumes the existence of a modest structure of effective empirical difference as intrinsic to human historical existence. She further argues that wherever there is significant empirical difference, different human categories, there will be some effort by each to impose itself on its other, or, if that fails, find an enduring form of either mutual recognition or enmity. Beauvoir's genealogy of patriarchy, her reconstruction of the master-slave analytic, begins with the account of pre-agricultural human life in which "the burdens of reproduction represented for them [women] a severe handicap in the fight against a hostile world" (72). Apart from those women who rejected maternity – Beauvoir notes the practices of Amazon women – ordinary women faced a regime of hardship: "pregnancy, giving birth, and menstruation diminished their capacity to work, and condemned them to long periods of impotence" (72), of, literally, powerlessness, hence helplessness, hence dependency on men for both food and protection. With no effective means of birth control and no biologically ordained periods of relief from pregnancy and child birth, women became systematically dependent on men for the feeding and protection of their children. Beauvoir is hypothesizing that there must have been a fundamental sexual division of social labor from the beginning of human social existence.[24] However, in itself a sexual division of social labor does not imply hierarchy, inequality, or domination. Nonetheless, from this modest starting point, Beauvoir begins to leverage the central terms of patriarchy.

She contends that nomadic hordes, operating in a generally hostile environment, were "barely interested in posterity" (73). Even for women, this state of affairs made the status of providing new life equivocal. And later, when "more value was attached to children" (73), the value of life-giving was bracketed:

> But in any case, to give birth and to breast-feed are not *activities* but natural functions; they do not involve a project, which is why the woman finds no motive there to claim a higher meaning for her existence; she passively submits to her biological destiny. Because housework alone is compatible with the duties of motherhood, she is condemned to domestic labor, which lock her into repetition and immanence; day after day it repeats itself in identical form from century to century; it produces nothing new.[25]
>
> (73)

These few sentences announce the themes of woman's domination and othering that recur throughout *The Second Sex*. Beauvoir knows that there is nothing necessary about this configuration of female life; and indeed, there is evidence that some indigenous peoples in fact do place a higher value on life and life-giving than those societies that bequeathed Western modernity. Hence, it is sufficient for Beauvoir's argument that some peoples responded to the predicaments of living in a hostile natural world by conceiving of women's lives as the *continuation* of nature in society, and that so-doing is thus to be categorized as submitting to the logic of natural necessity that ordains repetition as the order of natural functioning within social life; not just pregnancy and procreation, but everything surrounding biological and social reproduction: feeding, clothing, cleaning, caring for the ill and elderly, educating, and so on. Hence, rather than contrasting cyclic time with linear time, Beauvoir differentiates *immanence* as a structure of natural repetition bearing the burden of species life and species reproduction in society from *transcendence* which is the futural movement of sociality itself as the work of valuing and value-creation.

For Beauvoir's philosophical anthropology, if the woman-slave is consigned to provisioning living nature in society, the man-master is categorized as an axiological animal. One could conceive of male hunting and fighting as social versions of all-too-familiar animal routines; and again, in some cultures this has been the case. But we cannot explain the advances of civilization that occurred after the agricultural revolution without positing an anterior axiological self-understanding:

> Man's case is radically different. He does not provide for the group in the way the worker bees do, by a simple vital process, but rather by acts that transcend the animal condition. *Homo faber* has been an inventor since the beginning of time … [man] annexes the world to himself.… Through such actions he tests his own power; he posits ends and projects paths to them: he realizes himself as existent. To maintain himself, he creates; he spills over the present and opens up the future.
>
> (73)

In Hegel's philosophy, it is the slave who in forming nature learns to form himself, to conceive of himself as originating form, and hence as self-transcending praxis beyond nature. But from here, Hegel's argument leaves nature behind as it canvases modes of self-consciousness – Stoicism, Skepticism, and the Unhappy (religious) Consciousness – that adamantly repudiate their dependence on natural life (as well as other subjects). In actuality, there came to be a divergence between the role and value of nature in society and a self-conscious valuing of valuing through which meaning and value is explicitly assigned to nature. Incrementally, Beauvoir argues, this division of labor became a hierarchy. Even in nomad life, she argues a gap appeared between men risking their life in dangerous hunting and warrior expeditions, and women who merely give life. If life must be risked, then life cannot be the highest good: learning to value life appropriately became necessary for the intelligibility of life in a world reproduced through social practices. Hence, Beauvoir's bitterly ironic statement of the value system that emerges with axiological self-conscious humanity: "it is not in giving life but in risking life that man raises himself above the animal; this is why throughout humanity superiority has been granted not to the sex that gives birth but to the one that kills" (74).[26]

It is at this juncture that Beauvoir first claims precedence for the relation of man to woman over Hegel's version of the relationship of master to slave. Here is her first line of argument for man-woman precedence:

> The Master's privilege … arises from the affirmation of Spirit over Life in the fact of risking his life: but in fact the vanquished slave has experienced this same risk, whereas the woman is originally an existence who gives *Life* and does not risk *her* life; there has never been combat between the male and her; *Hegel's definition applies singularly to her*. "The other [shape of consciousness] is non-self-sufficient; for it, [animal] life, or being for an other, is the essence."
>
> (74; italics JMB)[27]

Hegel's pronouncement of slave consciousness in the quoted sentence literally identifies the double-dependency structure of *life* and *being for an other* as its essence. That double-dependency structure, Beauvoir is underlining, is constitutive of woman as other. Women, but not the Hegelian slave, or indeed, other versions of enslaved peoples – colonization, racial slavery, class slavery – identify life with being for another in this way. It is this way of instantiating the double-dependency structure, namely, as a decline from male sovereign oneness, that immures woman in patriarchal domination. On this account, it makes complete sense for *all* the other categories of enslaved peoples to seek unconditional recognition of equal standing from the master (class, race, nation); nothing in the make-up of, say, a colonized people prohibits them from forming a "we" that seeks equal recognition from the colonial master – and they did. What distinguishes woman from slaves as such is not only that free women have always existed, but that the position of woman in society as bearing the burden of species existence positions her as the sustainer of a *necessary but altogether subjugated domain and value: life*. Race difference, national difference, religious difference, cultural difference, and ethnic difference are all either illusory differences or merely contingent differences from the masters, differences that cannot plausibly ground difference in status or worth. Uniquely, woman's way of being human invokes a manner and power of the human that is necessary for the human, and which patriarchal male consciousness repudiates as less than fully human and represses as constitutive of themselves as human – ever and again: "If I can risk my life for value beyond life, then I am not (essentially) life." Patriarchy is the persistence of the Master claim to not be essentially life, not essentially a living and dying being, and not in some emphatic sense a dependent being. *Patriarchy, in its most achieved formations, has been the strict denial and repudiation of the double-dependency structure of self-consciousness.*

But this patriarchal work of denial, repression, and repudiation is complex and difficult to sustain. Beauvoir contends that patriarchal self-consciousness carries out this work of reconciling its claim to transcend life while, at every moment, dependent on life structures and practices, through two mechanisms: the positing of ontological dualism, and the projection of the double-dependency structure onto Woman as Other (to Man). Man knows that he was born and will die, and as such he is part of nature, his *existence* as a self-consciousness depends on life/Being even as he distinguishes himself from life. Hence, "since the coming of patriarchy, life in man's eyes has taken on a dual aspect: it is consciousness, will, transcendence, it is intellect; and it is matter, passive, immanence, it is flesh" (163). Dualism – as enacted in the risk of life, in the claim of reason, in the primacy of the future over the present, and so on – enables a severance of self-consciousness from life. But that severance leaves the significance of life, the powers of procreating nature, the darkness of death unresolved. While Beauvoir is tempted by the thesis that woman represents for man an escape and reprieve from the toils of self-transcending praxis, that she is the "dream" (160) of a reprieve through absorption in the presence of flesh, the "dream" thesis understates what is projected onto woman. Nonetheless, Beauvoir pointedly captures what *must be projected onto woman in order to compensate for the impossible excision of life from self-consciousness*:

> She is the perfect intermediary between nature that is foreign to man and the peer who is too identical to him. She pits neither the hostile silence of nature nor the hard demand of a reciprocal recognition against him; by a unique privilege she is consciousness, and yet it seems possible to possess her in the flesh…
>
> (160)

> She is nature raised to the translucency of consciousness; she is a naturally submissive consciousness. And therein lies the marvelous hope that man has often placed in woman: he hopes to accomplish himself as being through carnally possessing a being while being confirmed in his freedom by a docile freedom…. Appearing as the Other, woman appears at the same time as a plenitude of being by opposition to the nothingness of existence that man experiences in itself; the Other, posited as object in the subject's eyes, is posited as in-itself, thus as being. Woman embodies positively the lack the existent carries in his heart, and it is by seeking to reunite himself [*se rejoindre*] through her that man hopes to realize himself.
>
> (161/FI, 242)

Patriarchal self-consciousness is haunted by and riveted to the model of the double-dependency structure of self-consciousness called woman: the necessity of life for self-consciousness becomes the necessary possession and submission of the Other who inhabits the double-dependency structure of self-consciousness by being a singular life destined for possession by man.[28] Beauvoir is explicit in the opening paragraph of "Myths" that the resolution of the conflict between competing self-consciousnesses seeking recognition of their sovereign authority would be a "free recognition of each individual in the other, each one positing both itself and the other as object and as subject in reciprocal movement" (159). Operating from slightly differently premises, Hegel and Beauvoir agree that till now (patriarchal) man has defaulted on or suppressed this demand; the master-slave relation between man and woman is the practical – ethical, moral, religious, metaphysical – replacement or stand-in for mutual recognition. Patriarchy is the continual displacement of the necessity of mutual recognition into heteronormativity, into the normative/metaphysical structuration of the relation between man and woman as ontologically necessary.

Bearing the Burden of Reproduction

> History has shown that men have always held all the concrete powers; from patriarchy's earliest times they have deemed it useful to keep woman in a state of dependence; their codes were set up against her; she was thus concretely established as the Other. This condition served males' economic interests; but it also suited their ontological and moral ambitions. (159)

These are the opening sentences of "Myths." The "Myths" chapter explicates precisely how woman as Other serves man's ontological and moral ambitions. Volume II of *The Second Sex*, "Lived Experience," provides existential phenomenological depictions of the stages and possibilities of modern woman's life. In often uncanny third person descriptions that take-on the standpoint of first-person experience, Beauvoir provides an account of what it feels like to be othered, how such othering is lived, survived, suffered, evaded, and challenged. Although writing twenty-five years previous to the emergence of ecofeminism, it is in the opening chapters of Volume II that Beauvoir anticipates the structural economic analysis of patriarchal capitalism provided by ecofeminism by demonstrating how woman's life is emphatically shaped by having to bear the burden of biological and social reproduction within a society whose central ambitions are located outside the sphere of reproduction on which it nonetheless economically and ethically depends. In this depiction, woman is destined for man; her life is "waiting" for "Man" (341, 370, 381, 649ff.); her life is destined to be his sexual object; her life is destined to bear children into the world; her life is destined to be a mother; her life is destined to be responsible for the repetitions of domestic labor; her life is destined to experience the signatures of reproductive womanhood – menstruation, pregnancy, birthing, breast feeding, and so on – as fulfilling a species need rather than an individual project; and even if she chooses the pathway of independence now available to modern women, that pathway cannot be lived without contradiction and self-repudiation (723). In all this, I am claiming, Beauvoir has perfectly in sight the governing thesis of ecofeminism as it applies to women in the Global North, even as she lacks the framework to translate that critique into a critique of capitalism's patriarchal economy.

For the purposes to hand, some exemplary instances of experiences of bearing the burden species reproduction will suffice. In eliciting these, it is important to remember that what Beauvoir is portraying is not her conception of woman or the truth of woman, but the formation of woman in the condition of patriarchal society which she has internalized as her self-understanding from infancy forward. Again, Beauvoir accepts as a biological fact that in "higher forms of life, reproduction becomes the production of differentiated organisms; it has a twofold face: maintenance of the species and creation of new individuals; this innovative aspect asserts itself as the singularity of the individual is confirmed" (33–4). This bland statement guides Beauvoir's critical theory: patriarchy entails the division of labor where woman is consigned to species maintenance and man to the assertion of individuality, as if the dual condition, species maintenance and individuality, did not inscribe all human organisms equally, however differently. Because the human body "is a situation, it is our grasp on the world and the outline for our projects" (46), then the cleavage between species maintenance and individuality is experienced through the body, through its activities and functions. Hence, woman's consignment to reproduction is felt in patriarchal society with the first signs of mature womanhood.

> Just as the penis gets its privileged value from the social context, the social context makes menstruation a malediction. One symbolizes virility and the other femininity: it is because femininity means alterity and inferiority that its revelation is met with shame. The girl's life has always appeared to her to be determined by this impalpable essence to which the absence of the penis has not managed to give a positive image: it is this essence that is revealed in the red flow that escapes from between her thighs … the menstrual stain [inclining] her toward disgust and fear.
>
> (328)

In this instance, unlike almost everywhere else in the book, Beauvoir immediately reminds us that in "a sexually egalitarian society, she would envisage menstruation only as her unique way of acceding to an adult life…menstrual periods inspire horror in adolescent girls because they thrust them into an inferior and damaged category" (329). There is nothing about the development of breasts (320ff) or menstruation that requires the surrender of individuality; the depositing of the young girl into "an inferior and damaged category" is how she is gendered, how she is being made into a woman, how she becomes a woman (282), how she becomes the one who will take on the burdens of species reproduction in society.

After introducing the chapter on "The Married Woman" with a dark, deflationary account of sexual life in marriage that traces the disintegration of "erotic magic" into "a kind of mutual masturbation" (467), Beauvoir turns to housework. She introduces woman's investment in the home with man's minimization of its significance; he has "only middling interest in his domestic interior because he has access to the entire universe and because he can affirm himself in his projects" (470). Beauvoir comprehends the strategy of investing in the home as the same dialectical strategy she has employed in sexual life: "she takes [man] by becoming [his] prey; she liberates herself by abdicating herself; by renouncing the world, she means to conquer the world ... The home becomes the center of the world and even its own one truth" (470, 471). But while this investment projects a domain of creative action, what it comes to, finally, is domestic labor.

> By administrating her home, she achieves her social justification; her job is also to oversee the food, clothing, and care of the familial society in general. Thus she too realizes herself as an activity. But, as we will see, it is an activity that brings her no escape from her immanence and allows her no individual affirmation of herself.
>
> (472)

Beauvoir's inaugural gesture in describing domestic labor is to survey the poetry of housework in which the satisfactions of cleaning, ironing, the smells of clean laundry and dinner being cooked are given an almost sublime character. But this quickly gives way to an account of the sense of domestic labor when it is separated from work, when immanence is disconnected from transcendence, when reproduction is severed from world-making. In this situation, domestic labor becomes a Sisyphean torment: "day after day, one must wash dishes, dust furniture, mend clothes that will be dirty, dusty, and torn again. The housewife wears herself out running on the spot" (474). Or again, after describing the endless washing, ironing, sweeping as a negative work that not only holds "away death but also refuses life," Beauvoir underlines how the Manichean structure of housework is one in which she is never called upon to build a better world: "she attacks the dust, stains, mud, and filth; she fights sin ... But it is a sad destiny to have repel an enemy without respite instead of being turned toward positive aims" (476). In this setting, in the world, reproduction is given a reduced structure of life maintenance in utter separation from work and world-making and world-building.[29]

Reproductive labor is pure immanence and repetition *through* its severance from work and transcendence,[30] where the patriarchal model stretches into slave labor, colonial exploitation, and racist domination. Although Beauvoir is hyper-aware that the activity of domestic labor is part of woman's work of bearing the burden of species reproduction generally in a manner that leaves her outside male world-making, and further knows that reproductive labor generally provides a necessary economic benefit to the community at large, and although her understanding of patriarchy drew upon the model of American racial domination that has its roots in chattel slavery,[31] nonetheless, she never asks what apart from the operation of patriarchy itself could be responsible for consigning woman to species maintenance and depriving her of the opportunities of individuality in a world in which, *formally*, women do have access to education, political participation, and work outside the home.

Beauvoir's critical theory comprehends patriarchal domination, but, in a deeply puzzling and exasperating way, leaves invisible and uninterrogated the social and economic structures and practices that are responsible for and have a stake in woman *continuing* to bear the burden of reproductive labor, or into the role of those practices in *convergent* slave, race, and colonial practices of domination. Beauvoir perceived precisely why patriarchal domination's depositing of woman permanently into the role of reproductive labor and servicing species needs was the privileged exposition of the master-slave dialectic, but she failed to track back to those other master-slave structures, namely, the brutally dehumanizing subordination of racialized bodies, colonized bodies, enslaved bodies in order to demonstrate that the patriarchal model was not merely privileged because of its explicit binding of woman to living nature, but exemplary for and convergent with those other forms of domination and binding of bodies to reproductive labor. Patriarchal domination is not only the privileged form of the master-slave domination, but models the expropriation of reproductive labor generally.

Ecofeminism's patriarchal analysis of capitalism answers the explanatory puzzles and absences of Beauvoir's critical theory at it reveals the full critical force of Beauvoir's theory.

From Beauvoir to Ecofeminism

> If the central concern of all economic and social activity is not the accumulation of dead money but the creation and maintenance of life on this planet, nothing can remain as it is now.[32]

In Hegel's master-slave dialectic, it is the slave who must transform himself from his petrified attachment to life; and he achieves this detachment from life-nature through his coerced labor on the world in the service of the master. And generally, we imagine slave revolts as involving the slave overcoming past life-preserving fears for the sake of future freedom. But the Hegelian dialectic is radically misleading in one fundamental aspect: not only does the master deny his dependence on nature and life as essential to his self-consciousness, but equally the slave takes the master as his ego-ideal, as who he wants to be, and in his own processes of social learning achieves freedom through detaching himself, first, from the natural world through the suppression of natural desires, and then through the transformation of given nature in accordance with mental forms, thus coming to forge the idea of stoic freedom of mind as his essence. Life suddenly drops out of the dialectic at this juncture, despite the fact that from the outset the slave was thought to embody the structure of *double dependency*, of being dependent on life and on being for an other; stoics concede neither form of dependency – which is the point of Hegel's argument. Beauvoir's implacable achievement is to explicate the long and enduring history of patriarchy as made possible by a master-slave structure of domination and subjugation constituted by woman being the bearer of that double-dependency structure. Being the bearer of the double-dependency structure enables man to accomplish his illusory moral and ontological goal of leaving nature behind, of morally and ontologically subordinating nature and life to freedom, reason, mind, soul, culture. Beauvoir does not mean to deny the significance of the distinction between self-consciousness and life; on the contrary, she insists upon it. This leads directly to her equivocal paean to transcendence at the expense of immanence:

> It is because humanity puts itself into question in its being—that is, *values reasons for living over life*—that man has set himself as master over woman; man's project is not to repeat himself in time: it is to reign over the instant and to forge the future. Male activity, creating values, has constituted existence itself as a value; it has prevailed over the indistinct forces of life; and it has subjugated Nature and Woman.
>
> (75; ital. JMB)

Nonetheless, and Beauvoir has only scant resources for grasping this thought, it is woman who bears the burden of both the relational structure of self-consciousness (being for an other) and self-consciousness being a formation of life. The privilege of patriarchal consciousness, existential transcendence, is simultaneously its severe pathology: it conceives of individuality as distinct from constitutive sociality, and it conceives of freedom and transcendence as ontologically distinct from life and immanence. Sociality and life are externalities for patriarchal self-consciousness. Which is to say, the very idea of a woman's revolt is equivocal: women need to revolt against male power but not for the sake of male ontological privilege; it is male consciousness that requires radical transformation, a conversion to another form of living will be necessary if man, the male-gendered subject, is to surrender both the ideas of sovereign individuality independent from constitutive sociality and transcendent existence outside species life. Man's ontological and social pathology is what deforms woman's integral existence.[33]

But this necessity for a moral conversion overlooks what every experience of reproductive labor today signifies: the entire formation of late liberal capitalism now operates *through* the radical severance of economic production for the sake of profit from imposed and backgrounded formations of reproductive labor in its diverse forms – domestic labor, service industries, neocolonialism, raced based distributions of opportunity, and so on. More emphatically and importantly, over the past century, increasingly, the exploitation of female, colonial, racial, imperialist reproductive labor has been coordinated with the ever-expanding systemic destructive exploitation of living nature, the destruction of biosphere and ecological integrity. Capitalism is now *essentially* the operation of this double exploitation: the forces of reproduction, human and nonhuman, are suffering direct systematic exploitation, expropriation, and destruction. When the exploitation and destruction of the forces of reproduction displace the exploitation of wage labor as the primary contradiction of capitalism, capitalism has morphed into what I am calling *patriarchal capitalism*. To claim patriarchal domination is now the primary form and contradiction of capital is to claim that only through resolving the severance between production and reproduction, between transcendence and immanence, between master-man and slave-woman can the contradictions of capitalism be overcome. At the beginning and end of "The Married Woman," Beauvoir ferociously underlines the necessity for the joining of transcendence and immanence, self-consciousness with living nature.

> Woman is destined to maintain the species and care for the home, which is to say, to immanence. In truth, all human existence is transcendence and immanence at the same time; to go beyond itself, it must maintain itself; to thrust itself toward the future, it must integrate the past into itself; and while relating to others, it must confirm itself in itself. These two moments are implied in every living movement...
>
> (443)

> So the wife's work within the home ... becomes meaningful and dignified only if it is integrated into existences that go beyond themselves, toward the society in production and action.
>
> (484)

Although Beauvoir plainly thought that only a socialist society could be an egalitarian one, she did not foresee how the severance of reproduction from production that defined patriarchal society would come to define the structure of capitalism itself.

That severance is precisely how D'Eaubonne inherits Beauvoir, recruiting the feminism Beauvoir bequeathed for a new incarnation: ecofeminism. D'Eaubonne's allegiance to Beauvoir begins in her Introduction with an unattributed quote claiming "woman was a slave before slavery existed" – from Engels' *The Origin of the Family, Private Property, and the State*. Then, on the very first page of the text proper, D'Eaubonne declares, first, implicitly in place of the "*cogito, ergo sum*," "I am a woman,"[34] only to immediately qualify it by saying woman is other, "this zombie, this negativity, this hole"; which is in turn qualified: "We are not born as such; we become it."[35] I am a woman (so announcing the unavailability the absolute standpoint that has already been appropriated by man); woman is thus the absolute other of the male one; but this fatality is not final because it is a construction, one is not born but becomes a woman – three theorems of Beauvoir feminism. The fourth theorem, that women are condemned to the bear the burden of species life, comes later: "it is a question of chaining the woman to her own species, of 'preventing her from running,' to make her feel concretely her weight as an animal, an inferior."[36] In

D'Eaubonne the species doctrine is given its first explicit ecofeminist sounding; yet, it is only in the final pages of the book that D'Eaubonne announces that phallocracy is not only the domination of women, consigning her to the role of life-creation and life-maintenance, but the double destruction of the planet: through the destruction of natural resources and through overpopulation. While the overpopulation thesis was common at the time,[37] D'Eaubonne does begin sounding those theses that will become the hallmark of mature ecofeminism: feminism concerns not only the liberation of women, but the life or death of the planet; beginning with agricultural production, the male of the species was intent on ever-increasing productivity, hence there is a line between ever more efficient and expansive agricultural production and ever more efficient and expansive industrial production; and that both those modes of exploiting the earth's bounty have been pursued through means that have become increasingly destructive of the ecological integrity of the Earth System. Hence, her resounding thesis: "Do you want to live or die? If you refuse planetary death, you will have to accept the revenge of women, for their personal interests join those of the human community."[38]

D'Eaubonne underlines the doctrine of the severance of production from reproduction within capital,[39] which together with her announcement of women as an emergent revolutionary class is sufficient to claim that she has the idea of patriarchal capitalism fully in view. But it would take the next generation of ecofeminists to develop this thesis in line with the thought that it was not only women themselves but also feminized bodies – racialized bodies, enslaved bodies, and colonized bodies – that have collectively borne the burden of social reproduction.

Bennholdt-Thomsen and Maria Mies's *The Subsistence Perspective* is distinctive in ecofeminist literature for its attempt to map an economic alternative to patriarchal capitalism in a manner that would answer to the demands of the double-dependency structure of self-consciousness. Their premise is that food security cannot be achieved through current economic practices, and, more emphatically, the very idea of the rest of the world achieving the standard of living of the wealthy Global North is not viable – in part because this standard of living was achieved at the expense of those in living the Global South, and in part because it was achieved through devastating its ecological conditions of possibility. *The devaluation of women and nature constitutive of patriarchal modernity simultaneously devalues the subsistence perspective*: it cannot be seen or appreciated from the value premises of liberal capitalism. Their contention is thus that a future economy must take its orientation from the subsistence *perspective*, from the orientations and ethical world-view of the victims of capitalism, from the orientation of those remaining forms of life in which agricultural *subsistence practices* define relatively self-sustaining, life-maintaining village societies. What we have been designating through the conception of reproduction and species life, they contend is visible in subsistence-oriented forms of social life.[40]

While Bennholdt-Thomsen and Mies half concede that the subsistence perspective ill consorts with a world in which half the population live in large conurbations doing industrial and service industry labor, they insist that the subsistence perspective premised on the work of those now dominantly pursuing reproductive labor, labor for the sake of life and future lives, is necessary as a *paradigm* of an alternative to liberal capitalism.[41] The premise of the subsistence perspective is that *life*, life today and the lives of future generations, will be the main goal of work, not a side-effect of extended accumulation. *This is equivalent to saying that immanence must become our transcendence*, that the concentric circles constituting reproductive life must be embraced as a life-form to be achieved through creative activity, thereby overcoming the dualism between immanence and transcendence. The subsistence perspective discloses a world in which the formations of reproduction, the structures of human immanence, can become themselves a new formation of transcendence, as reasons for living and the values of life should not be transcended but themselves lifted into a mode of transcendence. Only through subliming immanence, making reproduction a form of *collective work* can the double dependency structure of self-consciousness be satisfied without a sexual division of social labor.

Bennholdt-Thomsen and Mies thus make the subsistence perspective a conception of economic life that simultaneously secures mutual recognition and collective dependency on living nature. The subsistence perspective, they argue, involves a form of life premised on: (i) relatively or maximally self-sufficient regions; (ii) a restoration of the commons under the control of a local community – there cannot be a commons without a community whose commons it is; (iii) decentralization of authority, regulations, and problem-solving practices; (iv) policymaking from below; (v) reciprocity versus mechanical solidarity. (vi) "the priority of use-value production";[42] (vii) being able to actively perceive and experience the relation between work and consuming as a condition for a new form of economic rationality; (viii) as opposed to the male contract-and-wage-labor model, making each worker an independent producer, coming to embrace mutuality among workers as the orienting model; entailing (ix) that the bond connecting workers, traders, and consumers is to be primarily ethical not contractual. Hence, the goal is a "new moral economy" which "every individual brings about by assuming responsibility."[43]

Neither technology nor economic growth can resolve the crises of patriarchal capitalism. The subsistence perspective, however otherwise limited, projects a form of life in which the severance of production from reproduction is overcome by subsuming production under reproduction, an overcoming that thus makes it possible to conceive of a life form no longer determined by a gendered master-slave structure of domination, no longer determined by humanity being divided between the master-man and the slave-woman.[44]

Notes

1 Ariel Salleh, "Ecological Debt: Embodied Debt," in *Eco-Sufficiency and Global Justice: Women Write Political Ecology* (London: Pluto Press, 2009), 3.
2 Salleh, "Ecological Debt," 11.

3 Stefania Barca, *Forces of Reproduction: Notes for a Counter-Hegemonic Anthropocene* (Cambridge: Cambridge University press, 2020). Ecofeminism has always forcefully underlined the alignment of gender domination with slavery, racial domination, colonization, and their continuations in contemporary capitalism. While slavery has existed full apart the logic of gendered nature-human dualism, and colonization and racism logically could operate with distinct logics, because capitalism has become defined by the severance of economic production from reproduction, there has been set in place a gendered constitutive structure whereby human domination is aligned with the separation of economic production from reproduction, depositing dominated humanity into the sphere nature.

4 The concept of *patriarchal capitalism* is an intensification of Veronika Bennholdt-Thomsen and Maria Mies' notion of a "patriarchal—capitalist economy" (*The Subsistence Perspective: Beyond the Globalised Economy*, trans. Patrick Camiller, Maria Mies and Gerd Weih (New York: Zed Books, 1999), 5), where the lead thought is that reproductive labor is subsumed under the demands of the primary market economy. What was for Bennholdt-Thomsen and Mies an incremental process of subsumption and destruction that was incapable of resolving the problems of food security and poverty has become the explicit, catastrophic contradiction of capitalism: a radical severance of the conditions of production from reproduction such that the latter are not merely used and backgrounded but continually violently exploited to the level of outright destruction.

5 Françoise D'Eaubonne, *Feminism and Death*, trans. Ruth Hottell (New York: Verson, 2022 (1974)), 194, 195. As widely noted, while D'Eaubonne powerfully prefigures the argument that capitalism is a formation of patriarchal domination, she problematically places equal weight on overpopulation as a potential cause of planetary death.

6 This thesis patently requires refinement when between 30% and 40% of executive positions are held by women in advanced industrialized societies. A knock-on effect of more women *actively* belonging to the capitalist class is that domestic and non-domestic reproductive labor in some wealthy countries is being off-loaded onto women and men from the Global South. For the purposes of my argument here, one can say that under patriarchal capitalism the *logic* of the severance between the productive economy and social reproduction is itself subject to demographic realignments and adjustments.

7 This thesis that overexploitation of the natural and social conditions of reproduction to the point of collapse forms the primary contradiction of capitalism is now widely accepted by Marxist critics. See, for example, James O'Connor, "The Second Contradiction of Capitalism" in his *Natural Causes: Essays in Ecological Marxism* (New York: The Guilford Press, 1998), 158–177; and John Bellamy Foster, *Marx's Ecology: Materialism and Nature* (New York: NYU Press, 2000). My contention is that the precise form of this over-exploitation is mistaken without comprehending it through the lens of patriarchal domination. I have a sketched my own version of this thesis in "Adorno and Ecofeminism," in *Adorno*, edited by Martin Shuster and Henry Pickford (New York: Oxford University Press, forthcoming).

8 For a wider analysis see Tanella Boni, "Why is Woman Other?" in *A Companion to Simone de* Beauvoir, edited by Laura Hengehold and Nancy Bauer (Hoboken, NJ: Wiley Blackwell, 2017), 174–184.

9 The original plan for this essay was to argue that Beauvoir's taking over Hegel's master-slave analytic for the purposes of explaining sex/gender domination *anticipates* ecofeminism, but only achieves its exemplary statement in the writings of later ecofeminists, especially Val Plumwood, *Feminism and the Mastery of Nature* (London: Routledge, 1993). Marie Loslier Simon, in class and conversation, convinced me that Beauvoir intended her reworking of Hegel to demonstrate that

sex/gender domination is presupposed by the master-slave relation since only it properly articulates the double-dependency structure of human self-consciousness. Loslier Simon's intervention forced me to reconceive the argument into the one being presented here. Loslier Simon would press the argument differently than I do, reading Beauvoir as pursuing an Adorno-style genealogy rather than reconstructing Hegel's master-slave directly.

10 I am borrowing the idea that Beauvoir's philosophy is best conceived as a version of critical theory from Stella Sanford, "Beauvoir's Transdisciplinarity: From Philosophy to Gender Theory," *A Companion to Simone de Beauvoir*, 15–27: "*The Second Sex* is the historical meeting point of critical theory and feminism, via philosophy" (24).

11 For authoritative developments of this claim see Eva Lundgren-Gothlin, *Sex and Existence: Simone de Beauvoir's 'The Second Sex'*, trans. Linda Schenck (Hanover: Wesleyan University Press/University Press of New England, 1996); and Nancy Bauer, *Simone de Beauvoir, Philosophy, and Feminism* (New York: Columbia University Press, 2001). For a useful overview of Beauvoir's relation to Hegel, see Kimberly Hutchings, "Beauvoir and Hegel," in *A Companion to Simone de* Beauvoir, 187-197. For my own reconstruction of Beauvoir's Hegelian-Lacanian reading of child development at the commencement of Volume II, see "Sexuate Being: Foundations for a Feminist Legal Philosophy?" in Renée J. Heberle and Benjamin Pryor (eds.), *Imagining Law: On Drucilla Cornell* (Albany: SUNY Press, 2008), 81 –102. My reading of Beauvoir is indebted to Bauer's.

12 Georg Wilhelm Friedrich Hegel, *The Phenomenology of Spirit*, trans Terry Pinkard (Cambridge: Cambridge University press, 2018), §177.

13 Hegel, *Phenomenology*, §175.

14 A necessary condition for Hegel's philosophical anthropology is the insistence upon replacing Kant's constitution of all nature as Newtonian, mechanical nature with living nature. Hegel, probably following Schelling's lead, comprehended human life as another formation of species life within the planetary whole of living nature. For a full and compelling consideration of this conception of nature in Hegel's philosophy generally see Karen Ng, *Hegel's Concept of Life: Self-Consciousness, Freedom, Logic* (New York: Oxford University Press, 2020).

15 Hegel, *Phenomenology*, §187.

16 Hegel, *Phenomenology*, §187.

17 Hegel, *Phenomenology*, §189.

18 Hegel's resolution of the role of life in the *Phenomenology* is fragmentary: through the self-defeat of asceticism in "the Unhappy Consciousness"; through the primacy of active rational self-consciousness over observing reason; through the critique of the separation of happiness from moral virtue in "The Moral Worldview"; and so on. The final acknowledgement of nature is more implicit than fully thematic; for Hegel, at this juncture in his thought, dependence on living nature and embodiment simply requires the overcoming of stoic and religious repudiation, not systematic analysis.

19 Putting the thesis this way is not meant to deny that Hegel *intended* his account to provide a *general* form of consciousness that could be realized in variety of settings and formations from ancient slavery to feudalism (lord and serf) to modern chattel slavery.

20 All unmarked pages references in brackets in the body of the text are to Simone de Beauvoir, *The Second Sex*, trans. Constance Borde and Sheila Malovany-Chevallier (New York: Vintage Books, 2011). References to the original French, *Le Deuxième Sex I*, and *Le Deuxième Sex II* (Paris: Gallimard 1976 (1949)) will follow the English translation, separated by a slash, /, and marked FI or FII (French volume I or French volume II) followed by a comma and page number.

21 Patently a partial generalization with numerous exceptions – queen bees, and black widow and redback spiders for openers!

22 For a reading of Hegel that underlines this aspect of his thought, see Andreja Novakovic, *Hegel on Second Nature in Ethical Life* (Cambridge: Cambridge University press, 2017).

23 In Hegel's austere syllogistic argumentation, borrowing Anna Katsman's formulation, women functions as the middle term that permits the mediation: she is both human and nature at once, and therefore can serve as the screen for both repudiation and reconciliation.

24 For a strong claim that woman's reproductive role is in fact responsible for the depths of human sociality, see Sarah Blaffer Hrdy, *Mothers and Others: The Evolutionary Origins of Mutual Understanding* (Cambridge MA: Harvard University Press, 2009). Although implicit in Beauvoir, Hrdy pointedly underlines the long helplessness and vulnerability of children, which equally reverberates onto the community of women as the solder of gender hierarchy. Giving birth without the continuation of female reproductive labor would not have yielded the actual and familiar formations of gender hierarchy. It is not just biological creation but the formations of sociality necessary to respond to the ongoing biological vulnerability of the human, from infancy to old age, that makes gender difference the bearer of an incipient ontological/existential contradiction. Beauvoir's immanence and transcendence distinction that contrasts repetition with creation is too agency or will based to capture the underlying stakes and logic.

25 Beauvoir is here almost certainly overly deflating women's reproductive labor. As she must have known, the evolution of reproductive labor and care work is historically and culturally variegated. One might thus want to say that it has been the historical failure of the development and variegation of reproductive labor, its repression, that spells out the significance of the severance of produce from reproduction as a formation of domination and repudiation.

26 Beauvoir's ironic tone is meant to underline that this is an ideology and not an argument. Even the sacrifice of life is, finally, for the sake of a formation of life. It is for this reason that Beauvoir sees the nomad moment as only an anticipation of what becomes solidified during the agricultural revolution.

27 In the quote from Hegel, *Phenomenology*, §189, I have followed Pinkard rather than the inaccurate translation of Beauvoir's original.

28 For a differently inflected reading of this stretch of argument see Bauer, *Simone de Beauvoir*, 189-90.

29 For a parallel analysis of labor in its grueling and alienating separation from work, bizarrely projected back into ancient Greek society, see Hannah Arendt, *The Human Condition* (New York: Anchor Books, 1959 (1958)), Chapter III.

30 As Bennholdt-Thomsen and Maria Mies, *The Subsistence Perspective* argue, 30–8, 98–101, capitalism inaugurated a process if "housewifeisation" (*Hausfrauisierung*) of female labor, separating domestic labor from other forms of reproductive labor, as part of industrializing agriculture and making it a component of the commodity producing economy. It is this process that is reflected into Beauvoir's and Arendt's distinguishing between labor and work.

31 For a critical engagement with this feature of Beauvoir's thought, see Kathryn T. Gines, "Sartre, Beauvoir, and the Race/Gender Analogy: A Case for Black Feminist Philosophy" in *Convergences: Black Feminism and Continental Philosophy*, edited by Maria Davidson, Kathryn T. Gines, and Donna Dale Marcano (New York: SUNY Press, 2010), 35–51; and Gines "Comparative and Competing Frameworks of Oppression in Simone de Beauvoir's. The Second Sex," *Graduate Faculty Philosophy Journal* 35 (1–2), 2014: 251–73.

32 Veronika Bennholdt-Thomsen & Maria Mies, *The Subsistence Perspective*, 7.

33 If this argument is right, it follows that *The Second Sex* not only re-deploys the master-slave model, but its driving orientation to demonstrate the repressed foundational role of the double-dependency structure of self-consciousness is the same

as the primary structural argument of the *Phenomenology of Spirit*. It is also the case that both works have the almost identical explanatory lacunae: liberal capitalism's role in enforcing the repression of the double-dependency structure of self-consciousness, its eviscerating elision of human sociality and the placement of human self-consciousness as a formation of living nature, as a living part of a living world.

34 See Bauer, *Simone de Beauvoir*, Chapter 2.

35 D'Eaubonne, *Feminism or Death*, 4, 9. From this opening Beauvoirian chorus, it would not be difficult to substantiate how D'Eaubonne is mostly concerned with giving Beauvoir's feminism a more angry, urgent, and political tonality, leading to her three part "For a Planetary Feminist Manifesto."

36 D'Eaubonne, *Feminism or Death*, 85.

37 For a critique of D'Eaubonne's handling of the population question, see Myriam Bahaffou and Julie Gorecki's Introduction to the New French Edition, xxxi.

38 D'Eaubonne, *Feminism or Death*, 219.

39 D'Eaubonne, *Feminism or Death*, 164.

40 I will take as given that accepting a subsistence perspective in the first instance may itself *first* require the moral overcoming of gender hierarchy.

41 Bennholdt-Thomsen and Maria Mies, *The Subsistence Perspective*, 14-15, open their argument by listing the range of critiques of their position; most of the critiques concern how generalizable their model is, especially in relation to industrial production in urban settings. My view is that while the subsistence perspective is not fully generalizable, it is easily compatible with the wider economic framework of, e.g., Kate Raworth, *Doughnut Economics: Seven Ways to Thinks Like a 21st-Century Economist* (White River Junction, Vermont: Chelsea Green Publishing, 2017). They have also been criticized for their apparent essentialism and primitivism by, among others, Maxine Molyneux and Deborah Lynn Steinberg, "Mies and Shiva's 'Ecofeminism': A New Testament?," *Feminist Review* 49 (1995), pp. 86–107. For a critique of Mies' primitivism see Victor Wallis, "Vision and Strategy: Questioning the Subsistence Perspective," *Capitalism Nature Socialism* 17:4 (2006), pp. 38–43. For formulations of core ecofeminist doctrines in this area that avoid even the look of essentialism and primitivism, see Barca's *Forces of Reproduction*. Thanks to Agnese di Riccio for urging this clarification.

42 Bennholdt-Thomsen and Maria Mies, *The Subsistence Perspective*, 58.

43 Bennholdt-Thomsen and Maria Mies, *The Subsistence Perspective*, 121.

44 My fulsome gratitude to Alice Crary, Agnese Di Riccio, Anna Katsman, Marie Loslier Simon, Esther Neuhann, and Gabriel Ricci for invaluable comments and criticisms that have made this essay better. Its inevitable faults are mine alone.

3 Nature, Art and Gender in Renaissance Italy

A Counter Narrative*

Mary D. Garrard

Where did we come from? Where are we headed? These questions take us back to the beginning of life on our planet, and forward into our future. Today, the second question is more pressing, as climate change rapidly transforms our idea of nature – from a benign, stable, dependable entity, like a benevolent nourishing mother, into a dangerous, capricious and frightening force, an indifferent or malevolent mother provoked into punishing us. Nature imagined as mother has a long history. Throughout the Middle Ages, nature was identified with the Earth, as the generator and mover of organic life. In medieval manuscripts, Terra, or Mother Earth, is shown surrounded by trees and hills, interacting with animals and people. In earlier figurations, Nature was a powerful goddess, as the Greco-Roman Diana of Ephesus and Gaia, back to the Egyptian Isis and Mesopotamian Inanna/Ishtar.[1]

Alongside those ancient cultures came patriarchy; with Christianity came Nature's subordination to the patriarchal God. While Nature survived in the Middle Ages as Mother Earth, limited to the geosphere, God was depicted as a creator inscribing the universe with a compass. The distinction had important philosophical implications. In ancient Mediterranean origin myths, the female creator emerged from Chaos to be born and simultaneously to give birth to the universe, a creator at one with her creation. But the male Christian deity preexisted and stood apart from the universe he created. Their difference was hierarchic: female nature's material realm is earthbound and physical, subordinate to the male god's realm of mind and idea.

Art joined this equation in the later Middle Ages, at first on the feminine side. In fourteenth-century manuscripts, Art is personified as female, like the

* Author's note. This chapter is drawn from Mary D. Garrard, *Brunelleschi's Egg: Nature, Art, and Gender in Renaissance Italy* (Berkeley: University of California Press, 2010). A lot has changed since I started writing that book. The climate change problem is becoming a crisis. Technology has created AI, which threatens to steal back the inventive creativity that Renaissance man claimed to steal from Mother Nature. Art historians today are skeptical of master narratives, especially those constructed by white men. But the issue of the human-nature relationship, which Renaissance people struggled to define, remains very much alive today. In the spirit of past-as-prologue, I've created this short version of a long book, hoping to make it more useful.

DOI: 10.4324/9781003385202-4

celestial Natura before whom she kneels. Art was subordinate to Nature, and Nature was subordinate to God, whose creative powers transcended their own. The notion of God as a designing artist was a foundation for the much-celebrated divine genius of the Italian Renaissance male artist. Throughout the Middle Ages, however, Nature's creative superiority over human art remained ascendant, sustained by the ontological superiority of her creations.

The Western European conception of Nature changed radically with the reinvigoration of patriarchy in the period we call the Renaissance. Philosophies of nature moved from a view of the earth as a living, intelligent organism imbued with spirit to the scientific conception of the world as a mindless machine, created and maintained by God. By the seventeenth century, the ancient concept of Mother Earth as a powerful generative force had been supplanted by the scientific view of nature as passive matter controllable by man. The divine *natura naturans* was gradually replaced by *natura naturata*, reduced from a creative power to trees and rocks, the inert products of creation.

This was considered a model of progress before modern feminist scholars revealed its gender bias. Sherry Ortner challenged the inferiorizing equation of women with nature and men with culture. Carolyn Merchant demonstrated that the transition from an organic to scientific worldview was assisted by the gender metaphor and men's supposedly natural dominance of women, which lent support to the exploitation of nature. Evelyn Fox Keller exposed a gender ideology at the root of modern science, with its dependence on masculinist sexual imagery for self-definition.[2]

What Merchant called the "death of nature" tends to be presented by historians of science as an abrupt paradigm shift in Western thought. But the history of art shows a more nuanced transition. In *Brunelleschi's Egg: Nature, Art and Gender in Renaissance Italy*, I showed how Italian Renaissance artists assisted this transition, by visually modeling Nature's gradual disempowerment and demise, sometimes in advance of science itself. Long before scientific investigations cohered into a discipline whose keynote was the mastery of nature, Italian Renaissance art theorists conceived of art and nature as gendered competitors in a power struggle, whose victor was proclaimed to be art. The theme was sounded first in fifteenth-century Florence, by male artists who sought to raise the status of the visual arts, from manual to liberal arts. The intellectual elevation of painting, sculpture and architecture was fostered by figuring nature as a female Other who was said to be threatened by their art. An escalating claim by artists to have rivaled or defeated nature was sustained in theoretical texts and images.

The visual arts have an especially close relationship with nature, most obviously because the natural world is a primary subject of the painter's art. But the art-nature relationship long preceded landscape painting, and art's capacity to model or anticipate changing concepts of nature also deserves attention. Basic elements of painting correspond to key concepts of philosophy, such as the form-matter binary that was famously gendered and hierarchized by Aristotle, as the superiority of active masculine form to passive feminine matter. Aristotle himself explained procreation through the analogy of art, equating male insemination with the artist "who works upon inert matter according to a design."[3]

In both philosophy and painting, matter was the inferior, feminine element to be transcended. For the Early Renaissance painter, pigment served the creation of form, and the goal was to neutralize its presence. But in Renaissance painting, unlike philosophy, the form-matter relationship became increasingly dynamic over time. Leonardo da Vinci's innovative *sfumato*, which merged forms with their environment and energized matter,[4] led to pictorial fields such as the late paintings of Titian, in which matter was in constant motion and the form-matter binary was finally dissolved – *before* the work of Kepler and Galileo confirmed it scientifically. Because the painters followed progressions inherent in their craft, it is not surprising that art might precociously develop models that apply to, yet evolve within, two separate spheres.

Renaissance visual imagery also modeled, and could have affected, how people felt about the changes they experienced or intuited, as technology and science gained footing and Nature lost ground. Over the fifteenth and sixteenth centuries, a drama unfolded, as artists mediated cultural man's separation from Mother Nature. The goddess Natura was no longer figured in art; she became conflated with other female divinities. As Mother, she became the Virgin Mary; her procreative and erotic dimensions were channeled into Venus, while Diana symbolized Nature's protection of her secrets. Imaged in these forms, concepts of nature entered cultural consciousness directly.

In the arc of change over two centuries, we can see patterns. Two distinctive perspectives on the art-nature relationship emerged in the major cultural centers of Florence and Venice. Florentine painters and theorists offered a narrative of mastery, boasting of man's transcendence over nature's constraints, his perfection of her imperfections, and men's priority in both art and human reproduction. Venetian painters, by contrast, depicted the poetic beauty of the natural world, with humans comfortably situated within it – a model of immanence rather than transcendence. In the Venetian perspective, Nature was the artist's mysterious yet benevolent guide and partner, and her slow demise was met with regret, even separation anxiety.

Another pattern emerges. Feminist scholars have identified the cultural transition from the organic medieval universe to the masculinization of thought in seventeenth-century philosophy and science as a progression of gender. Susan Bordo described it as a "flight from the feminine," invoking a modern psychological model, in which male ego development depends upon successful separation from the mother, expressed in changing representations of the female.[5] Such a progression is affirmed by Renaissance art, as male artists imaged their separation from the original mother (nature) to attain cultural independence and dominance. Key examples illustrate the trend: for Quattrocento artists Brunelleschi and Masaccio, nature was mother; for Botticelli, she was a youthful bride. In the Cinquecento, Giorgione and Titian posited art's equal partnership with nature, like a good marriage, while in the later sixteenth century, Cellini cast nature in erotic terms, as a slave or whore to be dominated.

Histories of art do not address the role of the visual arts in mediating the large cultural transition from the age of theology to the age of science. Nor, more surprisingly, do they historicize art's changing dialogue with nature. Many rubrics and story lines make up the narratives and metanarratives of art history, but not this one. In *Brunelleschi's Egg*, I present a counter-narrative, to show what happens when the story is told from nature's point of view. From this vantage point, many familiar, much-studied monuments of Renaissance art can be seen to engage a common primal theme. And they yield fresh meanings, manifest in these new interpretations of Renaissance classics.

Brunelleschi's Egg

The architect Filippo Brunelleschi has been heroized for his technical achievement in constructing the great dome of Florence Cathedral (Figure 3.1). Yet

Figure 3.1 Filippo Brunelleschi. Dome, Cathedral of Florence (Santa Maria del Fiore). 1420–36. Florence, Italy. Photo: Scala / Art Resource, NY.

his achievement was an act of appropriation. Long before Brunelleschi gave the church its crowning feature, Santa Maria del Fiore had been a major site of imagery dedicated to the Virgin Mary. In this monument, a civic symbol of nature's maternal and protective power was transformed into a symbol of the new and formidable power of art.

In 1430, two Florentine humanists sat on the Capitoline Hill in Rome, reflecting on the once majestic city. To Poggio Bracciolini and Antonio Lusco, the ruined buildings of ancient Rome were a tragic sign of Natura's destructive power, the calamitous triumph of a nature run amok.[6] It was a time of ecological desperation. Europe had been struck by the Black Death, a series of pandemic plagues that recurred throughout the fourteenth and fifteenth centuries, sweeping away nearly half its population and bringing chronic food shortages and economic crisis.

The humanists' laments bespeak an agonistic drama: it was man's heroic task to purge Rome, the symbolic cultural site, of the signs of decay, ruin and death brought by female Nature. The cultural "Renaissance" of the fifteenth century was, I suggest, partly spurred by the perception of nature as disorder, which required the supplement of art, to reassert human control over nature's dangerous forces in a time of acute need.

The chief site of culture's defense against nature's unraveling was not Rome, but Florence. When the plague hit Florence, the city's largest buildings were just completed or nearly so. In a time of social disarray, stable architectural structures can hold great psychological importance. Anxiety about the durability of the new buildings was expressed by the humanist Coluccio Salutati, who wrote in 1381 that "our Cathedral has developed a fissure and seems about to end in a state of hideous ruin."[7]

A hero soon came to the rescue. The Cathedral, or Duomo, was Florence's major building, and Filippo Brunelleschi's design and construction of the Cathedral dome was the largest creative project of the fifteenth century. Brunelleschi was the first artist to be celebrated for his *ingegno*, a term that embraced both artistic genius and engineering. Contemporaries compared Brunelleschi to Daedalus, the master artificer of ancient Greece. Like Filippo, Daedalus was an inventor of tools, including the wax wings by which he and his son Icarus escaped from Crete. Daedalus's invention of a device to transcend nature's limits effectively symbolized the triumph of technology over nature. And this was Brunelleschi's mythic achievement when, in 1418, he found a way to vault the dome of Florence Cathedral.

For fellow architect Leon Battista Alberti, Brunelleschi symbolized progress in the arts because he invented linear perspective and many technological devices. In treatises, Alberti affirmed the masculinity of the progressive enterprise. He described the family as a hierarchy ruled by a *pater familias*, the absolute master of his wife, children and servants. In Alberti's explanation, one-point perspective was a power hierarchy with gender dimensions, for it established the (male) artist's control of the entire visual field (metaphorically, the whole of nature).[8] These patriarchal hierarchies are, however, totally absent in Alberti's characterization of female Nature. For Alberti and the period in general, Nature was still more like a mother than a wife, and her powers were greatly revered. Alberti cautioned against open disobedience:

> For so great is Nature's strength that … she will always overcome and destroy any opposition or impediment…. We ought to be careful, then, to avoid any undertaking that is not in complete accordance with the laws of Nature.[9]

The Florentines had long intended to crown their new Cathedral with a dome, but lacked the technical knowledge. Why a dome? Scholars suggest the influence of the Pantheon in Rome, a major domed structure surviving from antiquity. Yet domes carried associations that have been insufficiently linked to Brunelleschi's cupola. Like many earlier Tuscan churches with domes, the Cathedral of Florence is dedicated to the Virgin Mary. In theological tradition, the body of Mary was understood as a house, or domus (Italian, *duomo*) for divinity. Mary herself was identified with Mater Ecclesia, the Mother Church, embodiment and protector of the congregation. Early Church Fathers defined the supportive love of Mater Ecclesia in earthy biological terms, such as the "sustaining bounty of her breasts."[10]

It cannot have escaped Renaissance Florentines that the shape of the Cathedral's dome resembled an emblem of the mother's nourishing power, the female breast. Paintings of the *Madonna lactans*, Mary nursing the Christ Child from one exposed breast, were prevalent in that period, a phenomenon that Margaret Miles connected with the recurrent plagues, when instability and fertility deficits drew people to symbolic expressions of nourishment.[11] In a period when images of the nursing Virgin Mary were ubiquitous, it would have been natural to imagine the Cathedral's dome as a nurturing female breast. Conceived in the midst of social crisis, the completed dome would provide both shelter and the psychological comfort of the nurturing and protective Mother Goddess.

Brunelleschi's contribution was to redirect the dome's significance from biological to technological. An art myth gives a clue to his agenda. In the account of Renaissance biographer Giorgio Vasari, Brunelleschi won the architectural competition to create the dome by producing a simple egg, which he challenged his competitors to stand upright. When each architect tried and failed, Filippo smashed the egg on the table, standing it on its broken end. When the others protested his trick, Brunelleschi replied that they would know how to construct the dome when they saw his design. Most writers consider this a typical Vasarian embellishment. But Alessandro Parronchi suggested nearly fifty years ago that Vasari might have told a garbled version of the truth. Noting that the unusual elliptical shape of Brunelleschi's dome resembles an egg, a natural form known for its exceptional strength, Parronchi proposed that Brunelleschi presented the egg to the committee as a model of organic structure.[12]

Parronchi's proposal has not dented Brunelleschi scholarship, yet the dome's unusually subtle shape supports his idea. This shape can be expressed mathematically – as a combination of a parabola and ellipse, or as something called a pointed fifth – but you see it more directly in an egg, which may have been the architect's model for the principle of dynamics needed to solve his problem. Effectively, he had stolen a secret from Nature. Brunelleschi said it himself in a sonnet: "Only the artist, not the fool / Discovers that which nature hides."[13]

An egg appears in Piero della Francesca's Brera *Madonna* of c.1472, held in dramatic suspension over the heads of the Virgin Mary and saints. Piero's egg alludes to the contemporary custom of hanging eggs within churches (including Florence Cathedral). In folk tradition, eggs were associated with fertility; when juxtaposed with Mary, it refers to the virgin birth and Christ's resurrection. But the egg's symbolic scope is broader. In ancient origin myths, the cosmic egg contains the universe in embryo, and it is a prime symbol of the divine creator, *natura naturans*. Piero suspended this egg in an architectonic cage: a captive specimen of Nature's order, on view in the brave new world of geometry and perspective. Captured and contextualized, the natural egg is claimed by Piero for art, nature's powers both tapped and tamed through representation.

The form of Brunelleschi's dome was doubly tied to emblems of female power – the nurturing breast of Mother Church, and the potent egg of Mother Nature. As Brunelleschi set out to rival Nature's creative power, his ambition conflicted with the dependency implied in the Mother-son relationship. Nature's manifestation of advanced physics in the lowly egg was kept a secret. Like Daedalus, who made wings for human flight, Brunelleschi imitated nature in order to defy her.

Brunelleschi's miraculous construction of the dome was instantly mythologized because the story matched men's aspirations: to raise the great dome unsupported from the ground and make it self-sustaining was to escape the gravity that is the earth's force. Quattrocento humanists and art theorists united to celebrate man's creative transcendence of physical nature, in what we might call "the aesthetics of upwardism."[14] Anticipating the dream of modern technology and science fiction, Marsilio Ficino described man as ascending in the air, like Daedalus, now on the brink of being able to live "without the aid of the body" that man's soul despises. Brunelleschi himself voiced the dream: "We rise above corruptible matter / And gain the strength of clearest sight."[15] For these men, the arts ascend through their victory over nature and by *homo faber*'s projected escape from the unfortunate tethers of gravity and flesh.

The Reproduction of Life

In Renaissance art theory, artistic creativity is described in metaphors of birth and reproduction. Its higher levels are represented by terms like *concetto* (concept or idea) and *ingegno* (genius or superior intelligence), words etymologically linked with the conception and generation of human life. Indeed, a birth metaphor is central to the self-characterization of the period that called itself a Renaissance. Yet the definition of art and culture through metaphors of birth stands curiously at odds with the description in Genesis of God as a "divine potter," and Plato's description of the cosmic creator as an artisan-god.[16] Man's creativity is equated with biological reproduction, while God's primal creation is described as art.

This displacement of biological and technological functions points to a missing, suppressed term, which is the reproductive power of both women and nature, everywhere apparent in the world yet unmentioned in the texts. Patriarchal mythologies invert the natural process. Eve is born of Adam's rib; Pallas Athena is born full grown from the head of Zeus. In the *Eumenides* of Aeschylus, Apollo declares that the mother is only a nurse of the seed that the father has planted. Aristotle echoed this dictum: "having no seed, she is without the *causa formalis* that can generate her own essence out of herself." She is thus subject to the male, who has both seed and essence.[17] Improbable as it seems, Aristotle's version of male-sponsored conception long remained Western civilization's chief paradigm for the roles of the sexes in generation. Not until the nineteenth century was the role of the female egg in reproduction scientifically confirmed.

Masculine denial of the female part in generation is all the more remarkable as it ignores the physical evidence: babies emerge from and are nourished by the mother's body. Is this simply a wishful reversal? Psychologist Karen Horney thought so, naming the phenomenon "womb envy."[18] To follow that line of thinking, men may have claimed both creativity and procreativity for themselves because they deeply feared that women might have both: the power to bring forth life from their bodies and creative ideas from their minds. Accordingly, men co-opted the paradigm to produce gender imbalance in their favor, claiming creativity for males while assigning female procreation a lower status.

To support the shaky claim that it was *they* who had it all, male artists touted their ability to create lifelike images. The fourteenth-century painter Giotto's breakthrough was announced by his contemporaries. Giovanni Boccaccio described Giotto's achievement as a competitive triumph, praising his ability "to make each thing resemble nature, intending through this, that these things have the very same effects as do those produced by nature."[19] The increasingly lifelike depiction of things and people is dramatically visible from Giotto through the next generation, stalled by pandemics in the later fourteenth century, then resumed in the fifteenth century.

A litany of progress in the imitation of nature reached its most developed form in Vasari's *Lives of the Artists* (1550), which presents art's rise from crude origins to modern perfection (in Michelangelo). The progressivism of artistic style was poised to claim success by its own standards. Nature can endlessly replenish the supply of living things in the world, but Art can produce an increasingly more lifelike succession of images. Nature's specimens not only are imperfect, they do not improve (pace Darwin). Progress thus ratchets up art's side in the competition with nature, since to make a *better* example of a living being is something nature cannot do.

To fifteenth-century eyes, art's miraculous advance on nature was led by Masaccio. The new standard set by this young painter (who died at twenty-seven) was recognized immediately, and he soon took his place as the next star after Giotto in the trajectory of progressivism. On the entrance piers of the Brancacci Chapel, Masaccio's image of Adam and Eve expelled from the Garden (Figure 3.2) directly faces Masolino's image of the Temptation.

Figure 3.2 Masaccio. *The Expulsion from Paradise*. Fresco. c.1424–27. Florence, Santa Maria del Carmine, Brancacci Chapel. Photo: Public domain, via Wikimedia Commons.

Masaccio's giant step beyond Masolino's timid and phlegmatic figures did not lie in greater anatomical naturalism, but in form and expressive impact. With their enhanced physical reality and the recognizable human emotions they so persuasively experience, these actors appear almost literally to have come to life.

By endowing his Adam and Eve with such marvelous humanity, Masaccio would seem to have echoed God's act of creation. But divine creativity was not yet ascribed to the painter, for the visual arts lagged behind other art forms in this respect. In 1482, the philosopher Cristofano Landino praised a certain poet as divinely inspired, creating *ex nihilo* like God, and contrasted his art with arts that merely transformed matter.[20] Because they worked with physical materials, painters and sculptors could not easily detach themselves from the inferior part of the form-matter dyad.

Upwardly mobile early Renaissance painters found a powerful strategy for negating the materiality that so stubbornly clung to art. With the new tool of linear perspective, an artist could create the illusion of a spatial environment on a wall or panel, seemingly dissolving its material surface. The painter's new creation, empty space, was a magic realm in which lifelike bodies seemed to move freely, and it diverged radically from the flat, tapestry-like surfaces of Late Gothic painting in Florence. Now space is put into a binary relationship with form. Like matter, space is dependent upon form to give it substance; otherwise, it is literally nothing.

Masaccio's bulky figures stand out against the recessive background, thanks to another powerful tool. Light and shade give the figures dimension and life, like God's inspiriting breath. Eve covers her genitals and breasts, while Adam's hands go to his head, the seat of consciousness. She is carnal, he is spiritual. Adam was the model of human free will and God-like creativity for Quattrocento humanists, who traced human arts and industry back to Adam's labors. They made an equal case for man's achievements in human reproduction. Gianozzo Manetti asserted that men "profit future ages" by constructing buildings, writing works of art and science, and by their "continuous production of children."[21]

Masaccio himself touted male reproductive power, as was revealed when the foliage that had covered Adam's genitals since the eighteenth century was removed by late-twentieth-century conservators. He turned out to be exceptionally well endowed, leading people to wonder why his genitalia were so flamboyantly displayed. Clearly, it was to convey the point that Adam would initiate human generation after the Expulsion. This idea had been forcefully expressed in the twelfth-century *Cosmographia* of Bernard Silvestris, which concludes in a celebration of the male sexual organs. These "unconquerable armies of procreation fight with death, renew nature and perpetuate the species."[22] Boccaccio built on Bernard's text, writing in the *Decameron* that Nature has brought the plague that damages men's genitals, undermining the only means by which life can be perpetuated on earth.[23] In a time of epidemic disaster, when nature was unusually adversarial to human fortune, the phallus was claimed to restore the natural order.

The theme of male responsibility for generational succession was conspicuously advanced in art. Jacopo della Quercia reiterated Adam's genital power in his Genesis relief sculptures at San Petronio in Bologna (c.1429–34). In the "Paradise Doors" created by Lorenzo Ghiberti for the Florentine Baptistery (1425–52), ten scenes representing the male ancestors of Christ are framed by the border figures of Adam and Noah, who conspicuously expose their genitals, undoubtedly to express their shared role in perpetuating the species – Adam's prime generative act and Noah's salvation of his people following the flood – as we are reminded by exegeses on Genesis.

The heroizing of male procreative power in public art was complemented by the minimizing of female presence in prominent imagery. In the twelve narrative scenes of the Brancacci Chapel (some completed by Filippino Lippi), not a single female appears on the entire left wall, Eve apart, and only a handful of women appear on other walls, solely in passive roles. In the Brancacci Chapel, men dominate the public spaces, by contrast to earlier scenes such as the *Good Government* fresco in Siena (1338), where women and men interact equally in public spaces. This renowned fresco cycle joins Brunelleschi's dome as a charged site of the displacement of women and nature by men and art.

Nature into Art

By the later fifteenth century, Venus had ousted Eve in Florentine art, most conspicuously as the leading character in Filippo Botticelli's *Primavera* and *The Birth of Venus* (Figures 3.3 and 3.4). Venus paralleled Eve in her association with sexuality and depiction as nude, yet she came to present a principle of pleasure and good rather than sin and evil. With genealogical continuity now less threatened, and with communal wealth vested increasingly in powerful families, art was commissioned by wealthy patrons who furnished their houses with imagery that supported the marital and birth economy.

Previously, the artist had challenged Nature in the reproduction of life; now he competes with her in the creation of physical beauty. Created within the intellectually refined ambient of the Medici family, Botticelli's monumental *Primavera* and *Birth of Venus* present a goddess philosophically charged with ancient dualities. For Plato, Venus (Aphrodite) was both a celestial and terrestrial goddess, whose dual dimensions represented intellectual and sensual love. Plato's interpretation of Aphrodite's duality was popularized in late Quattrocento Florence through Neoplatonist writings, particularly Marsilio Ficino's *Commentary* on Plato's *Symposium*.[24] Together, Botticelli's two paintings articulate a dialectic on the relation between Nature, Art and Beauty that radically advances the case for Art's superior powers.

The *Primavera* embodies springtime fertility, imaged as an orange grove carpeted with flowers, populated by mythological characters. The right-to-left progression presents a translation of the female from the natural to symbolic realm, and from nature into art. Chloris, a nymph pursued by the wind god Zephyris, undergoes metamorphosis, signaled by the real flowers flowing from

Figure 3.3 Sandro Botticelli. *Primavera.* Tempera on wood. c.1476–78. Florence, Galleria degli Uffizi. Photo: Public domain, via Wikimedia Commons.

Figure 3.4 Sandro Botticelli. *Birth of Venus.* Tempera on canvas. c.1484–86. Florence, Galleria degli Uffizi. Photo: Public domain, via Wikimedia Commons.

her mouth that turn into painted flowers on Flora's dress. She becomes Flora, goddess of spring and, here, a married woman.[25] It is a highly artful way to encourage marriage and reproduction, now by way of female beauty rather than masculine virility. The exquisite beauty of the female figures masks the purpose of marriage at that time, which was, as Luce Irigaray described it, a patriarchal convention in which women "represent a form of wealth circulated among men, whose reproductive use value underwrites the symbolic order of the fathers."[26] Viewing Botticelli's *Primavera*, a bride could envision herself, not as an asset to be passed from one man's house to another, but as an eternally youthful beauty in a chivalric romance.

Venus stands at the center of the transition from natural to artificial. Nominally a personification of Natura's generative powers, she neither generates nor affects the action around her. Rather, Botticelli's Venus is an elegant cultural artifact, whose adornments – a crescent moon pendant, pearls from the sea – allude to nature through the materials of art. As a virginal maiden, younger than Flora, Venus is subtly disempowered (the Virgin Mary was similarly juvenilized by painters in this period), and the Graces are also maidens. Mercury too is young, yet in the *Primavera* youth marks disempowerment only for women. Though outnumbered by females, the males initiate the significant movements: Zephyris assaults Chloris, Mercury pokes the clouds, Cupid shoots his arrow. Cupid is the keystone, both of the implicit arch behind him and of the painting. Aiming his arrow straight at the Graces, this little metaphoric stand-in for the painter transforms nature into art. The creation of female beauty is a masculine achievement.

Through gaze and gesture, Venus directs the viewer to join the left-moving progress, as the imaginary connoisseur of the three beautiful Graces. While the right half of the painting concerns the terrestrial world of *Venus vulgaris*, the left half is devoted to the higher realm of *Venus celestis*. Ficino similarly claimed in his Platonic commentary that while the Earthly Venus moves us to propagate, the Heavenly Venus instills the desire to contemplate beauty. Botticelli's three Graces exemplify the feminization of beauty within a masculine discourse. The progression ends with cloud-poking Mercury, who was for Neoplatonists "the ingenious god of the probing intellect." Botticelli's Mercury is from top to toe a work of art. With helmet, sword and shoes made of finely tooled metal, he is the antithesis of lusty Zephyris, the monochromatic, lightly draped wind god. Yet, together, Mercury and Zephyris drive both intellectual inquiry and the inseminating life force. Between them stand six women who do little but receive their energies.

In the *Birth of Venus*, painted several years later, the celestial Venus has become the central theme. In accord with ancient depictions of the goddess's arrival at Cythera after her birth from the testicles of Uranus, this fully nude Venus stands on a massive scallop shell. Her pudenda-covering pose is a direct quotation of the Greco-Roman Medici Venus. Indeed, every element in the painting is a quotation from art – not only Venus but also Zephyrus and Flora, characters carried over from the *Primavera*. In the *Birth of Venus*, art is now omnipresent, and nature is rearranged as the painter wishes. Botticelli flaunts his deviations from natural and ancient models. The intricately interlocked limbs of the *zephiri* cockily repudiate both anatomy and logic, while the huge shell casts a strong shadow, as if to emphasize its descent from Platonic form into an insubstantial material world. Waves form neat patterns on the water's surface, and the highly stylized trees, with their unnatural leaf structure and patterned gold hatching, are a calculated abstraction of the trees and leaves in the *Primavera*, which seem quite organic by comparison.

In this deliberate move away from natural appearances, and away from his own prior creation, Botticelli initiated a process of abstraction we might call progressive unnaturalism. He positions himself firmly as "against nature," forcefully asserting the artist's freedom to improve upon her creations. The licenses that Botticelli took led to the capricious distortions of the Mannerist painters Pontormo and Rosso Fiorentino who, a half century later, similarly thematized the artist's growing creative independence and deviation from nature's guidance.

The early sixteenth century brought an escalation in the rhetoric of competition. The epitaph composed by Pietro Bembo on the death in 1520 of Raffaello Sanzio, the most celebrated artist in Rome, introduces the language of victory and defeat: "Here lies Raphael; while he lived, Great Mother [Nature] feared that she would be conquered, and when he died, feared that she would die."[27] Nature is now personified as a fearful competitor of the artist, whose very existence depends upon her preservation by his art. The theme of art conquering nature was further sounded in poetic tributes that poured out to Raphael: art "defeats" nature by preserving human beauty that must fade; art "wins" by offering a more perfect beauty than is found in nature. Other encomia to Raphael celebrated the artist's divine creativity; Antonio Tebaldeo maintained that Christ "was the God of nature, you the God of art."[28]

Raphael himself had contributed to the upping of art's ante when, in a letter of 1516, he explained to Baldassare Castiglione how he painted his *Galatea* (Figure 3.5): "In order to paint a beautiful woman … I make use of a certain idea that comes into my head."[29] The letter may actually have been written by Castiglione, yet it marks a new stage in the art-nature relationship, for the artist is now said to create perfect beauty from a concept in his mind. The Quattrocento model of improving on nature was surpassed by the more ambitious notion of *grazia*. The idea of artistic grace connects the ideal with the seemingly natural, invoking an ancient concept of art that disguises itself as nature, *ars est celare artem*. Raphael was praised for the flowing movement and lifelike vitality of his figures, which he simulated through an informed application of artistic devices such as contrapposto. His Galatea is a composite of many sculptural prototypes, from antique to Michelangelo, and to connoisseurs she might have seemed a statue come to life.

In the *Galatea*, we see a signature of the creator. Unrelated to the story, three flying putti shoot arrows at Galatea, inscribing her as beauty on display, while a fourth perches on a cloud, holding an arsenal of arrows. The notion

Figure 3.5 Raphael. *The Triumph of Galatea*. Fresco. 1512–14. Rome, Villa Farnesina, Sala di Galatea. Photo: Public domain, via Wikimedia Commons.

of the archer as a stand-in for the artist, implied in the Cupid-*arciere* of Botticelli's *Primavera*, became explicit in the early Cinquecento. The metaphor was also used by Michelangelo and Dürer, and the archer-with-bow became the attribute of *Ingegno*, the personification of artistic genius in Cesare Ripa's *Iconologia* (1603).[30]

The Goddess in Arcady

The turn of the sixteenth century brought a new concept, in which art and nature were cast, not as competitors, but as complementary partners. This paralleled an innovative emphasis on gender complementarity in influential treatises of the time, such as Baldassare Castiglione's *Il Cortegiano* (1528). As feminist scholars have pointed out, theoretical gender equality probably signified the opposite: a reaction to the growing power of women in the cultural sphere.[31] By muting the voices of female protagonists in *The Courtier* (who were real women at the court of Urbino) Castiglione may have sought to tame anxiety about the newly emergent female court rulers in Italy, culturally ambitious women such as Isabella d'Este, whose art patronage is legendary, and others such as Cecilia Gallerani, Caterina Sforza and Caterina Cornaro. These matriarchs posed a conceptual threat to male artists and courtiers because their identification with the realm of culture potentially displaced the men and confused the categories.

Quite unlike this strategy of resistance was the more indirect treatment of gender issues in art, which reflect and may have facilitated a deeper transformation in men's psyches. Central to the changing perceptions of nature was the art produced in Venice, an ecologically fragile city built on water, with an unusual relationship to its natural environment. The Venetian Republic's precarious access to her mainland possessions placed those lands at the center of Venetian art and poetry, filtered through the nostalgic imagination.

No picture better represents the Venetian perspective than the *Concert champêtre* (c.1509–10), painted by Giorgione and probably finished by Titian (Figure 3.6). Two clothed men and two nude women consort in a lush verdant setting – an idyllic ambient of music and poetry, shepherds and sheep. The figures are harmoniously integrated with the trees and hillocks by their echoing shapes, and a soft, moist atmosphere envelops the whole. The image exemplifies the new poetic genre of pastoral, seen in literary texts such as Pietro Bembo's *Asolani*, and Jacopo Sannazaro's *Arcadia* – a genre born at Caterina Cornaro's court in Asolo. The pastoral, in which nature is conceived as timeless

Figure 3.6 Giorgione (and/or Titian). *Concert champêtre*. Oil on canvas. c.1509–10. Paris, Musée du Louvre. Photo: Public domain, via Wikimedia Commons.

and restorative, offers escape from uncomfortable modern realities, but also a way to disguise and deflect them. It is suggestive that pastoral art and poetry emerged in North Italian courts at a time of unprecedented female political assertion: as women threatened to rule the sphere of culture, they had to be reinscribed in nature. But the *Pastoral Concert* articulates the art-nature relationship in a more nuanced way.

Seated on the ground in fine clothes is a modern man who is *in* nature but not *of* it. He plays a lute and sings his appreciation of nature's beauties, a cultural homage contrasted with the natural music of the shepherd's bagpipe. His urban outfit differs from the rough and plain garb of the adjacent rustic, barefooted like the nude women before him. Together, the three males present three positions in the culture/nature spectrum. The nude women in the *Concert* represent two aspects of the figure known in Venice as Venus Physizoa, the goddess of living nature. The generative Venus pours water into the fountain, sustaining and renewing human life. The tutelary Venus sits in dialogue with the men, their cultural guide in the realm of nature. Her reedpipe brings to mind the reedpipe awarded to victors in ancient pastoral singing contests – an allusion, along with the music-making shepherd, to rustic art forms.

In the Venetian perspective, art is not necessarily superior to nature; rather, they are interdependent. The lutenist teaches art's higher forms to the rustic, yet he also draws psychic refreshment from nature. In this place he can only visit, the lutenist's own song is the least enduring form of art. In Arcady, where nothing dies, the dry reedy voice of the shepherd's pipe endorsed by tutelary Venus will live on forever, but the modern lutenist has only the claim of his difference to stake. In singing his fragile high-art song, he takes the risk that divine Venus will turn her back on him, to sustain the process of endless generation that is entirely indifferent to art. The *Concert* poses a philosophical question about the relative powers of nature and art that is unanswered, but finely held in balance.

The painting represents a new stage of familiarity between cultural man and female nature. He now wants to know her on adult terms. This psychic maturing of the male artist, now a philosophical quester, operates on an entirely different level from the social desire to control women. Rather, it is a stage in man's evolving effort to define himself in relation to nature – in Jungian terms, to reconnect with his feminine anima. In this journey, as mapped by Venetian artists, Natura will be his guide to productive and mature artistic creativity.

The female body was a default signifier for Nature, but in the allusive imagery of Venetian painting, elements of landscape also took on symbolic meaning. Giorgione's *Three Philosophers* of 1507–09 (Figure 3.7) is a case in point. The philosophers are probably mathematicians or astronomers, as indicated by the set square, compass and tablet of astrological images in their hands. These attributes led scholars to suggest particular names, such as Pythagoras or Ptolemy. But we need not fix specific identities to recognize that a relationship is specified between the men's philosophical interests and the natural physical environment.

Figure 3.7 Giorgione. *Three Philosophers*. Oil on canvas. c.1507–09. Vienna, Kunsthistorisches Museum. Photo: Public domain, via Wikimedia Commons.

The seated man stares at the open mouth of a cave (more prominent before the picture was abridged on the left side). In the distance is a rising sun, its bright glow contrasted with the dark interior of the cave. This is a juxtaposition of gendered entities, the sun as masculine and the cave as the female earth. On the tablet held by the bearded man, emblems of the sun and crescent moon represent a variant polarity. In Renaissance astrology and alchemy, masculine Sol and feminine Luna are equal in value, the two perfect metals not reducible by fire. These entities were personified in the figures of Apollo the sun god and Diana the moon goddess. Though equal in principle, Apollo and Diana were depicted by Albrecht Dürer (1501–03) and Jacopo de' Barbari (c.1503–05) as the ascendant masculine triumphing over a waning and weak feminine.[32]

The privileging of the sun did not occur in a vacuum. Its unspoken corollary was the growing devaluation of the ancient figure of mother earth, and her displacement in the conceptualizing of nature. The dismantling of the old world order was forecast in philosophy and art before it was accomplished in science. Just as Brunelleschi's mythic displacement of Natura in the Cathedral dome helped to foster the rising power of technology, so art prepared the way for heliocentrism, by rendering the goddess's dethronement a thinkable thought.

At this historical moment, Copernicus was constructing his revolutionary proof that the earth revolves around the sun. *On the Revolution of Celestial Orbs* was written between 1507 and 1530, and published in 1543. The Copernican model, in which heliocentrism replaced geocentrism, was foundational for astronomy. Giorgione's *Three Philosophers* may refer to this radical new worldview, particularly since he could have known of Copernican thought through personal connections.[33] And it could have been the other way around. As scholars have noted, Copernicus arrived at his discovery through mathematical calculation, but he was buoyed by a contemporary interest in the sun expressed by modern philosophers and artists.[34]

The *Three Philosophers* presents new and old concepts of nature. The young man is contrasted by his active use of measuring tools with the bearded old philosopher, who holds, but does not use, a pair of dividers. Seated on a geometrically chiseled shelf of rock, the young man is the protagonist of the new direction of learning, in which truth is sought in a nature subjected to analytical measurement. The old philosopher's understanding of cosmological bodies is based in astrology, soon to be replaced by modern astronomy. In the diagram on his tablet, the moon is positioned above the sun, implying that the old man is biased toward the feminine, a worldview that was beginning to be challenged.

But the painting does not simply celebrate the new world order. In their rich warm colors, the older philosophers match the sun's luminosity. Behind them are leafy green trees; behind the young philosopher is a tree stripped bare, possibly dead. The older men's venerable presence upholds a worldview long revered, in which the elements of nature, terrestrial and celestial, are interactive and affect each other. Advanced by contemporary alchemists and astrologers, this worldview was later sustained by philosophers such as Paracelsus, Bernardino Telesio, Giordano Bruno and Tommaso Campanella, whose cosmologies embraced living, organic bodies that influence one another.[35]

Nature's laws proved to be more complex than the organic philosophers realized, yet they were not wrong in viewing the earth as an organism. In the new intellectual framework of emerging science, personified by Francis Bacon, Thomas Hobbes and René Descartes, the organic concept was replaced by a mechanical model, in which nature was an impersonal machine, to be objectively studied by scientists who saw themselves as detached from it.[36] Modern ecology has challenged the mechanistic underpinnings of early modern science, and now emphasizes the primacy of interactive processes in a nature that includes us. The interdependency of Earth's systems is today a felt reality, as global warming causes icebergs to melt and oceans to rise, and heat waves produce wildfires that claim lives.

In the older, alchemical system, the womb of the earth was a living spirit. The shadowy cave on which the young philosopher's gaze is fixed could represent nature as the object of analysis or, equally, as the source of life and wisdom. In antiquity, caves represented the Source – the mythic point where the earth's rivers converge, the source of nature's mysteries and the origin of life. An allusion to this myth is evident in a little stream of water at the mouth of the cave, a small rivulet that flows a few feet and then ends. Giorgione's visual metaphor of a Source whose waters are drying up echoes Jacopo Sannazzaro's *Arcadia*, a contemporary pastoral poem, which ends in the disintegration of the pastoral vision. Mourning the loss of a nature no longer accessible, the shepherd Sincero laments, "Our Muses are extinct; our laurels are dried up, our Parnassus is ruined; the woods are all mute; the valleys and the mountains are grown deaf with sorrow."[37]

In *The Three Philosophers*, Giorgione acknowledges both the rising new sun of science and dying old nature's wisdom, maintaining the artist's right to juxtapose positions without taking sides. His ambivalence is more pronounced in the *Tempesta*, a cryptic painting that has generated over a hundred different interpretations. The *Tempesta*'s larger cultural signification, like that of the *Three Philosophers*, is clarified in the metanarrative of nature and art. Giorgione images the male artist-quester's dramatic encounter with a modern Natura, his anima – an image of art and nature joined by a transformative thunderstorm that brings about their seemingly unachievable psychic connection.[38] These paintings express an uncertainty, a nostalgic ambivalence, about man's gradual separation from Natura in the rise of technology and science.

In his last paintings, the aged Titian resumed the discourse of art and nature, recombining many earlier formulations. The *Nymph and Shepherd* (Figure 3.8) features an ivy-crowned, flute-playing man and a reclining nude woman, together in a murky pastoral realm. The precise subject has baffled everyone, but within the Giorgione-Titian continuum, they take on larger, composite identities. He is the artist-quester who pursued nature's secrets through the changing identities of musician, shepherd, pilgrim, satyr, and goat. Her identity is more constant: with her right hand she appears to perform the generative act of self-stimulation first seen in Giorgione's *Sleeping Venus*, which identifies her as the goddess Natura.[39] Their intimacy hints at union between these archetypes, now in poignant terms. Behind them, a goat, ancient companion of the

Figure 3.8 Titian. *Nymph and Shepherd.* Oil on canvas. 1570s. Vienna, Kunsthistorisches Museum. Photo: Public domain, via Wikimedia Commons.

earth goddess Terra, ruts at a regenerating tree stump – a metaphor for the artist's productive interaction with nature.

We see the pair as vibrating, shimmering, insubstantial forms, as if through a veil. In Titian's late paintings, the broken brushstroke dissolves objects and unites form and space as a continuum. As in twentieth-century quantum mechanics, there are no such things as solids and voids, only a field of molecular structures constantly in motion. As if to dramatize the atomic vision of Lucretius,[40] Titian produces a vision of instability and change. Leonardo da Vinci had dissolved contours, initiating the merger of form and space, but Titian goes farther, instaurating movement as the driving force of the universe. He claims pictorially that space and matter might *matter* more than form, for the shifting brushstroke is all we see, and form constantly dissolves under its aggressive action. It is a metaphoric expression of nature as dynamic process, and its creative action upon art.

The *Nymph and Shepherd* offers a precise counterpart to the philosopher Giordano Bruno's description of a world in which nature "is in things,"[41] where matter is infused with light, air is thick with substance, and darkness glows resplendently. In Titian's vision, sky and horizon merge in a turbulent whirl, as if the earth were to burst into flames and dissolve. In this uncertain, darkening world, two figures are in intimate yet tentative communication. He hesitates, as if awaiting her response; she has turned away, seemingly inaccessible, yet she pauses to look back. She has heard his flute, but it's understood to be in parting. It's no longer that Natura is becoming obsolete, as Giorgione briefly boasted, but that she is leaving him. Titian's *Nymph and Shepherd* effectively expresses the extent to which the artist saw nature dying in his time. A world is coming to an end, and the last Renaissance figuration of a living Natura pays tribute to her departure.

Natura Bound

Elsewhere, Natura became an opponent to be conquered and subjugated. A conspicuous sign of change was the eroticizing of the female body. Marcantonio Raimondi's book of pornographic engravings, *I Modi* (1524–27), with erotic sonnets by Pietro Aretino, thematized modes of sexual intercourse. In paintings, rape scenes dramatically increased, as Jupiter the Olympian womanizer takes his female prey by stealth, disguised as a bull or a shower of gold. Lustful satyrs uncover graphically nude female bodies – an obscene inversion of the ancient idea, expressed by Macrobius (354–430), that Natura withholds her secrets from vulgar eyes by a protective veil, and her exposure to them is an offense to the goddess.[42] In these new tropes, the male is aggressively dominant, and holds his position by force. Art no longer supports a procreative or collaborative agenda; it now sustains the fantasy life of the amoral sexual adventurer.

Paralleling the eroticized female body was the invaded female body, as anatomical dissection and illustration began to employ female models. In a development spearheaded by Andreas Vesalius, nature's biggest secret, the female body's mysterious capacity to generate life, would be uncovered by scientific investigation. The title page of Vesalius's influential anatomical treatise, the *Fabrica* (1543), shows a female body in a frontal view that exposes its cavernous interior. The organ under scrutiny is a womb. As Katharine Park has shown, this image underscored the female body's symbolic identity as a representation of Nature, who held no mystery in the face of advancing science.[43]

The scientific demystification of the female body was accompanied by a rising rhetoric of art's triumph over nature. By the time of Michelangelo's funeral in 1564, the theme had become a topos. In multiple verses affixed to his bier, the divine Michelangelo was cast as Nature's heroic rival. For Vasari, Michelangelo surpassed not only others who surpassed Nature, but the ancients who also surpassed her. In a treatise of 1567, Vincenzo Danti claimed that architecture was superior to other arts because it defeated nature.[44]

A more sinister measure of the changed power balance appears in new genres of imaging Natura that emerged in the Cinquecento. Monstrous female-animal hybrids such as harpies, gorgons and chimeras pop up in the *grottesche* found in decorative borders: disabled, armless females trapped in bird bodies, sometimes shown in bondage. Fantasies of domination appear in winged boys riding fantastic dragons paired with female-headed serpents. As Michael Camille has shown, little-noticed marginalia often express hostile subconscious attitudes – in this case toward woman/nature.[45] To present a woman as animalized and bound renders her simultaneously dangerous and powerless.

Art's defeat of Nature is violently imaged in Benvenuto Cellini's bronze statue of *Perseus Slaying Medusa* (Figure 3.9), which stands in Piazza Signoria, the civic center of Florence. The Greek hero's slaughter of the Gorgon is shown as a gory decapitation. Perseus triumphantly displays Medusa's head and its dripping ligaments, his feet planted on her mangled body. The patriarchal Medusa, a dangerous woman whose gaze turns men to stone, inverts an archetype of Nature. In older legends, Medusa-Metis was a serpent-goddess who stood for female wisdom and death itself. To see her face – that is, to die – was to be turned into (funerary) stone.[46] Cellini would not have known the earlier myth, yet the savage excess of his statue connotes the presence of a latent female power that must perpetually be denied. His commission did not call for the body of Medusa at all; Cellini gratuitously supplied this mangled yet oddly eroticized body, modeled on that of his mistress.[47] The eruption of misogynist violence in this highly public image surely played a part in the desacralization of a once divine Natura.

Cellini's debasement of the feminine imaginary is relentless, down to the images of Diana of Ephesus on the statue's marble base. The ancient Temple of Artemis at Ephesus sustained a mother-goddess cult, renowned in antiquity. In her primordial image, preserved in two majestic statues in the museum at Ephesus, the goddess appeared with four rows of bull's testicles across her chest. In the Renaissance, Diana of Ephesus became a symbol of Nature, the raw fertility emblems now depicted as breasts. In Cellini's version, this iconic figure is radically disempowered, with no legs and virtually no arms, a trickle-down

Figure 3.9 Benvenuto Cellini. *Perseus Slaying Medusa.* Bronze, with marble base. 1548–54. Florence, Piazza della Signoria, Loggia dei Lanzi. Photo: Scala/Ministero per i Beni e le Attività culturali/Art Resource, NY.

vestige of the monster slaughtered above. With swelling multiple breasts shaped into ornamental arabesque curves, these stone-cold Dianas project an empty fecundity, emblems of a Natura drained of her throbbing life-blood.

Cellini's imaging of a captive and defeated nature anticipated by half a century the verbal debasement of Nature by Francis Bacon. In *The Masculine Birth of Time* (1602–03), Bacon described the objectives of modern science in terms of aggressive sexual imagery. Nature must be "hounded in her wanderings," "taken by the forelock," and "put in constraint." Through metaphors of rape, conquest and torture, he advised the newly virilized scientist to "to bind her to your service and make her your slave."[48] Feminist scholars have deplored the misogynist underpinnings of Bacon's model, but he did not invent the metaphors. The transition from an organic to a scientific worldview took place over several centuries, supported by visual and rhetorical models that gave concrete form to the gender ideology of masculinist science.

Dissolving the Binary I: Women Artists Break the Mold

Just when Cellini was completing his *Perseus*, some drawings by a young female artist, Sofonisba Anguissola, were shown to Michelangelo. One of them depicted a weeping boy who has been pinched by a crayfish, possibly Sofonisba's response to Michelangelo's comment on her earlier drawing of a laughing girl, that a crying boy would have been better. Michelangelo may have meant to direct the young artist to a worthier theme: male, not female; tragic, not comic. But Sofonisba's laughing girl was the more potent image, if it is identical with her drawing of a smiling girl teaching her old nurse to read. As mentor to her nurturer, the girl exults in a traditionally forbidden role, appropriating masculine cultural agency to reformulate the nature-culture construct on new terms – not an opposition but a continuum, expressed as a symbiotic interchange between two females.

In these drawings and in *The Chess Game* (Figure 3.10), Sofonisba undermined the gender construct that obliged women to play nature to men's culture. She depicted two of her younger sisters playing chess, their match closely followed by a nurse and their youngest sister. This famously intellectual game,

Figure 3.10 Sofonisba Anguissola. *The Chess Game*. Oil on canvas. 1555. Poznán, Poland, Muzeum Narodowe. Photo: Erich Lessing / Art Resource, NY.

stereotypically masculine, is now a natural and spirited activity for girls, to judge from the flourishing tree and foliage behind the giggling young Minerva. The sisters compete with each other – Lucia has won and Europa concedes – yet the flow of gazes connects all three sisters with the painter outside the frame, emphasizing their interdependence. The chess game here stands for female cultural achievement in general, especially in the art of painting. Both Lucia and Europa were trained in painting by Sofonisba, and all six girls in their upper-class household were well-educated. In this extraordinary image, Sofonisba ennobles the theme of a female community bonded through kinship and artistic aspiration.[49]

During the early modern period, women artists grew in number. Yet they were still relatively rare, and their uniqueness attracted patrons and collectors of rare treasures. Sofonisba Anguissola's invitation to the Spanish court of Philip II was replicated in other invitations to other women artists. Such isolated social prestige was offset, however, by broader masculine resistance to women's presence in art professions. Female artists were barred from life drawing classes; they didn't receive major commissions, and they were inferiorized in the critical language used to describe their art.[50]

Such proscriptive practices were misogynist, but they were also reactive. Women artists disrupted the gendered nature-art dyad by calling its very premise into question. Their very existence presented a radical challenge, because they combined the two entities, female nature and male art, so long held in opposition. Natura was no longer tenable as an abstraction but, more threateningly, was now personified in living women.

Some women artists challenged the system through their art, presenting new models of female behavior. Sofonisba was one, Artemisia Gentileschi was another. In Artemisia's paintings, women perform heroic deeds with virile agency; her Judith's decapitation of a helpless male was terrifying (and still is) for its radical transgression of gender norms. Tropes of masculine art and feminine nature continued, as did defenses of men's natural sovereignty in art. Yet once female artists escaped Pandora's box, the confining metaphor no longer carried the force of an unquestioned truth.

Dissolving the Binary II: The Feminizing of Art

As science and technology advanced, the binary was destabilized in another way. Representative is the Studiolo of Francesco I de' Medici (Palazzo Vecchio, 1569–72), which thematizes the relationship of nature and art. The Grandduke's collection of pearls, coral and precious stones were stored in cupboards faced with paintings that corresponded to his treasures – scenes of gold and diamond mining, a bronze foundry. In a ceiling painting, Natura passes a torch to Prometheus, symbol of man's power to transform nature for his purposes. As capitalism expanded to the New World, Nature was widely regarded as an exploitable resource. In the Studiolo, natural elements acquire value from their conversion into art. Precious stones are displayed in gold and silver mounts.

This demonstration of art's perfection of nature's rarities anticipated the *Kunstkammers* and *Wunderkammers* popular in the next century.[51]

Paintings in the Studiolo similarly contextualize nature's products within luxurious beauty. In *Perseus Liberating Andromeda*, Vasari sublimated *naturata* into *maniera*, the ascendant style in Florence. Blood streaming from the gorgon's decapitated head turns into hard, spiky coral, admired by elegant nymphs. Both sexes appear as androgynous bodies, dainty in movement and affected in gesture. Perseus's feminized elegance is an upstaging response to Cellini's virile counterpart. So anxious were the *maniera* artists to exhibit grace, they were willing to risk effeminacy.

Ironically, Mannerist art began to look like that feminine Other its proponents claimed to have defeated. This is the logical end of a course that began with Botticelli's feminization of beauty, and the emergence of *grazia* as an artistic ideal. To explain why, after the Florentine artists' long-standing effort to detach male artists from feminine nature, they wound up feminizing art, we might invoke a Freudian psychological concept: the return of the repressed. From the artists' collective subconscious came real nature's correction of a sociopsychological disorder, recognizable even in the perverse image of femininity in which they presented her.

The feminizing of art by artists was trumped by the advocates of science. As Lorraine Daston and Katharine Park demonstrated, *Wunderkammer* treasures that expressed the fictive collaboration of nature and art inspired Bacon and Descartes to postulate that nature and art were effectively the same.[52] Both nature and art were gendered feminine in this conflation, and contrasted to virile science. When these philosophers spoke of art, they usually meant "mechanical" arts such as dyeing or glassmaking. They were not talking about the arts of painting, sculpture and architecture. Yet through irrelevance-by-omission, newborn science also feminized the arts we moderns call "fine art," as the passive female partner in a dyad that was no longer art vs. nature, but science vs art/nature. A new power structure emerged, in which feminized art played nature to masculine science, a hierarchized binary that reappeared in nineteenth-century France, as Norma Broude has shown.[53]

Dissolving the Binary III: Form and Matter in a New Key

At the turn of the seventeenth century, Italian painting was newly virilized with the advent of a vigorous naturalism in the art of Caravaggio. In his closely observed depictions of humans in their unimproved natural appearance, untouched by *grazia*, Caravaggio forged an innovative connection with nature on new terms – he both privileged nature and de-gendered it. The formerly female Natura has vanished, to return as a pervasive substantive reality. Ordinary things – a loaf of bread, a rotting apple – appear not as degraded *naturata* but as humble carriers of philosophical or religious truths.

A new form-matter relationship is staged in Caravaggio's chiaroscuro. The shadow cast by an object is not a thing at all, but intangibly affirms the thing's presence. Painting the shadow's shape, Caravaggio reifies both the object and its "negative" counterpart, whose substance is shrouded. One might say that Caravaggio invented the concept of dark matter – modern physicists' hypothesis of an undetectable yet powerful form of matter operative in the universe – three centuries before its time would come in science. As in modern physics, the tangible and intangible, visible and invisible, are interactive parts of material reality.

The formative role of Caravaggio's shadow is comparable to that of Titian's late-work brushstroke, which interacts with form to generate motion. Michelangelo developed a similar vision of the form-matter relationship in the so-called "unfinished" sculptures of his later years. Having earlier described the block of marble as containing a living essence waiting to be discovered by the artist, a conception of matter as generative, Michelangelo left many statues unfinished – arguably, because in that state they expressed the creative participation of unworked stone, a birth-giving matrix that is consubstantial with the form it contains.[54]

These new form-matter relationships anticipated major shifts in philosophy and science. Aristotelian form-matter was finally replaced by atomic theory, and the view of matter as passive and inert was proven to be wrong. The dynamic images of form and matter united by motion in time by Leonardo da Vinci, Titian, and Michelangelo prefigured the new science of motion developed by Galileo and Newton, in which planets and stars, electrons and neutrons, swirled in a universe newly conceived as dynamic.[55]

Conclusion

Two models of the human-nature relationship, one of detachment and the other of connectedness, are visible in the histories of both art and science. In one model, the Earth is a utilitarian spaceship; in the other, it is co-extensive with ourselves. These two positions remain gendered, not by the sex of their proponents, but by mind-sets that have been ascribed to the sexes. The masculinist model of man's conquest of nature dominated the early history of science and it underlay the subtle privileging of Florentine art in twentieth-century art history. But now, in the twenty-first century, in growing realization that the masculinist exploitation of nature has brought us to the brink of ecological disaster, climate change scientists have parted ways with Earth's rapists and, sometimes in collaboration with artists,[56] they seek to better understand Nature's complex operations to prevent Armageddon. A new model has emerged that puts nature and culture, nature and science, in partnerships of mutual self-interest.

Notes

1 For a fuller history of Nature's pre-Renaissance imaging, see Garrard, 2010, chap. 1.
2 Sherry B. Ortner," Is Female to Male as Nature is to Culture?," in Michelle Z. Rosaldo and Louise Lamphere, eds., *Woman, Culture and Society* (Stanford: Stanford University Press, 1974), 67–87; Carolyn Merchant, *The Death of Nature:*

Women, Ecology, and the Scientific Revolution (San Francisco: Harper and Row, 1980); Evelyn Fox Keller, *Reflections on Gender and Science* (New Haven: Yale University Press, 1985).

3 Aristotle, *Metaphysics*, book δ, 1029a20; Garrard, 2010, 12–13.

4 See Garrard, 2010, chap. 5, for Leonardo's multi-dimensional position in the art-nature discourse, which is too complex to represent in this essay.

5 Susan R. Bordo, *The Flight to Objectivity: Essays on Cartesianism and Culture* (Albany: State University of New York Press, 1987), and 6–7 for examples from developmental psychology.

6 See Garrard, 2010, 33, and 322nn1, 2 for these citations.

7 Garrard, 2010, 34 and 323n14.

8 L. B. Alberti, *The Family in Renaissance Florence (I libri della famiglia)*, trans. Renée Neu Watkins (Columbia: University of South Carolina Press, 1969), 207–8; and Alberti, *De pictura* 2.30ff (Grayson ed., 1972, 67ff).

9 Alberti, *On the Art of Building in Ten Books (De re aedificatoria* 2.2), trans. Joseph Rykwert, Neil Lynch, and Robert Tavernor (Cambridge MA: MIT Press, 1988), 35–36.

10 St. Cyprian, *De unitate*, 5; see Garrard, 2010, 37, for this and other Church Father descriptions.

11 Margaret R. Miles, "The Virgin's One Bare Breast: Nudity, Gender and Religious Meaning in Tuscan Early Renaissance Culture," in Norma Broude and Mary D. Garrard, eds., *The Expanding Discourse: Feminism and Art History* (New York: HarperCollins, 1992), 27–39.

12 Alessandro Parronchi, "L'enigma della cupola: Perché 'sanza armadura,'" in Alessandro Parronchi and Eugenio Casalini, eds, *Le due cupole* (Florence: Convento SS Annunziata, 1977), 7–27.

13 Quoted in Isabelle Hyman, *Brunelleschi in Perspective* (Englewood Cliffs NJ: Prentice-Hall, 1974), 33.

14 Garrard, 2010, 51–52.

15 Marsilio Ficino, *Theologia Platonica* 2:224–5 and 226, quoted by Charles Trinkhaus, *In Our Image and Likeness: Humanity and Divinity in Italian Humanist Thought* (Chicago: University of Chicago Press, 1970) 483, 485; and Hyman, *Brunelleschi in Perspective* (Englewood Cliffs NJ: Prentice-Hall, 1974), 33.

16 See Garrard, 2010, 13, 54.

17 Aristotle, *On the Generation of Animals* 729.22; Garrard, 2010, 55.

18 Karen Horney, *Feminine Psychology* (New York: W. W. Norton, 1967).

19 Giovanni Boccaccio, *Decameron* 6.5; trans. by Joan Kelly Gadol, *Leon Battista Alberti: Universal Man of the Renaissance* (Chicago: Chicago University Press, 1969), 100.

20 Landino, commentary on the *Ars poetica* (for full citation see Garrard, 2010, 329n44).

21 Gianozzo Manetti, as quoted and translated in Trinkhaus, *In Our Image and Likeness: Humanity and Divinity in Italian Humanist Thought* (Chicago: University of Chicago Press, 1970), 1:240–41.

22 Bernard Silvestris, *Cosmographia* 2.14.153–80; Garrard, 2010, 61 and 330nn76, 77.

23 See Giuseppe Mazzotta, *The World at Play in Boccaccio's 'Decameron'* (Princeton: Princeton University Press, 1986) 26–30.

24 Marsilio Ficino, *Commentary on Plato's Symposium on Love (De Amore)*, trans. Sears Jayne (Dallas TX: Spring Publications, 1985).

25 *Brunelleschi's Egg*, 96, for Lilian Zirpolo's reading of the *Primavera* as patriarchal marriage ritual.

26 Luce Irigaray, *The Speculum of the Other Woman*, trans. Gillian C. Gill (Ithaca NY: Cornell University Press, 1985), 170–91.

27 John Shearman, *Raphael in Early Modern Sources, 1483–1602* (New Haven: Yale University Press, 2003), vol. 1, no. 1520/74.

28 John Shearman, *Raphael in Early Modern Sources, 1483–1602* (New Haven: Yale University Press, 2003), vol. 1, no. 1520/85.
29 John Shearman, *Raphael in Early Modern Sources, 1483–1602* (New Haven: Yale University Press, 2003), vol. 1, no. 1522/16.
30 Garrard, 2010, 240 and Fig. 180.
31 Valeria Finucci, *The Lady Vanishes: Subjectivity and Representation in Castiglione and Ariosto* (Stanford: Stanford University Press, 1992); Pamela Benson, *The Invention of the Renaissance Woman* (University Park PA: Pennsylvania State University Press, 1992).
32 Garrard, 2010, figs. 129, 130.
33 Garrard, 2010, 173–73, and 352n110 for Jeremi Wasiutynski's tracing of connections between Giorgione and Copernicus.
34 Frances A. Yates, *Giordano Bruno and the Hermetic Tradition* (London: Routledge and Kegan Paul, 1964), 151–56.
35 Merchant, *Death of Nature*, chap. 4.
36 Merchant, *Death of Nature*, chap. 7.
37 Quoted in David Quint, *Origin and Originality in Renaissance Literature: Versions of the Source* (New Haven: Yale University Press, 1983), 67.
38 On the *Tempesta, see* Garrard, 2010, 180–86.
39 Garrard, 2010, 169.
40 Garrard, 2010, 154–55, on Lucretius's importance in the Renaissance, especially for Leonardo.
41 Giordano Bruno, *The Heroic* Frenzies *[De gli eroici furori]*, trans. Paul Eugene Memmo Jr. (Chapel Hill: University of North Carolina Press, 1964), 185; Garrard, 2010, 224, 233.
42 Garrard, 2010, 20 and 320n86.
43 Katharine Park, *Secrets of Women: Gender, Generation and the Origins of Human Dissection* (New York: Zone Books, 2006), 150–59 and chap. 5.
44 Garrard, 2010, 275 and 370n2.
45 Michael Camille, *Image on the Edge; The Margins of Medieval Art* (Cambridge: Harvard University Press, 1992), esp. 10 and 26.
46 Garrard, 2010, 288–89, and for literature, 372n59.
47 Garrard, 2010, 286, and 371nn54, 55, for important essays on this statue by Yael Even and Margaret Carroll.
48 Francis Bacon, "The Masculine Birth of Time (1603)," in Benjamin Farrington, ed. and trans., *The Philosophy of Francis Bacon* (Liverpool: Liverpool University Press, 1964), 59–72.
49 On the works discussed here, see Mary D. Garrard, "Here's Looking at Me: Sofonisba Anguissola and the Problem of the Woman Artist," *Renaissance Quarterly* 47, no. 3 (1994), 556–622.
50 On the last point, see Fredrika H. Jacobs, *Defining the Renaissance Virtuosa: Women Artists and the Language of Art History and Criticism* (Cambridge: Cambridge University Press, 1997).
51 Garrard, 2010, 305–08, with references.
52 Lorraine Daston and Katharine Park, *Wonders and the Order of Nature* (New York: Zone Books, 2001), 290–96.
53 Norma Broude, *Impressionism, A Feminist Reading* (New York: Rizzoli, 1991; HarperCollins, 1997).
54 Garrard, 2010, 310–12 (Caravaggio), and 266–74, esp. 269–71 (Michelangelo).
55 Ernan McMullen, *The Concept of Matter in Modern Philosophy* (Notre Dame IN: University of Notre Dame Press, 1978), introduction, esp. 27.
56 E.g., *Diane Burko: Seeing Climate Change*, exh. cat., with essays by Mary D. Garrard and Norma Broude (Washington, DC: American University Museum, 2021).

4 Universal Application

The Natural World as Metaphor and Phenomenon in Melville, Thoreau, and Dickinson

Brian Yothers

Introduction

In his last published volume of poetry, Herman Melville reflected on the inevitable conflict between human aspiration and the material world in which human plans play themselves out. In his poem "Fragments of a Lost Gnostic Poem of the Twelfth Century," published in *Timoleon*, Melville's imagined medieval gnostic reflects:

> Found a family, build a state,
> The pledged event is still the same:
> Matter in end will never abate
> His ancient brutal claim.
>
> (Melville, *Published Poems* 284)

That Melville's inspiration for the poem is closer to home in terms of chronology becomes evident in the second titular fragment:

> Indolence is heaven's ally here,
> And energy the child of hell:
> The Good Man pouring from his pitcher clear
> But brims the poisoned well.
>
> (Melville, *Published Poems* 284)

Although the poem is attributed to a twelfth-century gnostic (perhaps a Cathar/Albingensian, given the timing of the poem),[1] the poem evidently reflects the influence of the British Romantic poet William Blake, whom Melville read as part of his self-education as a poet, acquiring a copy of Alexander Gilchrist's 1863 volume *Life of William Blake "Pictor Ignotus" with Selections from His Poems and Other Writings* in 1870. The second stanza clearly mirrors the language of Blake's *Marriage of Heaven and Hell*, with its connection of "energy" to "hell," but what links to the two fragments is a wry observation of the ironic subversion of human beings' intentions

DOI: 10.4324/9781003385202-5

when they try to control the world around them. Whether the assertion of human agency is small-scale and common ("Found a family"), or large-scale and historic ("build a state"), Melville suggests that humans cannot control the consequences of their actions. This suggestion is intensified by the image of the "Good Man," whose efforts at purification "But brim the poisoned well." This late poem by Melville points to something that shapes his wider career; namely, his skepticism about human efforts at imposing their own aspirations and intentions on the natural world. This skepticism is shared with two of Melville's contemporaries who are centrally concerned with the tension between human aspiration and natural phenomena, Henry David Thoreau and Emily Dickinson, and it shapes the careers of all three authors to a remarkable degree.

In one of his more frequently quoted letters to Nathaniel Hawthorne, Melville rebukes Goethe's advice to "Live in the All," wryly observing that

> Here is a fellow with a raging toothache. "My dear boy," Goethe says to him, "you are sorely afflicted with that tooth; but you must *live in the all*, and then you will be happy!" As with all great genius, there is an immense deal of flummery in Goethe, and in proportion to my own contact with him, a monstrous deal of it in me.
>
> (*Correspondence* 193–4)

Although Melville acknowledges the pull of the "all feeling," he objects to what he calls the "universal application of a temporary feeling or opinion." In this chapter, I consider how three authors broadly associated with Transcendentalist impulses relative to nature: Melville, Henry David Thoreau, and Emily Dickinson both make strategic use of nature as a unifying metaphor and refuse to let the natural world be reduced to a metaphor, urging the importance of concrete attention to ecological phenomena even as they value and critique the Emersonian impulse to understand nature through a web of correspondences with the human mind. In keeping with Hester Blum's bracing reminder in her work on Oceanic Studies that "the Sea is not a metaphor" (Blum 670), this chapter traces the ways in which Dickinson, Thoreau, and Melville both affirm the metaphoric power of nature and resist allowing natural phenomena to be subordinated fully to human consciousness.

While it would be possible to multiply examples at length, my approach in this chapter is to get at Melville's approach to the question of "universal" application through *Moby-Dick* and three volumes of poetry he was writing at the end of his life, *John Marr and Other Sailors, Timoleon*, and *Weeds and Wildings*, while referring glancingly to earlier moments in Melville's career, at Thoreau's approach through his wonderfully compact and suggestive essay "Walking," and at Dickinson's through two poems, one of her earliest and one of her latest, one set on land and one on sea.

Melville's Critique of "Universal Application," From *Moby-Dick* to *Weeds and Wildings*

Melville's status as a writer deeply engaged by the natural world has risen considerably in recent years, as critics have considered his work as part of the ecological turn in American literary studies and as part of a broader turn toward materialism in literary studies. Such scholars as Tom Nurmi, Gillian Osborne, Lawrence Buell, Geoffrey Sanborn, Merdith Farmer, Jonathan Schroeder, and Michael Jonik have all considered Melville in relation to the ecocritical and posthuman turns in literary studies, and Melville is increasingly acknowledged as a figure who speaks to ecological concerns.[2] My own essay on the animals in *Moby-Dick* stresses how central concrete details from the natural world are to Melville's work in the early 1850s (Yothers, "Whales").

Several of Melville's works seem to capture his resistance to reducing the natural world to a metaphor especially well, and it is striking that they range across his long career as a novelist, poet, and short story writer. In order to plot out the significance of this motif throughout Melville's career, I draw on his acknowledgment of the whale's unknowability in *Moby-Dick* before settling in to a discussion of the latter portion of Melville's career, when he toiled in relative obscurity as a poet even as he began the work that would ultimately become *Billy Budd*.

In *Moby-Dick*, Melville repeatedly satirizes human efforts to domesticate, with one of the most striking moments coming in a chapter called "The Prairie," where Ishmael writes about the difficulty of understanding the whale through a consideration of his physiognomy:

> Champollion deciphered the wrinkled granite hieroglyphics. But there is no Champollion to decipher the Egypt of every man's and every being's face. Physiognomy, like every other human science, is but a passing fable. If then, Sir William Jones, who read in thirty languages, could not read the simplest peasant's face in its profounder and more subtle meanings, how may unlettered Ishmael hope to read the awful Chaldee of the Sperm Whale's brow? I but put that brow before you. Read it if you can.
>
> (Melville, *Moby-Dick* 347)

Melville's treatment of the body of the whale as an unreadable text seems a direct reflection of his skepticism toward the "all" feeling on his letter on Goethe. Melville sees human character as fundamentally mysterious, but on a more profound level, he regards the otherness of the natural world as a characteristic that prevents it from simply being absorbed into human consciousness. The whale presents us with a visage that, if we are to read it at all, requires more learning than the most famous linguist in recent history, and the idea that the sperm whale can simply serve as an analog for our own soul or some portion thereof is simply absurd to Melville. Indeed, Ishmael offers here a foil to Captain Ahab's crucial mistake, which is to view the natural world through the anthropomorphic lens of his own experience rather than trying to understand the whale as a creature separate from himself.

A series of moments in Melville's marginalia to his copies of Emerson's *Essays*, which he purchased in the early 1860s, captures Melville's resistance to the idea that the human mind can fully encompass the natural world, or more specifically, that we can come to know the natural world by looking within. This presumption, of course, is rather crucial to Emersonian transcendentalism. In "The American Scholar," Emerson argued strenuously that self-knowledge and knowledge of the natural world went hand in hand:

> He shall see, that nature is the opposite of the soul, answering to it part for part. One is seal, and one is print. Its beauty is the beauty of his own mind. Its laws are the laws of his own mind. Nature then becomes to him the measure of his attainments. So much of nature as he is ignorant of, so much of his own mind does he not yet possess. And, in fine, the ancient precept, "Know thyself," and the modern precept, "Study nature," become at last one maxim.
>
> (Emerson 86–7)

For Emerson, the "All feeling" about which Melville expresses skepticism in his reading of Goethe is actually an epistemological method. The human mind and the natural world correspond on a fundamental level, and so self-knowledge becomes knowledge of the world, and vice versa. Emerson's position elevates nature, of course, but it also circumscribes it within the scope of human consciousness. As Lawrence Buell pointed out, Emerson "sacralized nature as humankind's mystic counterpart," but this association of nature with the human mind could work against the impulse to know nature in its complexity and in the ways in which it is more than a mirror for our souls (117). Buell captures something of this paradox when he quotes "one of Harvard's natural historians" as saying to Bronson Alcott that "if Emerson had not spoiled him, Thoreau would have made a good entomologist" (118).

Melville's reactions to Emerson's writings are typically harshest on exactly this point. When Emerson writes in "Spiritual Laws,"

> As in dreams, so in the scarcely less fluid events of the world, every man sees himself in colossal, without knowing that it is himself. The good, compared to the evil he sees, is as his own good to his own evil.
>
> (*Melville's Marginalia Online*)

Melville responds on the page top and bottom with two biting reactions. In a note at the top of the page cued to an "X" in the margins, Melville writes:

> A perfectly good being, therefore, would see no
> evel [*sic*].—But what did Christ see?—He saw what
> made him weep.—Howard, too, the "Philanthropist"

> must have been a very bad man—he saw, in jails,
> so much evel [*sic*].
>
> (*Melville's Marginalia Online*)

At the bottom of the page, in a comment cued to an asterisk, Melville is still more direct:

> To annihilate all this nonsense
> read the sermon on the mount,
> and consider what it implies.
>
> (*Melville's Marginalia Online*)

In both cases, Melville suggests that the idea that the natural world provides a reflection of human consciousness is not only incorrect, but productive of immoral conclusions as it guides us away from compassion and concern for others.

Perhaps no moment in *Moby-Dick* more thoroughly captures Melville's skepticism that there can be any absolute correspondence between the human mind and the natural world than the conclusion to his chapter "The Mast-Head."

> "Why, thou monkey," said a harpooneer to one of these lads, "we've been cruising now hard upon three years, and thou hast not raised a whale yet. Whales are scarce as hen's teeth whenever thou art up here." Perhaps they were; or perhaps there might have been shoals of them in the far horizon; but lulled into such an opium-like listlessness of vacant, unconscious reverie is this absent-minded youth by the blending cadence of waves with thoughts, that at last he loses his identity; takes the mystic ocean at his feet for the visible image of that deep, blue, bottomless soul, pervading mankind and nature; and every strange, half-seen, gliding, beautiful thing that eludes him; every dimly-discovered, uprising fin of some undiscernible form, seems to him the embodiment of those elusive thoughts that only people the soul by continually flitting through it. In this enchanted mood, thy spirit ebbs away to whence it came; becomes diffused through time and space; like Wickliff's sprinkled Pantheistic ashes, forming at last a part of every shore the round globe over.
>
> There is no life in thee, now, except that rocking life imparted by a gently rolling ship; by her, borrowed from the sea; by the sea, from the inscrutable tides of God. But while this sleep, this dream is on ye, move your foot or hand an inch; slip your hold at all; and your identity comes back in horror. Over Descartian vortices you hover. And perhaps, at midday, in the fairest weather, with one half-throttled shriek you drop through that transparent air into the summer sea, no more to rise for ever. Heed it well, ye Pantheists!
>
> (Melville, *Moby-Dick* 159)

This is among the most frequently discussed passages in Melville's oeuvre, and with good reason. It is at once lyrical, erudite, and profoundly funny, and it provides a compelling affirmation of the ways in which the natural world resists the shaping power of human consciousness. "The blending cadence of waves with thoughts" is a gorgeous phrase, but it is also a comic jab that the inefficacy of approaches based on the idea that waves and thoughts are somehow identical for finding actual, material whales. The "uprising fin[s]" that seem to correspond to human thought are, on the physical plane, either a sign of the presence of whales that the philosophical sailor is ignoring, or more ominously, a sign that sharks are on the prowl. Reading the passage, Melville's audience has to at once contend with the emotional power of the idea that nature and the human mind share a profound correspondence, and the sense that this idea can be seductive in ways that are ultimately destructive. Failing to take the natural world on its own terms may well be a fatal error, Melville suggests.

If Melville grappled with the tension between phenomena and idealism in *Moby-Dick*, as a writer coming into his own in his early thirties, this tension reasserts itself in the very last works that he wrote: three collections of poetry entitled *John Marr and Other Sailors*, *Timoleon*, and the posthumous *Weeds and Wildings*. As Tom Nurmi and Wyn Kelley have both pointed out, the "Pebbles" sequence at the end of *John Marr* offers a sharp critique of the idea that the universe is centered on humans or is responsive to human desires. The fifth of these fragments personifies the "old, implacable sea," and the seventh offers a particularly severe rebuke to human pretensions to controlling nature:

> Healed of my hurt, I laud the inhuman sea—
> Yea, bless the angels four that there convene,
> For healed I am even by their pitiless breath
> Distilled in wholesome dew named rosemarine.
>
> (Melville, *Published Poems* 249)

What the sea offers to the speaker here is defined precisely by the degree to which it is "inhuman"—to which it resists our impulse to imagine and shape it according to our own needs.

Timoleon, meanwhile, offers another set of poems that focus on both the beauty of the natural world and its resistance to human desire and to being easily absorbed by the human mind. As Cody Marrs has observed, in *Timoleon*, "even the poems that are ostensibly about the epic monuments of earlier civilizations are primarily about the natural shapes and materials out of which such monuments were composed" (30). "The Ravaged Villa" shows the transience of art in the face of natural processes, as it envisions "sylvan vases" "in shards," "their links of dance undone," and reflects on the way that the "weed displaces the flower"—meaning that domesticated nature and human preferences for vegetation are ultimately not determinative. As observed above, the poetic fragments that Melville attributes to a twelfth-century gnostic imagines human volition as subordinate to natural inevitability. Most strikingly, in "Venice," which Michael Jonik has described as "an ode to the creative power of non-human life" (220), Melville compares the human architects responsible for Venice to coral polyps, and appears to regard the coral polyps as superior artists.

Weeds and Wildings is perhaps the least well-known of Melville's volumes of poetry, but it actually functions as perhaps the fullest realization of his attraction to and repulsion from the "all feeling" that he noted in Goethe years before. To quote a characteristic moment in *Weeds and Wildings*, in a poem called "The Rose Farmer," Melville offers a reflection on the ways in which, even in a garden of domesticated roses, the natural world presses back against human efforts to control and define it:

> Adore the Roses; nor delay
> Until the rose-fane fall,
> Or ever their censers cease to sway:
> "To-day!" the rose-priests call.Without Price
> Have the Roses. Needs no pelf
> The blooms to buy,
> Nor any rose-bed to thyself
> Thy skill to try:
> But live up to the Rose's light,
> Thy meat shall turn to roses red,
> Thy bread to roses white.
> Grain by Grain
> Grain by grain the Desert drifts
> Against the Garden-Land:
> Hedge well thy Roses, heed the stealth
> Of ever-creeping Land.
>
> (Melville, *Billy Budd* 126)

Melville's Rose poems only intensify the emphasis on the otherness of nature that he had expressed in *Moby-Dick* nearly four decades earlier, and part of this intensification is due to the fact that he is no longer writing about the untamable sperm whale, but rather about the roses that so many people raise in their own gardens. In this poem, Melville uses the language of theology to capture something of the independence and elusiveness of the natural world.

The roses are seen as being gratuitous in a way that recalls famous formulation of divine grace and mercy in the Hebrew and Christian scriptures. Biblically literate readers of these poems, certainly including Elizabeth Shaw Melville, to whom the poems in *Weeds and Wildings* were dedicated, would recognize that Melville is echoing the language of Isaiah 55. 1–2 in the King James Version of the English Bible:

> Ho, every one that thirsteth, come ye to the waters, and he that hath no money; come ye, buy, and eat; yea, come, buy wine and milk without money and without price.

> [2] Wherefore do ye spend money for that which is not bread? and your labour for that which satisfieth not? hearken diligently unto me, and eat ye that which is good, and let your soul delight itself in fatness.

The evident biblical allusion here comes to be tied in with the wide range of encounters that people have with roses, encounters that are not limited to those who cultivate them. Roses appear here both as signs of a gratuitous splendor in the natural world and as evidence of the fragility of the natural world. What matters most in this case is that the roses are both natural phenomena and domesticated plans. The rose garden exists because it has been planted by human beings according to a human design, but the roses themselves exist in excess of humans' capacity to shape them, and their broader natural context far exceeds the scope of human agency. For Melville, early and late in his authorial career, an ecstatic sense of union with nature is seductive, but never provides sufficient grounds for a true comprehension of the complexities of the natural world and our relationship to it.

Thoreau's "Walking" and the "Migratory Instinct"

Thoreau's careful attention to natural phenomena is well known, as the remark on his affinity for entomology quoted by Lawrence Buell above indicates. What is striking is how this attention exists in tension with the Transcendentalist tendency to find correspondences between the natural world and a universal mind.[3] Although *Walden* provides us with the most comprehensive overview of Thoreau's views on the natural world, his understanding of the relationship between world and mind is articulated with special clarity and compression in his essay "Walking," which follows the path that we take when we move out into the natural world in one of the most direct ways possible.

In "Walking," Thoreau writes a manifest of sorts for how he wishes to approach the task of understanding the natural word:

> I wish to speak a word for Nature, for absolute freedom and wildness, as contrasted with a freedom and culture merely civil,—to regard man as an inhabitant, or a part and parcel of Nature, rather than a member of society. I wish to make an extreme statement, if so I may make an emphatic one, for there are enough champions of civilization: the minister and the school committee and every one of you will take care of that.
>
> (Thoreau 627)

Thoreau's argument here is for an understanding of the relationship between the human and natural world that does not merely treat the natural world as a backdrop, but imagines human beings as fully integrated with the world that surrounds them. Regarding humans as "part and parcel of Nature" means rejecting the idea that humans can see themselves as separate from nature, which is required by the idea that the natural world can provide a mirror of the human mind. For Thoreau, the vantage point that Emerson imagines that can allow humans to survey nature confidently in order to extract self-knowledge is simply not available. Humans are themselves embedded in the natural world. This might seem to take us near to the "all" feeling that Melville describes in his reflection on Goethe, but it differs from the "all" feeling precisely in that humans lack the capacity to absorb nature, but rather are inevitably absorbed into it, whether they desire such absorption or not, a significantly different phenomenon.

This call for absolute wildness on Thoreau's part leads him into his explanation of what walking should mean, and it takes him in unexpected directions relative to both human consciousness and natural phenomena. Regarding the practice of walking, Thoreau writes:

> I have met with but one or two persons in the course of my life who understood the art of Walking, that is, of taking walks,—who had a genius, so to speak, for *sauntering*: which word is beautifully derived "from idle people who roved about the country, in the Middle Ages, and asked charity, under pretense of going *à la Sainte Terre*," to the Holy Land, till the children exclaimed, "There goes a *Sainte-Terrer*," a Saunterer, a Holy-Lander. They who never go to the Holy Land in their walks, as they pretend, are indeed mere idlers and vagabonds; but they who do go there are saunterers in the good sense, such as I mean. Some, however, would derive the word from *sans terre* without land or a home, which, therefore, in the good sense, will mean, having no particular home, but equally at home everywhere. For this is the secret of successful sauntering. He who sits still in a house all the time may be the greatest vagrant of all; but the saunterer, in the good sense, is no more vagrant than the meandering river, which is all the while sedulously seeking the shortest course to the sea. But I prefer the first, which, indeed, is the most probable derivation. For every walk is a sort of crusade, preached by some Peter the Hermit in us, to go forth and reconquer this Holy Land from the hands of the Infidels.
>
> (Thoreau 627)

Thoreau's reflections on walking as a journey to the Holy Land has been interpreted in various ways, including as a reflection on the role that nineteenth-century Ottoman Palestine played in the literary imagination of the United States (see, for example, Hilton Obenzinger's *American Palestine* and my own *The Romance of the Holy Land in American Travel Writing, 1790–1876*). Here, I want to press less on the connection with the Holy Land of biblical lore and nineteenth-century geography, and more on the relation to the natural world suggested by Thoreau's remarks. Thoreau seems concerned here to consider how humans might purposively make themselves "part and parcel" of the natural world—not so much by absorbing the natural world into the human mind, but rather by recognizing a divinity in nature that cannot be contained by human expectations. "Go[ing] to the Holy Land in [a] walk," while it points toward a desire to "go forth and reconquer this Holy Land from the hands of the Infidels," is most centrally concerned with an infidelity to nature, a failure to acknowledge each human being's profound embeddedness in the natural world. The desire to use and transform the natural world without contending with it on its own terms figures as a form of infidelity for Thoreau, and a walk becomes a way of retrieving a faith the value of confronting nature on its own terms that can otherwise be lost.

A central part of Thoreau's understanding of the practice of walking as an instance of humanity's embeddedness in nature appears in his discussion of how the human desire to walk can be compared to the "migratory instinct" in birds and mammals. He considers that humans who are walking are not so much "significant" and "singular" in their actions as they are engaging in activity that reminds us of human continuity with the natural world, as he writes:

> I know not how significant it is, or how far it is an evidence of singularity, that an individual should thus consent in his pettiest walk with the general movement of the race; but I know that something akin to the migratory instinct in birds and quadrupeds,—which, in some instances, is known to have affected the squirrel tribe, impelling them to a general and mysterious movement, in which they were seen, say some, crossing the broadest rivers, each on its particular chip, with its tail raised for a sail, and bridging narrower streams with their dead,—that something like the *furor* which affects the domestic cattle in the spring, and which is referred to as a worm in their tails,—affects both nations and individuals, either perennially or from time to time. Not a flock of wild geese cackles over our town, but it to some extent unsettles the value of real estate here, and, if I were a broker, I should probably take that disturbance into account. (Thoreau 639)

Here again, Thoreau resists positioning his walker as a Transcendentalist "transparent eyeball," in Emerson's terms, or as a Romantic poet comprehending a cosmic visions, but rather imagines humans as acting instinctively in the way that other species do. Humans become in this matter less exceptional, and more like squirrels or cattle or geese, than many of Thoreau's contemporaries might imagine them as being. Moreover, Thoreau calls into question the ideal of the Romantic individual (typically gendered male) who finds an ideal correspondence to nature in his own soul, but suggesting that human experience in general is enmeshed in the animal world. In "Walking," Thoreau suggests that humility is as essential to our relationship to the human world as is exaltation.

Emily Dickinson's Fraudulent Summers and Undulating Rooms

Emily Dickinson's poetry draws upon the same complicated relationship to Emersonian Transcendentalism that we find in Melville and Thoreau, and like Melville and Thoreau, Dickinson resists the attempt to establish too close a correspondence between the natural world and the human mind. Dickinson finds aspects of the natural world compelling, and she is often inclined to find analogies to human experience in nature, but as with Melville and Thoreau, she resists the impulse to subordinate the natural world to human experience and desire.[4]

One of the few poems Dickinson published during her own lifetime captures this tension particularly well, as she writes about the phenomenon of "Indian Summer," a phrase which provides the title for early published versions of the poem. She imagines the plight of birds who are deceived by the apparent return of warm weather in the midst of autumn and who find their way back to an illusory summer:

These are the days when Birds come back—
A very few—a Bird or two,
To take a final look—
These are the days when skies resume
The old—old sophistries of June—
A blue and gold mistake.
Oh fraud that cannot cheat the Bee,
Almost thy plausibility
Induces my belief,
Till ranks of seeds their witness bear,
And swiftly thro' the altered air
Hurries a timid leaf—
Oh Sacrament of summer days!
Oh last Communion in the Haze—
Permit a Child to join—
Thy sacred emblems to partake—
Thy consecrated bread to take—
And thine immortal wine—

(Dickinson J130)

I quote this poem at length because, despite its simplicity, it captures many of the complications that drive Dickinson's engagement with the natural world and with questions of faith and meaning. On one level, this poem is a classic Victorian faith/doubt poem. On another, it stresses the distance between nature as we understand and interpret it and nature as it functions on its own terms. Dickinson begins with a curious phenomenon: the way that birds can appear to interrupt their migratory patterns in response to the return of warm weather in the midst of autumn. As with Thoreau, she finds a correspondence to such migratory movements in the human world, in this case in the nineteenth-century crisis of faith in religious revelation. If for Thoreau walking is like the migration of birds, for Dickinson migratory patterns can correspond to intellectual and religious movements as well. The "fraud that cannot cheat the Bee" in Dickinson's poem is the appearance of the warm weather that leads the birds to return. Dickinson's lines "Almost thy plausibility/ Induces my belief," echo the language of Agrippa's response to the apostle Paul in the King James translation of the biblical book of Acts: "Almost thou persuadest me to be a Christian." Biblicism continues to invest the lines that follow: the "ranks of seeds" that "witness bear" recalls the "great cloud" of "witnesses" from the first verse of Hebrews 12, and all this leads toward the sacramental imagery of the concluding two stanzas. Most readers of this poem have rightly understood it as a skeptic's nostalgic expression of desire for a faith that seems no longer tenable. Dickinson's poem, however, is not only metaphoric, but rather, it is engaged with the actual phenomena of the natural world itself.

Like Thoreau, Dickinson in this poem is fascinated by the migratory instinct that drives the movement of some species, and she is concerned to make fine distinctions between how different species respond to the weather, drawing a clear line even between birds and bees. If the poem is about how it feels to live in a disenchanted cosmos where the possibility of religious faith seems remote, it is also about the affective power of natural phenomena to stir contemplation of something out side of the human mind.

Like the language of faith, the language of sacrament here intersects interestingly with Thoreau's leading metaphor of walking to the Holy Land in "Walking." A natural phenomenon that can appear confusing or even deceptive becomes charged with profound religious and moral significance: the "Sacrament of summer days" is not summer itself, but it is conveying a gratuitous truth about the summer. The careful, attentive observation of nature in which Dickinson is engaged becomes its own sort of act of faith here, one that involves engagement with a reality that exceeds the limits of the human mind and human wishes. Thoreau suggests that the act of taking a walk can carry us out of ourselves, and Dickinson finds that when she "keep[s] the Sabbath staying at Home," to invoke another of her best-known poems, she also is compelled to engage with a world that at once compels her attention and operates according to its own principles.

Dickinson's commitment to reading the ways in which nature resists absorption into human consciousness appears in very different settings from the garden setting imaging for "These are the days when Birds come back." As I have argued elsewhere, Dickinson's poetry of the sea offers an especially stark picture of the distance between the natural world and its human interpreters (Yothers, "Going to Sea"). In a late poem, Dickinson writes:

Water makes many Beds
For those averse to sleep—
Its awful chamber open stands—
Its Curtains blandly sweep—
Abhorrent is the Rest
In undulating Rooms
Whose Amplitude no end invades—
Whose Axis never comes.

(Dickinson, J1428)

Here Dickinson envisions the all feeling as a nightmare of sorts. The scope of the ocean is such that it is immeasurable on any standard human scale, and Dickinson's concluding couplet recalls the medieval understanding of God as a circle or sphere "whose center is everywhere and whose circumference is nowhere," but here this vision of infinitude seems charged with an awe verging on horror. The poem invokes absorption into nature, but here the absorption is not the pleasure of allowing nature to fill one's mind, but rather the terror of drowning in an expanse too vast for human comprehension. The opening lines seem suggestive of suicide: those "averse to sleep" may find that suicide provides them with a chance to rest, but this sleep of course is eternal. Dickinson then extends the conceit of watery beds to create imagined bedchambers in the sea. Dickinson's poem then becomes a sort of ghost story, as the rest experienced at sea is said to be "abhorrent," and the "undulating rooms" of a watery world are sinister rather than inspiring in their sublimity. Dickinson reminds us that we may indeed experience a union with the cosmos, but it may be at the cost of our own annihilation.

So what is Dickinson's view of the "all feeling" in this poem? On the one hand, she certainly appears to acknowledge it, as she offers a version of the ocean that is formed in awe. On the other, that awe seems to shade into the appalled realization that a natural world that so far exceeds the human scale as does the sea cannot easily be domesticated as a metaphor for the human mind: the sea engulfs us, rather than vice versa. We may use metaphor to help us to fathom the scope of the sea, but we forget that our metaphor is a metaphor at our own peril.

This acknowledgment by Dickinson draws together the broader critique that she, Melville, and Thoreau each expresses of an idealist reading of nature as a mirror for the human mind. As critical as Melville could be of Goethe and the "flummery" that he saw in both Goethe and himself, Melville also acknowledged that there was "something in" the all feeling that he associated with Goethe. Reading Thoreau and Dickinson, as well as Melville's own later work, suggests that the "all feeling" continued to exert an appeal and elicit resistance from American writers in the nineteenth century, with substantial implication for how they understood the human relationship to the natural world. Rather than simply embracing a conventional piety towards nature, Melville, Thoreau, and Dickinson all asked their contemporaries, and, by extension, us, to reflect critically on the place of human beings in a natural world in which they and we inevitably participate. The world generates powerful emotion for all three authors, but the material reality of the world operates independently of human aspiration and desire, and casts a critical light on human attempts to manipulate and appropriate the natural world for their own ends. The self-critical reflection on our relation to the world around us continues to be intellectually and morally essential, and Melville, Thoreau, and Dickinson have offered us a model for undertaking such a reflection.

Notes

1 Thanks to Paul Hurh for raising the question in conversation of just who might be the sort of Gnostic to whom Melville is referring in twelfth-century Europe.

2 Nurmi argues that Melville's work is characterized by an emphasis on natural science across his career. Farmer and Schroeder's collection offers a view of *Moby-Dick* that is centered on the "New Materialism" in literary studies and regards Melville's most famous novel as being centrally concerned with the material world. Sanborn shows that Melville's significance derives in no small part from his engagement with the nonhuman world, a viewpoint amplified in Jonik's book-length study of the posthuman in Melville. Otter's *Melville's Anatomies* regards Melville as a writer profoundly concerned with the material aspects of bodies, but human and animal, as well as with questions of literary form. Kelley suggests that the poems in the "Pebbles" sequence points toward our contemporary interest in the posthuman. Marrs finds that reading Melville in relation to natural beauty re-arranges his canon, placing *Timoleon* and *Weeds and Wildings* in an unexpectedly prominent position. Osborne considers the gendered and ecological dimensions of the rose poems in *Weeds and Wildings*.

3 For considerations of the relationship between Thoreau's intellectual biography and his treatment of the natural world, see Robert D. Richardson and Laura Dassow Walls.

4 For widely influential considerations of Dickinson's intellectual development, see Richard Sewall and Alfred Habegger. For a consideration of Dickinson's ecological consciousness, see also the special issue *Lyrical Ecologies* on the *Dickinson Electronic Archive*, edited by Marta Werner and Eliza Richards.

Bibliography

Blum, Hester. "The Prospect of Oceanic Studies." *PMLA* vol. 125, no. 3, 2010, pp. 670–677.

Buell, Lawrence. *The Environmental Imagination: Thoreau, Nature Writing, and the Formation of American Culture*. Cambridge, MA: Harvard University Press, 1996.

Dickinson, Emily. *Complete Poems*. Ed. Thomas Johnson. New York: Little, Brown, and Company, 1960.

Emerson, Ralph Waldo. "The American Scholar." In *The Complete Works of Ralph Waldo Emerson*. Concord Edition. Boston and New York: Houghton Mifflin Co., Riverside Edition, 1903. pp. 79–116.

Farmer, Meredith, and Jonathan S. Schroeder. *Ahab Unbound: Melville and the Materialist Turn*. Minneapolis, MN: University of Minnesota Press, 2021.

Gilchrist, Alexander. *The Life of William Blake, "Pictor Ignotus," with Selections from His Poems and Other Writings*. London and Cambridge: MacMillan and Company, 1863.

Habegger, Alfred. *"My Wars Are Laid Away in Books": The Life of Emily Dickinson*. New York: Modern Library, 2002.

Jonik, Michael. *Herman Melville and the Politics of the Inhuman*. New York: Cambridge University Press, 2018.

The Bible. *King James Version*.

Marrs, Cody. *Melville, Beauty, and American Literary Studies: An Aesthetics in All Things*. Oxford and New York: Oxford University Press, 2023.

Melville, Herman. *Billy Budd, Sailor, and Other Uncompleted Writings*. Ed. Alma A. MacDougall, George Thomas Tanselle, Harrison Hayford, Hershel Parker, Robert A. Sandberg. Evanston and Chicago: Northwestern University Press and The Newberry Library, 2017.

Melville, Herman. *Correspondence*. Ed. Lynn Horth. Evanston and Chicago: Northwestern University Press and The Newberry Library, 1993.

Melville, Herman. *Moby-Dick; Or, the Whale*. Ed. Harrison Hayford, Hershel Parker, and G. Thomas Tanselle. Evanston and Chicago: Northwestern University Press and The Newberry Library, 1988.

Melville, Herman. *Published Poems*. Ed. Alma MacDougall, Robert C. Ryan, Harrison Hayford, Hershel Parker, and Robert A. Sandberg. Evanston and Chicago: Northwestern University Press and The Newberry Library, 2009.

Melville's Marginalia Online. Ed. Steven Olsen-Smith and Peter Norberg.

Nurmi, Tom. *Magnificent Decay: Melville & Ecology*. Charlottesville: University of Virginia Press, 2020.

Obenzinger, Hilton. *American Palestine: Melville, Twain, and the Holy Land Mania*. Princeton, NJ: Princeton University Press, 1999.

Osborne, Gillian. "Herman Melville, Queen of the Flowers." *Leviathan: A Journal of Melville Studies* vol. 18, no. 16, 2016, pp. 129–148.

Otter, Samuel. *Melville's Anatomies*. Berkeley and Los Angeles: University of California Press, 1999.

Sanborn, Geoffrey. "Melville and the Nonhuman World." In *The New Cambridge Companion to Herman Melville*. Ed. Robert S. Levine. New York and Cambridge: Cambridge University Press, 2014.

Sewall, Richard. *The Life of Emily Dickinson*. Cambridge, MA: Harvard University Press, 1972.

Thoreau, Henry David. "Walking." In *Walden and Other Writings*. Ed. Brooks Atkinson. New York: Modern Library, 2008, pp. 625–664.

Richardson, Robert D. *Henry D. Thoreau: The Life of the Mind*. Berkeley and Los Angeles: University of California Press, 1988.

Walls, Laura Dassow. *Henry David Thoreau: A Life*. Chicago: University of Chicago Press, 2009.

Werner, Marta, and Eliza Richards. *Lyrical Ecologies*, a Special Issue. *DEA-2: Dickinson Electronic Archive*. Fall 2017. https://www.emilydickinson.org/emily-dickinson-lyrical-ecologies-forays-into-the-field

Yothers, Brian. "Going to Sea in Emily Dickinson's Poetry: Decentered Humanism and Poetic Ecology." *Lyrical Ecologies*, a special issue edited by Marta Werner and Eliza Richards. *DEA-2: Dickinson Electronic Archive*. Fall 2017. http://www.emilydickinson.org/emily-dickinson-lyrical-ecologies-forays-into-the-field/going-to-sea-in-emily-dickinson-s-poetry-decentered-humanism-and-poetic-ecology

Yothers, Brian. *The Romance of the Holy Land in American Travel Writing, 1790–1876*. Aldershot: Ashgate, 2007.

Yothers, Brian. "Whales, Mother Carey's Chickens, and a Heart-Stricken Moose." In *Animals in the Classics*, Ed. John Gruesser. Texas A&M University Press, 2022. pp. 88–110.

5 The Raging Torrent

Myth, Metaphor and Technology

Patricia Likos Ricci

Looking Back

Having reached a point in history when the artificial habitat that humans have constructed overwhelms and threatens our original natural habitat, the myths of an earlier age before modification of the Earth accelerated may provide perspective. Myths grew out of an animistic worldview in which the natural environment, organic and inorganic, was sentient and responsive to human interaction. While myth-making is a universal activity that exhibits cross-cultural patterns, similar motifs and recurrent themes, some myths are culture-bound and originate in and record the perceptions of a specific human group in a particular geographic location. The Greek poets maintained an isomorphic relationship between natural phenomena and poetic symbolization that made their myths plausible in the pre-scientific era. The geographic sites they anthropomorphized in myth were not always fictional; Olympus was appropriately designated the home of the gods because it was the highest mountain in Greece. On the other hand, the Garden of Eden is a symbolic place that exists in the imagination. Unlike the Greek Arcadia, the biblical paradise cannot be found on a map.

Ancient Greek and Roman poets transmuted nature into narrative so that myths continued to live in a parallel reality not bound by time or space. This chapter provides a cross-cultural analysis of a myth in Hesiod's *Theogony* that has migrated and evolved over two millennia: the Origin of the Oath in the Falls of the Styx. The correspondence between a topographic feature and a legal principle may seem obscure to us in the twenty-first century, but as its history unfolds, Hesiod's message becomes apparent: the natural order and the moral order are one.

Styx, Dread River of Oath

According to Hesiod's *Theogony*, when Zeus decided to overthrow the Titans, he called the immortals to Olympus promising to preserve the rights and honor those who fought with him. The River Styx, eldest daughter of Okeanos, was the first to answer the call. Zeus rewarded her with the creation of the Oath of the gods and promised that her children, Emulation, Nike, Power and Strength, should dwell with him forever (Th. 383–385). As one-tenth of her father

DOI: 10.4324/9781003385202-6

Okeanos, Styx flowed underground to the border of Hades. In a literal depiction of the myth by the Dutch engraver Philip Galle in 1587, a classical female personification of Styx sits in a grotto dangling her fingers in the river that bears her name while resting her foot on a nautilus shell that identifies her as an Oceanid. A short distance way, Charon waits in his boat for souls to ferry into the infernal realm of Hades where flames rise behind its arcaded walls (Figure 5.1).

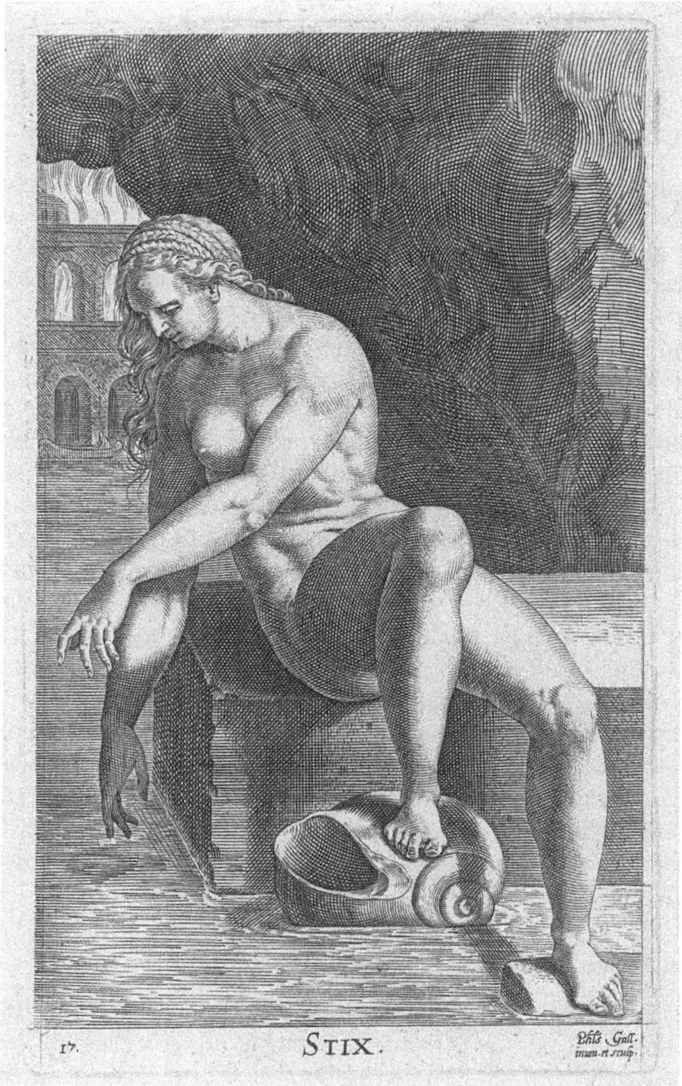

Figure 5.1 Philip Galle (1537–1612), *Styx*, engraving, Pl.17 from *Icons of the Nymphs, Oceantides, Ephydrids, Potamids, Naiads and Lynads*, 1587.

Styx will surge back to the surface of Earth and plunge off a precipice in an icy waterfall to provide the libations on which the gods swear oaths:

> There dwells the goddess loathed by the immortals,
> dreadful Styx, eldest daughter of Okeanos, whose stream
> flows back on itself; she dwells apart from the gods
> in a stately palace roofed by lofty rocks and ringed
> by silver pillars that tower into the sky.
> Seldom does fleet-footed Iris, the daughter of Thaumus,
> Roam on the broad-backed sea to bring her message
> when strife and quarrel arise among the immortals
> and a dweller of Olympos is guilty of lying.
> Then Zeus sends Iris far away to fetch in a golden jar
> the legendary cold water by which the gods swear great oaths,
> water that tumbles down from a steep and soaring rock.
> This water flows through the black night
> from sacred river, far below the earth of the wide paths.
> It is a branch of Okeanos allotted one-tenth of the water;
> The other nine parts wind round the earth and the broad-backed sea
> And, silver-swirled, cascade into the briny deep,
> but this one branch——an affliction for the gods—runs off a cliff.
> If one of the gods who hold the peaks of snowy Olympos
> pours a libation of this water and then swears a false oath,
> he lies breathless for no less than a full year's course;
> he cannot come close to ambrosia
> for nourishment but, no longer able to speak or breathe,
> lies in bed, wrapped in the shroud of an evil coma.
> When the illness is over at the long year's end,
> Another, even harsher, trial is in store for him.
> For nine years he is an outcast to the eternal gods
> And does not mingle with them at council or feast
> for nine full years, but on the tenth he joins again
> the meetings of the gods who call Olympos their home.
> Such is the oath the gods made of the primaeval and immortal
> Water of Styx that gushes through a rugged place.[1]

For Hesiod (ca. 750–625 BCE), the oath is a divinely ordained ritual to compel individuals, even the deathless gods, to keep their word.[2] Without mutual trust, a viable community cannot survive even among the Olympians. By tracing the course of the waters of Styx through the topography of the Earth, Hesiod defines the oath as an ineluctable part of a recurring natural cycle. Suzanne Lye has observed that

> most scholars largely ignore the locations of Styx in Hesiod's account and downplay her physical nature. This is in keeping with a tendency to view the earlier deities as primarily personifications of moral concepts, while downplaying their roles as physical entities in the universe.[3]

In his description of Styx in *Theogony*, Lye sees Hesiod "mapping" Zeus's new values onto the natural world like spatial coordinates. "Although Gaia [Earth], Ouranos [Sky], and Tartarus [Underworld] become the physical frame for the world," she argues, "the world does not gain final stability and order until Styx appears, since she embodies the physical and moral connections between the three realms."[4]

By the time that Homer (ca. eighth to seventh centuries BCE) composed his epic poems, he assumed that the ritual of gods swearing oaths on Styx was understood by his audience. In the *Iliad*, when Zeus accuses Hera of conspiring with Poseidon to support the Achaeans in the war with the Trojans, Homer tells us, "Hera shuddered" (a visceral reaction that alludes to the etymology of Στύξ) and declared her innocence: "I swear, by Earth and the spacious Sky above, and by the waters of Styx below, which is the greatest and most fearsome oath for all the gods who live at ease"[5] (15.34–36). The amused listener already knows from previous verses that Hera is equivocating but confident that Zeus will not test her oath with a libation of Styx waters. In the *Odyssey*, when Calypso agrees to send Odysseus home, he "shudders" with mistrust and requests a guarantee of safe passage. The goddess promises her protection by swearing "Let the earth be my witness in this, and the high sky above, and the waters of Styx that flow below, which is the greatest and most terrible oath for the blessed gods, that I shall not plot any more arduous troubles for you"[6] (5.184–187). With the statements of these goddesses, Homer elucidates the two temporal aspects of the Greek oath. Hera's oath attests to knowledge of an act that occurred in the past; it is an assertory oath. Calypso's oath is a vow to perform an action in the future; it is a promissory oath. Consisting of only words, both types of oath are deemed credible by the willingness of the oath-takers to call upon a higher authority to punish them if they lie. But what power is higher than an immortal god? "The combination of Gaia (Earth), broad Ouranos (Heaven) above, and the dripping water of the Styx, is a formulaic triad of oath-guarantors invoked by divinities in epic," according to I.C. Torrance.[7] Since these gods are the primordial Titans, Hera and Calypso are swearing on their ancestors who generated the natural world.

Horkos, the Oath of Mortals

Throughout the *Iliad* and the *Odyssey*, Homer's mortals also swear oaths, but to the gods, rather than Styx. Unlike Styx, Horkos, the oath of mortals, is not a physical feature of the natural environment. He was spawned by the violent disruptions in the social order that Hesiod described in *Theogony* (226–232):

> Then loathsome Strife bore Ponos, the bringer of pains,
> Oblivion and Famine and the tearful Sorrows,
> The Clashes and the Battles and the Manslaughters, the Quarrels and the Lies and Argument and Counter-Argument,
> Lawlessness and Ruin whose ways are all alike,
> And Oath, who more than any other, brings pains to mortals
> Who of their own accord swear false oaths.[8]

Horkos plays an important role in Hesiod's *Works and Days*, a didactic poem on the necessity of honest labor and Justice in an agricultural community. The fatherless son of negative Strife, the struggle for dominance, as opposed to the positive Strife of competition, Horkos is a male personification of oath-taking identified with the civil order.[9] His birth was attended by the Furies (WD. 802–804), avenging female deities who are "an embodiment of the act of self-cursing contained in the oath."[10]

The poem is addressed to Hesiod's brother Perses whom he accuses of collaborating with corrupt officials to defraud him of his patrimony. Yet his belief that Justice is ordained by Zeus allows him to predict that Perses and his ilk will eventually suffer retribution: "The Oath Demon follows the trail of crooked decrees; Justice howls when she is dragged about by bribe-devouring men whose verdicts are crooked when they sit in judgment" (WD. 219–221).[11]

According to J. Donald Hughes, Hesiod's pessimism is a form of environmental determinism that derives from the biological model of life-cycles. "Hesiod presented this idea in his myth of the Five Ages, in which Earth became less fertile and human life shorter as the Golden Age gave way to Silver, Bronze and Iron ages."[12] Yet if Hesiod accepts that entropy is inevitable, he also believes that humans have moral agency otherwise his diatribe would be pointless. He denounces the behavior of the people of the current Iron Age as a time of moral inversion when

> The man who keeps his oath, or is just and good,
> Will not be favoured, but evildoers and scoundrels
> Will be honored, for might will make right and shame will vanish.[13]
> (WD. 190–192)

"The perjurer and the wrongdoer harm not only themselves," Apostolos N. Athanassakis explains. "They are public menaces because their misdeeds can destroy whole cities. Actions against Justice can bring about infertility of crops and of women, thus causing death through famine and dwindling of the population through sterility."[14] These calamities would be attributed to divine retribution, but they might have been the result of mismanagement of agriculture. Hughes found "evidence of lower yields, farmers leaving land that can no longer support them, and deserted fields, not only in literature, but also in laws enacted to counter these trends."[15] Since less than one-fifth of Greece is arable land, Hesiod advised farmers to be attentive to the seasonal cycles and work on days when the biological cycles of their plants and livestock required timely action. Only then will

> Men who honor justice know neither hunger nor ruin,
> but amid feasts enjoy the yield of their labors.
> For them the earth brings forth a rich harvest.[16]
>
> (WD. 230–233)

Farming was not simply a private activity. "Agriculture was by far the paramount sphere of employment and investment."[17] The communal nature of agrarian society required honesty and cooperation to sustain life. Hence the essential role of the oath in promoting a cohesive society as Walter Burkert summarizes:

> In the institution of the oath, religion, morality, and the very organization of society appear indissolubly linked together. Its function is to guarantee that a statement is absolutely binding, whether it be a statement about something in the past or a declaration of intent for the future. In a culture without writing where there are no records to act as proof, no legal documents, this function is of unique importance … the written word made slow progress against the oath and never entirely displaced it.[18]

Arcadian Styx

Thus far Hesiod's narrative of the Oath of Styx has not been found in any other source, prompting speculation that it was his poetic invention.[19] While episodes of oath-taking in the works of Hesiod and Homer indicate that the custom was already a practice in Archaic Greek society, the oath may have lacked an origin story. Hesiod's detailed genealogy of the goddess Styx and her children at the inauguration of the Olympian regime corresponds to a typical aetiological strategy to justify a social innovation by treating it as an established tradition. Antje Wessels and Jacqueline Klooster found that "techniques of aetiological thinking are mainly employed during periods of fear, threat and danger" in order to exert control over a potentially threatening environment."[20] The importance Hesiod places on Horkos, the oath of mortals, in the unstable, corrupt society he laments in *Works and Days* supports their observations.

Was Hesiod's image of the Oath of Styx as a waterfall merely a poetic conceit or might he have found a model in nature? At least by the fifth century BCE, his description of Styx was identified with a waterfall in Arcadia, a mountainous wilderness region in the north central Peloponnese. The historian Herodotus (ca. 484–425 BCE) made the earliest known reference to a waterfall identified as Styx in relating the misadventures of the mad Spartan king Cleomenes I (d. ca. 490 BCE) who tried to compel the Arcadian leaders to swear an oath of allegiance to him on Styx.

> For it is here at Nonacris that the Arcadians believe the waters of the infernal river to be visible. It is, indeed, a fact that one can see a little water trickling out of rock into a basin surrounded by a circular wall.[21]

His perfunctory description suggests that he did not actually see the soaring waterfall that later impressed the geographer Pausanius (ca. 110–180 CE) near the ruins of Nonacris: "I know of none other that rises to so great a height. A water trickles down the cliff, called by the Greeks the water of the Styx."[22] In October of 1895, James G. Frazer, who translated Pausanius's *Description of Greece*, traveled to Arcadia to see the waterfall for himself. Assisted by a guide, he made the trek through mountainous terrain to reach the foot of the waterfall "where the water tumbles or trickles, according to the season, over the smooth face of an immense perpendicular cliff, the top of which is not far below the conical summit of Mount Chelmos (nearly 8000 ft. high)." The water came mostly from the snowfields on the summit and descended over 600 ft. to "the foot of the cliff where it formed a small stream flowing down a very steep rocky bed into the bottom of the glen far below."[23]

Frazer's report is a remarkably accurate description of the Arcadian Styx that exists today in a virtually pristine state. The waterfall is in fact a spring on the east face of Mount Chelmos at an altitude of 6,900 ft. that plunges 650 ft. into a pool where it feeds the river now known as Krathis. "There is no way of telling whether the traditional conception of the infernal river was influenced by knowledge of the Arcadian Styx and its falls," writes Robin Hand, "or whether, conversely, the Arcadian Styx was first given that name because its chilly falls resembled those of the Styx in Hesiod's description."[24] It is not implausible that Hesiod learned of a legendary waterfall in Arcadia and incorporated its features into his genealogy of the gods, particularly because the Arcadians had a mythical status in Greece as autochthonous people "born before the stars and moon," in the words of Statius (*Thebaid*, IV, 275–276).[25] The mountain spring that plummets from a great height into the earth and seemingly reappears at its source on the top of the mountain fits Hesiod's description of the water cycle of Styx. The icy water is caused by the snowpack that crowns Mount Chelmos each winter and seeps into the rock. Encased in ice and snow, the sheer grey limestone walls of Mount Chelmos might resemble Hesiod's description of Styx's "stately palace roofed by lofty rocks and ringed by silver pillars that tower into the sky" (Th.778–779).

Separating fact from fiction in ancient writers' descriptions of places was complicated by the permeable boundaries of classical literary genres. Greek rhetoricians classified all verbal and visual representations of places as *topographia*, the term for landscapes depicted by poets, geographers, historians, and scene-painters.[26] The extensive concept of the *topos* meant that customs, rituals, and legends could be accommodated by the topographic genre in addition to a variety of landforms. The tradition of aetological. myths about the natural environment had so pervaded the consciousness of even the more rationalistic Greek and Roman writers that some of them required empirical proof to separate natural phenomena from poetic invention. The Stoic philosopher Seneca (ca. 4 BCE–65 CE), for example, requested an eye-witness account of the legendary whirlpool Charybdis from a friend recently returned from Sicily: "I know very well Scylla is a rock not to be dreaded by mariners; but with regard to Charybdis I should like to have a full description, in order to see whether it agrees with the accounts of mythology."[27]

Once ancient travelers agreed that the Falls of the Styx was located in Arcadia, the question arose whether the water, which Hesiod described as coma-inducing in gods who lied under oath, was actually toxic. In *Natural Questions*, the Stoic philosopher Seneca (ca. 4 BCE–65 CE) not only accepted the lethality of the water as a fact, he also attributed malevolent intention to it. "In Arcadia, near Nonacris, the river called by the people there the Styx lures strangers to ruin, as its appearance and smell rouse no suspicion," he warned. "This is like the drugs of accomplished poisoners, which cannot be detected save by their fatal effects."[28] A rumor circulated for centuries that Alexander the Great had been murdered with a vial of water from the Styx conveyed to his final banquet in Babylon by the royal cupbearer Iolaus, but Plutarch (46 CE–119 CE) dismissed the story as a fabrication in the "Life of Alexander."[29] Pliny the Elder (23 CE–79 CE) asserted that "to drink of the Styx near Nonacris in Arcady causes death on the spot, although the river is not peculiar in smell or colour."[30] Pausanius reported that Arcadian Styx "once brought death even upon goats, which drank of the water first."[31] The dark appearance of the falls contributed to its sinister reputation but when Frazer investigated the phenomenon, he found that the water only appeared black because of a "dark incrustation which spreads over the smooth face of the rock wherever it is washed by the falling water."[32] Today the Styx River is called Mavronéri, or black water, by local residents who will not drink it.

Could there be some truth in Hesiod's claim that drinking from the Styx caused illness? Stanford University classicist Adrienne Mayor and toxicologist Anoinette Hayes set out to answer the question scientifically. Hayes concluded that the myth was substantiated by the presence of a bacterium that produces the dangerous compound calicheamicin which is extremely toxic if ingested.[33] This discovery made the ancient rumor that Alexander the Great was murdered by drinking water from Styx more plausible. Mayor conjectured that "such a sacred poison, used by the gods, would be appropriate for Alexander who was already being thought of as semidivine."[34]

Hesiod wrote that Iris, the personification of the rainbow, a phenomenon frequently seen arching over waterfalls in sunny weather, was sent by Zeus to collect the icy waters in a golden jug for libations. A legend arose that the waters of Styx could not be contained in ordinary vessels. *On water*, a lost book by the Peripatetic philosopher Theophrastus (ca. 371–287 BCE), included one of the earliest comments on Styx's extraordinary power to break all vessels except those of *keratina*, animal hoof.[35] The Roman architect Vitruvius (80 BCE–70 BCE) elaborated on this mysterious property. "No vessel, whether of silver, bronze, or iron, can stand it without flying to pieces and breaking up," he asserted. "Nothing but a mule's hoof can keep it together and hold it."[36] The Roman writer Apuleius (fl. ca. 155 CE) inserted this belief into his tale of Cupid and Psyche in *The Golden Ass or Metamorphoses*. Venus, angry with the mortal Psyche for breaking her promise never to look upon her mysterious lover, Cupid, forces her to submit to several trials designed to destroy her. Her broken vow explains her third ordeal, the impossible task of filling a crystal urn with the icy dark water that gushes from the summit of the Falls of the Styx. Unbeknownst to naïve Psyche, she has in essence been asked to pour an oath libation. As she is about to fall from the slippery serpent-infested crag, Omniscient Jupiter dispatches his royal eagle to warn Psyche not to "steal" or even touch the sacred water:

> You must have heard that this water of Styx is feared by the gods themselves, even Jupiter, and that the oaths which mortals swear by the power of the gods, the gods swear by the majesty of Styx.[37]

Snatching the vessel, the eagle swoops over the falls, fights off the serpents, and fills it with water to complete the task for Psyche. Although she did not keep her promise, she will not be punished because Justice is served by the spirit not the letter of the law. The resolution of Psyche's ordeal is not merely a happy ending for a charming fable; rather, it represents the exceptions that are justified to prevent abuse of power. Plato's explanation of exceptional circumstances regarding contracts in *The Laws* applies to Psyche's situation, inserted here in brackets:

> If someone agrees to a contract and fails to act in accordance with the agreement, except in cases where the laws or a decree prevent its fulfilment [retrieve sacred waters], or an agreement is made under unfair compulsion or force [Venus], or when someone is hindered, despite his intentions, by unforeseen circumstances [serpents], let there be, for all other unfulfilled agreements, a trial before the tribal courts, unless they are able to settle the matter previously before arbitrators or their neighbours.[38]

In this case, divine intervention, rather than retribution, ensures a just outcome and Psyche is reunited with Cupid and rewarded with immortality.

The Raging Torrent

"In spite of the local nature and uses of so many mythical narratives, and their significance in giving places meaning and status," Katherine Clarke explains, "it is travel, connections, and journeys from which the real power of myth in geography and geography in myth derives."[39] Both real and fictional travelers transmitted Hesiod's story of the Styx waterfall from Greek to Latin which entered the European vernacular languages with its essential features intact: a powerful waterfall protected by a deity or spirit that avenges violated oaths or crimes and flows into the underworld.

Hesiod's waterfall motif appears in Book VII of Virgil's *Aeneid* (ca. 29–19 BCE) to dramatize the outrage over a broken oath that shaped the course of Roman history. Latinus, King of Latium, breaks his promise to betroth his daughter Lavinia in marriage to Turnus, king of the Rutuli, and instead offers her to the Trojan Aeneas. To avenge the betrayal and thwart a permanent foreign colony on Italian soil, Hera summons the Fury Alecto from the underworld to foment strife and provoke war with between the Trojans and the indigenous Latins. When Alecto is finished wreaking havoc, she returns to the underworld as a raging torrent through an abyss at the foot of a cavern, a horrifying pit of swirling water:

> There is a place in the heart of Italy, beneath high hills, renowned and famed in many lands, the Vale of Ampsanctus. On either hand a forest's fringe, dark with dense leafage, hems it in, and in the centre a roaring torrent resounds o'er the rocks in swirling eddies. Here is shown an awful cavern, a breathing-place of horrid Dis; and a vast gorge, whence Acheron bursts forth, opens its pestilential jaws. Herein the Fury hid her loathed power, relieving earth and heaven.[40]

Virgil personifies the waterfall with the voice of an enraged Fury that reverberates throughout the region. Curiously, the conspicuous sound that waterfalls generate even at a distance was missing from other descriptions of the waterfall *topos*, perhaps because Hesiod did not include auditory effects in his myth.

In emulation of his mentor, Dante (1265–1321) revived the infernal waterfall in the *Divine Comedy* (1314). In *Inferno* Canto XVI, Virgil leads Dante to the edge of a precipice and informs him that they must go over a great cascade by holding onto the back of the monster Geryon in order to descend from the seventh circle of the Violent against Nature and Art into Malebolge, the eighth circle of the Fraudulent and Malicious. Like Hesiod, Dante considers fraud a worse offense than violence, even for the gods. Dante comments that the deafening sound filling the air reminds him of the falls of the Acquacheta River "as it cascades in one leap" near the monastery of San Benedetto dell' Alpe in the Appennines. "So did we hear that blackened water roar, as it plunged."[41]

By the eighteenth century, the *topos* of the raging torrent described in *Aeneid* VII was familiar to classically educated Northern Europeans and could serve as a metaphor for righteous conflicts. In 1791, the Austrian painter Joseph Anton Koch (1768–1839), a Jacobin sympathizer, abandoned his studies at the politically conservative Stuttgart art academy and went on a sketching trip in the Swiss Alps. In the mountain wilderness he discovered his passion for waterfall scenery which mirrored the political turmoil within him. He confided to his diary that the exhilarating sight of the Rhine Falls at Schaffhausen, the most powerful in Europe, "stirred his soul" with revolutionary fervor:

> It seemed as if the god of the Rhine were calling to me from the jagged rock: Stand up! Act! Bestir yourself with steadfast fortitude! Pit yourself

> forcefully against despotism! Rend the shameful bonds that bind you! Be unshakeable like the rock against which I fight in defence of the freedom of mankind! [42]

Koch made several drawings of the broad Rhine Falls, only 75 ft. high but 490 ft. wide, churning the water into billowing white foam as it crashed into the river. Over his career he would make other paintings of waterfalls, including an illustration of Dante and Virgil traveling down the cascade on the back of Geryon in Canto XVI of the *Inferno*.[43]

The waterfalls in the Swiss Alps also resonated with the torrential passions of the Romantic poet Lord Byron (1788–1824). In 1816, having fled England after scandalous rumors of an incestuous relationship with his half-sister Augusta Leigh became public, Byron sought refuge in Switzerland. There he composed *Manfred: A Dramatic Poem* about a Faustian nobleman living in the Bernese Alps who is tormented by guilt for the death of his beloved Astarte. Act II, Scene 2 takes place at a cataract in the lower Bernese valley. Conjuring up the Witch of the Alps, he confides to her that he feels like a stranger among men:

> My joy was in the Wilderness, to breathe
> The difficult air of the iced mountain's top,
> Where the birds dare not build, nor insect's wing
> Flit o'er the herbless granite; or to plunge
> Into the torrent, and to roll along
> On the swift whirl of the new breaking wave
> Of river-stream, or ocean, in their flow.[44]

Byron's self-described "metaphysical drama" internalized divine retribution. "My solitude is solitude no more," he writes, "but peopled with the Furies."[45]

Perhaps the most famous expression of the waterfall *topos* as a symbol of crime and punishment is the setting of *The Adventure of the Final Problem* in Sir Arthur Conan Doyle's Sherlock Holmes series.[46] On holiday in the Bernese Alps in 1893, Conan Doyle and his wife visited Reichenbach Falls, which was famous for its breathtaking scale and power, a cascade plummeting 820 ft. in seven steps through a deep gorge. The dramatic scenery of Reichenbach Falls seemed like a fitting location to end his fictional detective's life so that the author could move on to other literary projects. He imagined Holmes wrestling with his arch-nemesis Professor James Moriarity on the rocky ledge he had observed adjacent to the torrent. In the story, Dr. Watson relates the scene:

> It is, indeed, a fearful place. The torrent, swollen by the melting snow, plunges into a tremendous abyss, from which the spray rolls up like the smoke from a burning house. The shaft into which the river hurls itself is an immense chasm, lined by glistening, coal black rock, and narrowing into a creaming, boiling pit of incalculable depth, which brims over and shoots the stream onward over its jagged lip.[47]

Originally published in *The Strand Magazine* in Great Britain in 1893, the story was intended to conclude the series with Holmes's heroic decision to sacrifice himself to rid the world of a diabolical criminal. Sidney Paget's bold illustration depicts the moment Holmes accepts his fate as he watches his deerstalker cap blow off just before he and Moriarty tumble into the abyss. Captioned "The Death of Sherlock Holmes" (Figure 5.2), the image of their hero's demise so dismayed the magazine's readers that they compelled Conan Doyle to resurrect him. In 1991, the 100th anniversary of the fictional event, a

Figure 5.2 Sidney Paget, "The Death of Sherlock Holmes," illustration for Arthur Conan Doyle's "The Final Problem," *Strand Magazine*, December 1893.

plaque was placed on the rock wall above the ledge at Reichenbach Falls in English, German, and French that reads: "At this fearful place, Sherlock Holmes vanquished Professor Moriarty, on May 4, 1891."[48]

Horror and Beauty

Italy's largest waterfall was the result of the Roman consul Manius Curius Dentatus's (d. 270 BCE) plan to drain the wetlands in the Reate Valley in order to provide land for agriculture. In 271 BCE, he ordered the construction of the *cava curiana* (Curiano Trench), a channel cut into Lake Velinus (Lago di Piediluco), a natural reservoir in the Apennines, to divert its waters over a 540 ft. cliff into the River Nar below.[49] The modification of the natural environment successfully reclaimed land for farming and removed stagnant waters suspected of causing illness (probably malaria) as far as twenty miles south. However, the falls raised the water level of the Nar resulting in navigational hazards that jeopardized the commercial interests of the neighboring town of Interamna (Terni). In 54 BCE, Cicero and Aulus Pompeius attempted to arbitrate the conflict in the Senate but no resolution was reached.[50] The Falls of the Velino caused unintended consequences throughout the watershed and increased the frequency and the severity of floods in Rome, thirty miles downstream.[51] Livy, Tacitus, and the younger Pliny all recorded disastrous inundations of the city.[52]

Despite their engineering prowess in manipulating the natural environment, many Romans believed that rivers were ultimately ruled by gods who may disapprove of human intervention in their domain. Gregory Aldrete found that fear of "reducing the majesty of Father Tiber" was one of the reasons that proposals to divert the river's tributaries were defeated."[53] Consequently, catastrophic floods in the city of Rome were frequently interpreted as divine disfavor and portents of political upheaval. As Horace warned in an ode dedicated to Rome's first emperor, Augustus: "We saw the yellow Tiber, its waves hurled back in fury from the Tuscan shore, advance to overthrow the King's Memorial [Regia] and Vesta's shrines…"[54] Augustus (63 BCE–14 CE; reign from 27 BCE) was the first to establish a Curatorship of the Tiber channel, and subsequent emperors – notably Tiberius, Claudius, and Trajan – continued to manage flood control at the mouth of the Tiber at Ostia.[55]

Did the Romans consider the beauty of the Falls of the Velino as it thundered down a mountain side in Umbria? Would the citizens downstream menaced by the floods it caused respond to the waterfall aesthetically? Natural waterfalls are typically inaccessible, as Psyche discovered, and always treacherous at close range because the drenching spray of the torrent makes the surrounding surfaces precarious, as Sherlock Holmes realized. The scale and danger of waterfalls require distance and a position of safety from which to contemplate the site aesthetically. Though Roman engineers did not intend to achieve aesthetic effects when they cut a channel into a mountain lake, they had inadvertently created the conundrum of a man-made natural wonder and the largest artificial waterfall in the world. The Falls of the Velino drop 541 ft. from a precipice in three widening cataracts down the rocky steps of the Nar riverbed, producing a spectacle of immense breadth as well as height. As centuries passed, the calcium-rich waters of the Velino encrusted the slopes with white mineral deposits giving the rocky wall the appearance of marble; hence, the cascade became known as the marble waterfall, "La Cascata delle Marmore." In a letter of 1662, the painter-poet Salvator Rosa (1615–1673), whose pleasure in wilderness scenery was exceptional at the time, described the sight of the cascade four miles from the road as "a thing to excite every teeming brain with its horrid beauty (*orrida bellezza*), in seeing a river that falls from a mountain a half mile from the precipice and raises its spray about as much."[56] The oxymoron of horrid beauty is a rare expression of the contradictory emotions that would later be classified as the Sublime.[57]

Rosa's choice of the word *orrida*, from the Latin, alluding to Virgil's use of *specus horrendum*, or horrifying pit, attests to his association of the Cascata delle Marmore with the waterfall in the *Aeneid* (VII.563-571).

For those travelers adventurous enough to make the journey into the Umbrian wilderness, the waterfall did not disappoint. "I went out of my way to see the famous cascade about three miles from Terni," Joseph Addison (1672–1719) recorded in his travel diary:

> It is formed by the fall of the river Velino, which Virgil mentions in the seventh Aeneid ... I do not in the least question, notwithstanding the opinion of some learned men to the contrary, that this is the gulf through which Virgil's Alecto shoots herself into hell; the very place, the great reputation of it, the fall of waters, the woods that encompass it, and the smoke and noise that arise from it, are all pointed at in the description.[58]

A pilgrimage to the Falls of the Velino near Terni remained on the itinerary of grand tourists, writers and artists for more than 200 years and was recorded in numerous travel diaries, letters, poems, and paintings. "The Grand Waterfalls at Terni" (Figure 5.3) by the Swiss painter Franz Kaisermann (1765–1833), with two figures contemplating the cascade from below, is one of the finest examples of the topographical accuracy that artists could achieve before photography.

Whatever mythical or actual terrors waterfalls had evoked in Antiquity, by the nineteenth century their demons had been exorcised and their waters reconsecrated with the spirits of the Picturesque, the Beautiful, and the Sublime. [59]

What Is the Value of a Waterfall?

The parallel developments of industrialization and landscape aesthetics in the mid-eighteenth century made a collision of cultural values inevitable by the mid-nineteenth century. Industrialization valued natural resources as raw materials while for artists, writers and tourists they were sources of spiritual

Figure 5.3 Franz Kaisermann, *The Grand Waterfalls at Terni*, 1790s. watercolor, National Gallery of Art, Washington DC.

uplift and physical well-being. Although water power was a familiar technology used for mills and irrigation since antiquity, by the eighteenth century the scale and design of factories required hydraulic energy on a greater scale to meet the demands of industrial production. It was only a matter of time before waterfalls, free, abundant, and reliable sources of energy, were commandeered by industry. In 1896, water was diverted from the Curiano Trench above Cascata della Marmore to operate steel mills which undeniably benefited the local community with employment opportunities. The value of the falls was now measured with economic, rather than aesthetic, criteria. But in 1929, when the Galleto hydroelectric power plant reduced the flow over the falls thereby spoiling the magnificent cascade that brought tourists to Terni, suddenly there were two competing sources of revenue with mutually exclusive goals. By then, the national prestige of the magnificent waterfall, a feat of Roman engineering created over 2,000 years earlier, was too important as a cultural icon to be reduced to an industrial site. Eventually a compromise was reached between the economic needs of the hydroelectric plant and the cultural demands of the population by regulating the water flow on a schedule to accommodate tourism. The famous Reichenbach Falls faced a similar situation in 1906 when the hydroelectric plant turned the torrent into a trickle. The diminished waterfall jeopardized the attraction of the Reichenbach Funicular built across the gorge in 1899 for tourists and the increasing number of Sherlock Holmes's devotees. Ultimately, the power plant agreed to scheduling their use of the water flow. No such compromise was reached with the mighty Schaffhausen Falls. In 1951, the construction of a hydroelectric plant was vehemently opposed by 150,000 Swiss citizens who successfully petitioned to prohibit any future energy-generating projects at the falls. As for Arcadian Styx, the remote location and season-regulated flow have protected the waterfall from human interference.

A study conducted by R. Kerry Turner's team of environmental scientists and economists found that "society may also regard nature or some of its attributes as socio-culturally, historically or symbolically valuable; and for some people such value cannot be meaningfully expressed in monetary terms."[60] Geographer Jean-Paul Haghe points out that both aesthetic value and commercial value serve anthropocentric interests."[61] His case study of the physical impact that changes in values had on the Gimel waterfall in Corrèze, France, for over a century resulted in the realization that the environmental movement had introduced a new criterion: "we are witnessing the recognition of the non-human environment's right to be respected: the ecosystem has acquired a moral value."[62] Although the gods and goddesses are no longer visible, we seem to be revisiting Hesiod's world. "Can the value of a waterfall be analyzed solely from the point of view of its contribution to human well-being," Haghe asks, or "should that value also be recognized as being intrinsic and non-anthropocentric?"[63] However humans answer that question, we will still be deciding for the other lifeforms in the ecosystem. That is the paradox of living in the Anthropocene.

Notes

1 Athanassakis, Apostolos N. *Hesiod: Theogony, Works and Days, Shield.* pp. 30–31. © 2022 Johns Hopkins University Press. Reprinted with permission of Johns Hopkins University Press.

2 Hesiod's exact dates are not known. Athanassakis, *Theogony*, xi. places Hesiod between the second half of the 8th century and the first quarter of the 7th century BCE.

3 Suzanne Lye, "The Goddess Styx and the Mapping of the World Order in Hesiod's *Theogony*," *Revue de Philosophie Ancienne*, vol. 27 no. 2 (2009) 6–7.

4 Lye, "The Goddess Styx," 7–8.
5 Homer, *The Iliad*, trans. Emily Wilson (NY: W. W. Norton and Co., 2023) 250.
6 Homer, *The Odyssey*, trans. Anthony Verity and William Allan, Notes and Introduction (Oxford UK: Oxford University Press, 2016) 66.
7 I. C. Torrance, "The Oaths of the Gods," *Oaths and Swearing in Ancient Greece*, ed. Alan H. Sommerstein and Isabelle C. Torrance (Berlin: De Gruyter, 2014) 197.
8 Athanassakis, *Hesiod*, 16–17.
9 For an extensive history of the concept and its social applications see Alan H. Sommerstein and Judith Fletcher, *Horkos: the oath in Greek Society* (Exter: Bristol Phoenix Press, 2007).
10 Walter Burkert, *Greek Religion* (Cambridge MA: Harvard University Press, 1985) 198.
11 Athanassakis, *Hesiod*, 70.
12 Hesiod, *Works and Days* 110–201; J. Donald Hughes, *Pan's Travail: Environmental Problems of the Ancient Greeks and Romans* (Baltimore and London: Johns Hopkins University Press, 1996) 66–67.
13 Athanassakis, *Hesiod*, 70.
14 Athanassakis, *Hesiod*, 93n 213–285.
15 Hughes, *Pan's Travail*, 131.
16 Athanassakis, *Hesiod*, 71.
17 Hughes, *Pan's Travail*, 131.
18 Burkert, *Greek Religion*, 250.
19 Hesiod's invention of the myth is proposed by Hermann Fränkel in Suzanne Lye, "The Goddess Styx and the Mapping of the World Order in Hesiod's *Theogony*," *Revue de Philosophie Ancienne*, vol. 27 no. 2 (2009) 6 n3.
20 *Inventing Origins? Aetiological Thinking in Greek and Roman Antiquity*, Antje Wessels and Jacqueline Klooster, eds. (Leiden, The Netherlands: Brill, 6 Dec. 2021) 3–5.
21 Herodotus, *Histories* 6.74, trans. Aubrey de Sélincourt; introduction and notes by John Marincola (London and NY: Penguin Books, 2003) 385.
22 Pausanias. *Description of Greece* (8.17.6) trans. by W.H.S. Jones and H.A. Ormerod, Cambridge, MA: Harvard University Press, 1918. https://www.theoi.com/Text/Pausanias8B.html.
23 Pausanias's Description of Greece, trans. commentary by J.G. Frazer, Vol. 4 (Commentary Books VI-VIII (London and NY: Macmillan and Co., 1898) 250–251.
24 Robin Hand, *The Routledge Handbook of Greek Mythology*: based on H. J. Rose's *Handbook of Greek Mythology* (London: Routledge, 2004) 110.
25 Arthur O. Lovejoy and George Boas, *Primitivism and Related Ideas in Antiquity* (NY: Octagon Books, 1965) 380.
26 See the definition and use of *topographia* in J.J. Pollitt's *The Ancient View of Greek Art: Criticism, History and Terminology* (New Haven, 1974) 332–334. For the use of the Greek root *topos* in Latin terminology, see Pierre Grimal, *Les Jardins Romains* (Paris, 1969) 89, 92–94.
27 Seneca, *Ad Lucilium Epistulae Morales*, trans. Richard M. Gunmere, NY: Loeb, 1920) II, LXXIX 5–7, 201.
28 Seneca the Younger, *Physical Science In The Time Of Nero, Translation Of The Quaestiones Naturales*, trans. John Clarke (London: Macmillan and Co. 1910) 137.
29 Plutarch, "Life of Alexander," *Lives of the Noble Grecians and Romans*, trans. John Dryden, rev. Arthur Hugh Clough (NY: Modern Library, 1938) 854.
30 Pliny the Elder, *The Natural History*, vol. 1 trans. John Bostock (London: Henry G. Bohn, 1855) Bk.2.106.35, 136.
31 Pausanias. *Description of Greece* (8.18.4) 1918.
32 Frazer, *Pausanius's Description of Greece*, 251.

33 Adrienne Mayor, *The Poison King: The Life and Legend of Mithradates, Rome's Deadliest Enemy* (Princeton: Princeton University Press, 2010) 392 n43.
34 Adrienne Mayor, *The Poison King: The Life and Legend of Mithradates, Rome's Deadliest Enemy* (Princeton: Princeton University Press, 2010) 392, 69; Adrienne Mayor and Anoinette Hayes, "The Deadly Styx River and the Death of Alexander," Princeton/Stanford Working Papers in Classics: Version 1.3, May (2011); https://www.historyoftheancientworld.com/2010/10/the-deadly-styx-river-and-the-death-of-alexander/.
35 Robert W. Sharples, *Theophrastus of Eresus: Life, Writings, Thought and Influence; Commentary*, vol. 3.1 *Sources on Physics* (Leiden: Brill, 1998) 200–201.
36 Vitruvius, *Ten Books on Architecture*, trans. Morris Hicky Morgan, 8.3.16 (Cambridge: Harvard University Press, 1914) 238.
37 Apuleius, *The Golden Ass or Metamorphoses*, trans. E.J. Kenney (London: Penguin Books, 2004) 84.
38 *Plato's Laws*, Bk. 11. 920d. trans. David Horn (The Foundation for Platonic Studies, 2021). https://www.platonicfoundation.org/laws/laws-book-11.
39 Katherine Clarke, "Walking through History: Unlocking the Mythical Past', in Greta Hawes (ed.), *Myths on the Map: The Storied Landscapes of Ancient Greece* (online edn, Oxford Academic 2017).
40 Virgil, *Aeneid*, Loeb Classics, 1986, trans. H. Rushton Fairclough (VII, 563–571) 40–43.
41 Dante Alighieri, *Divine Comedy*, trans. Mandelbaum *Inferno*, Canto XVI, 96–104. https://digitaldante.columbia.edu/dante/divine-comedy/inferno/inferno-16/.
42 Martin Warnke, *Political Landscape: The Art History of Nature* (Cambridge PA: Harvard University Press, 1995) 96. fig. 101 *Falls of the Rhine*, 1791, Graphische Sammlung, Staatgalerie, Stuttgart.
43 Joseph-Anton Koch, *Dante and Virgil Riding on the Back of Geryon*, National Gallery, Washington D.C., No. 2007.111.113.
44 Lord Byron, *Manfred: A Dramatic Poem*, Act II Scene II (London: John Murray, 1817) 34.
45 Byron, *Manfred*, 37.
46 Arthur Conan Doyle, "The Final Problem," *The Strand Magazine* 6 (December 1893) 559–570; Brian J. Hudson, *Waterfall: Nature and Culture* (London: Reaktion Books, 2012) 135–137.
47 Conan Doyle, "The Final Problem," *The Strand Magazine* 6 (December 1893) 568.
48 Christopher Redmond, *Sherlock Holmes Handbook* 2nd edition (Toronto: Dundurn Press, 2009) 302.
49 For the history of the Cascata delle Marmore, see Torquato Secci, *Disegni e Stampe della Cascata delle Marmore dal 1545* al 1976, Terni, 1980.
50 Cic. Att.4.15.5; Tac. Ann. 1.79.
51 See Gregory S. Aldrete, *Floods of the Tiber in Ancient Rome* (Baltimore: Johns Hopkins University Press, 2006).
52 Livy, XXXV, 9; Tacitus, *Hist.*, I, 86; Pliny, *Epistle*, VIII, 17.
53 Aldrete, *Floods of the Tiber*, 219.
54 Horace, *Odes* 1.2.13–16. Quoted in Aldrete, *Floods of the Tiber*, 22.
55 Seutonius, *Augustus* 37; Dio, *History* 57.14.7–8. See "Primary source descriptions of floods in Ancient Rome," Aldrete, *Floods of the Tiber*, 13–33.
56 Salvator Rosa to G.B. Ricciardi, in G.A. Cesareo, *Poesie e Lettere Edite e Inedite di' Salvator Rosa*, vol. II, cxi, p. 117, Napoli, 1892; for Rosa's role popularizing the Falls of Velino, see Patricia Likos, *The Landscape Paintings of Salvator Rosa*, Bryn Mawr College, ProQuest Dissertations Publishing, 1988; Helen Langdon, *Salvator Rosa* (London: Paul Holberton, 2010) 126–163.

57 For an introduction the British landscape aesthetics see Christopher Hussey, *The Picturesque* (Hamden, CT: Archon Books, 1967).
58 Joseph Addison, *Remarks on Several Parts of Italy, Etc., in the Years 1701, 1702, 1703* (Oxford, 1830) 74.
59 Hudson, "The Beautiful, the Sublime and the Picturesque," *Waterfall*, 63–75.
60 Turner, R. K., Paavola, J., Cooper, P., Farber, S., Jessamy, V., & Georgiou, S. (2003). "Valuing nature: Lessons learned and future research directions," *Ecological Economics* 46 (2003) 495.
61 Jean-Paul Haghe, "Do waterfalls have value in themselves? A metamorphosis in the values of the Gimel waterfall in France," *Policy and Society*, Volume 30, Issue 4 (December 2011) 249–256.
62 Haghe, Gimel waterfall, 255.
63 Jean-Paul Haghe, "Do waterfalls have value in themselves? A metamorphosis in the values of the Gimel waterfall in France," *Policy and Society*, Volume 30, Issue 4 (December 2011) 249.

6 The Ecology of the Color Purple in Greco-Roman Antiquity

Christopher Schliephake

The study of color and color perception in the ancient world has made quite a splash over the last few decades. Numerous book-length studies and edited collections have appeared from major presses and interdisciplinary projects have been launched that ask for the meaning(s) of color in a diachronic as well as a cross-cultural perspective. Although the respective research interests may vary, some common strands can be identified: they range from the semantics of color descriptions in ancient texts over what we may call the epistemology and experience of color to the study of the socio-political significance of colors in ancient cultures.[1] Despite the risk of oversimplifying the results of these diverse projects, it is safe to say that they have turned the long-held stereotype of a monochrome antiquity (the white-marbled world of historical paintings and Hollywood blockbusters) on its head. Instead, they present us with a colorful, multifaceted picture of ancient life and its lifeworld, whose many shades we are only beginning to understand. What holds true for the visual color of antiquity also holds true for what the ancients themselves said and wrote about colors (cf. Bradley 2009: xi): their "color *perception*" is no longer thought to have been markedly "different from ours, even if their color-related *conceptions* were often very different from ours" (Wharton 2021: 11, emphasis original), associated with a "flux of sensory experience" (Clarke 2004: 133).

There has been a rich tradition of studies devoted to the different dimensions of the subject (in practice, these dimensions often overlap and, as we shall see, can hardly be disentangled from one another). Johann Wolfgang von Goethe's *Farbenlehre* (2016, orig. 1810) and the *Studies on Homer* by the British statesman and devoted Hellenist William Gladstone (1858) are often referenced as important forerunners of the current research paradigm (cf. Bradley 2009: ix; Grand-Clément 2011: 11–20; Wharton 2021: 10–13). Where Goethe had, for instance, observed that Greek terms for a single color could range across a wide area of the Newtonian spectrum, thus in fact encompassing many different colors that would be clearly perceived as distinct in later eras, Gladstone argued that Homer's color system was based on the differentiation of light and darkness, but that color vision was only poorly developed – an argument that would resonate with Darwinian theories of cultural evolution

DOI: 10.4324/9781003385202-7

by suggesting that color vision had been defective in the past, but had seemingly progressed over the ages.

While none of the modern studies would support many of these theses (at least not in their extreme versions) and have reverted the idea of the material and experiential poverty of color in the ancient Mediterranean, they agree with earlier approaches in that they underline the social and cultural differences between ancient and modern takes on color. Where modern color categories are used to differentiate between various hues, ancient terminology reflects a "multisensory" experience (Bradley 2013: 131) that includes luminosity and texture, but also smell and liquidity, for instance. Moreover, in some cases terms for colors cannot only be used to describe the surface of an observable object, but they carry connotations that refer to the very process and/or the materials from which colorants were made. For Bradley, ancient (and particular Roman) perceptions of color were therefore not solely attached to visible features such as light and hue, but to an ontological notion of color as an inherent property of an object and its place within the lifeworld. As he argues, "because of its close ties to objects in the environment (…) the ancient colour experience could tap into smell, touch, taste, and even sound. For Greeks and Romans", he continues,

> colour was a basic unit of perception, a source of information and knowledge, and a tool for accurately understanding the world around them. It was a primary index for describing an object, a person, a building or a landscape, as well as an evaluative category of 'character' and personality.
>
> (139–140)

Bradley's ontological take on the subject resonates well with approaches that define color as "a sensory experience that helps us navigate in our environment" (Mohr/Jonauskaite 2019: 347) and that understand color as a "complex phenomenon": "we *see* colors, but we also *talk* and *write* about colors" (348, emphasis original). Because the interpretation of color(s) could vary across time and cultures – and even between individuals – color was not "a static, objective *thing*: it was a fluid, subjective, interactive unit of value and meaning" (Bradley 2009: xi; emphasis original).

Starting from the environmental dimensions alluded to in Bradley's argument, this chapter will ask for what we could call the 'ecology' of ancient colors. As I want to show this does not only encompass cultural representations and systems of meaning, but also the material fabrics of colors and how they relate to the nonhuman world and the experience of this world. Taking purple as my main analytical example, I am concerned with reflecting on the ancient discourses and practices connected to this particular color and in how far they make visible what posthumanist philosopher Donna Haraway has referred to as "natureculture(s)" (2003) in the ancient world. Starting from a discussion of the 'agency' of purple as treated in ancient texts, the first part of this chapter will bring ancient sources in dialogue with material ecocriticism and the "prismatic ecology" developed by Jeffrey J. Cohen (2013) and others. It will ask for the way in which ancient color perception was an embodied and highly relational enterprise that focused on the many entanglements between humans and the nonhuman elements of their lifeworld and the vibrancy of that relationship. Moreover, I will analyze the sociocultural meanings and practices attached to the color purple, thereby illustrating the way that the extraction of resources, the discursive perception of matter, and the particular ecologies of the Mediterranean interacted.

The material properties and vibrancy of 'purple'

Purple has a history that stretches back 4,000 years (cf. Boesken Kanold 2017: 67). It "was arguably the most famous dye of antiquity" (Gleba/Berghe/Cenciaioli 2017: 131; cf. also Bradley 2009: 189 and 209) and "the color par excellence of Greek and Roman dress" (Llewellyn-Jones 2021: 92). It was also the "one color in antiquity that truly indicated power" (Olson/Wharton 2021: 51) and "the most long-lived status symbol of antiquity" (Reinhold 1970: 6). Representing a wide range of hues, "this dye could evoke the appearance of the sea, blood, gems and marbles, even swans and rainbows" (Bradley 2009: 189). Yet, as Bradley cautions,

> to understand these manifold references as abstract descriptions of things that shared similar wavelengths with sea-purple dye (as we might to with our color 'purple') is to ignore the underlying connections that existed between the color category and all the physical, economic and political properties of the material to which that category referred.
>
> (190)

When it comes to analyzing purple's – or in fact, any other color's – meanings, "materiality (…) matters a great deal" (Wharton 2021: 13). This is because in antiquity an immense variety of dyes were produced from vegetable, animal, and mineral sources: for red and purple, for instance, safflower, madder, lichen, henna, or whortleberry could be used (cf. Becker 2021: 36; Llewellyn-Jones 2021: 92–93; Plin. *NH* 16.31; 22.3).

There were different dyes and terminology for purple in Greek: *halourgos*, *porphyreos*, *hyakinthos*, *orphninos* (Brøns 2017: 109–11). By far the most prestigious dye was sea-purple, which was extracted from the bodies (or, to be more exact, from the hypobranchical glan) of mollusks (notably *Hexaplex*, or *Murex*, *truncullus*, *Bolinus brandaris*, *Stramonita haemastoma*) whose bodies contain chemically active color compounds, which were induced to combine with natural fibers, producing colors that ranged from red to various shades of violet to blue (cf. Olson/Wharton 2021: 51; Llewellyn-Jones 2021: 88; Gleba/Berghe/Cenciaioli 2017: 131). As modern-day experimental archaeology has shown, the color of the fabric soaked in the vat depended upon several circumstances: "the atmospheric conditions of the dye process, the quality of the vat, but also the species of the shell used" (Grand-Clément 2016: 124; cf. Koren 2005; Becker 2021: 40; Olson/Wharton 2021: 51). The value of purple dyes therefore was not only aesthetic or economic, "but also practical: they remained both a status symbol and, because of their origin, an important item of trade throughout Greek and Roman worlds" (Llewellyn-Jones 2021: 92–93). In addition, Grand-Clément highlights the special symbolic value of this process:

> True purple was credited with possessing special powers, and this was partly due to the dyeing process. The dyeing agent located in the snail was an almost colorless liquid. When the shell was crushed, contact with air and light changed the color of the juice and it took on a violet hue.

As she continues,

> The Ancients managed to keep the dye agent colorless in huge vats, and they dipped garments or wool yarns, it was only when they took them out that the color changed gradually, moving from white, to yellow, green, and then blue, violet, or red. This metamorphosis seemed to be magic.
> (125; cf. also Sassi 2015: 267)

That the color and its microstructure qualities were sensitive to variations in the local environment had already been noticed by ancient authors, who marveled at the fact and attempted different interpretations of the process that accounted for geographical as well as climatic influences (cf. Plin. *NH* 9.62 and on the different shades of purple 21.22.45–46). In Roman author's Vitruvius' interpretation, exposure to the sun is the primary reason for why purple in Gaul and Pontus almost appears black (because these regions are closer to the north) and why it is reddish in Rhodes and other regions in the South, which are closer to the sun's course (cf. Vitr. *De Architectura* 7.13; cf. Bradley 2009: 192–193).

But the seemingly magical properties of purple did not solely have to do with the variety of hues that the dye could bring forth. It had to do with the effect of the color: philosophers like Aristotle (who held that purple was not only the most 'attractive' color, but a mixture of the primary colors black and white, Aristot. *De sensu* 3.439b 33–440a 1 and 4.442a 20–25) commented on the effect of light and the additive synthesis involved in color perception (cf. Sassi 2015: 271). "The sensations of brightness and of movement display in the visual process" were important in this context and, as Sassi suggests, "the very name *porphura* is related to *porphurō* meaning 'swirling' besides 'growing/dying purple'" and evokes the very process by which the dye is made as well as "the several hue changes occurring in the dyeing process would determine the final effect of purple, which was seen shimmering under the light rays, with a side effect of movement" (271). In a similar vein, Gage too notes that purple was imagined to be "a bearer of light" and that *porphyrein* means to "heave, surge, or ripple", "often of the sea which may perhaps be accounted for by the many color-changes in the dyeing process but which is also of course the condition required for the perception of the luster itself" (Gage 1993: 25; Hom. *Od.* 2.428, *Il.* 16.391). The effect of purple and the way it interacted with light and perception could be seen in analogy to color contrasts in rainbows (cf. Aristot. *Meteorologica* 372a 7–8 and 375a 22–28; an image that could already be found in epic, cf. Hom. *Il.* 17.547–552; see Grand-Clément 2016: 125–126). The Hellenistic poet Meander described a cloth of interwoven purple and white threads, which shimmered according to the fall of illumination (Fr. 667 K 2 Kock; see Olson/Wharton 2021: 51; for a similar effect cf. Philost. *Imagines* 1.28.4 with Sassi 2015: 267). What is important to note here is the vibrancy that is ascribed to the effect of purple – and how ancient authors tried to account for it by drawing on what we might refer to as the environmental properties of this specific color: it brought spectator, atmospheric or geographical features as well as the material and its texture together.

As I want to argue, it is for this reason that ancient takes on purple resonate well with recent trends in ecocritical theory that try to account for the material dimensions of 'natural' and 'cultural' interactions by focusing, among other things, on a wide variety of colors. In a thought-provoking rereading of Goethe's *Farbenlehre*, Sullivan argues that studying color from an ecological vantage point is to "unsettle our basic assumptions regarding nature as a 'place' separate from the human realm and to posit it instead as natural-cultural processes continually occurring *all around*, *through*, and *in us*" (2014: 80; emphasis original). Grounding her argument on Goethe's conviction that colors do not so much denote "static states", but "are representative for nature's fluctuating processes continually fueled by solar energy" and light, Sullivan states that "the sensory perception of living beings functions as the site of natural-cultural interfaces. (…) The ecology of colors highlights the reciprocity of our bodily materiality with energy forms, discursive information, and other-than-human materiality of many species" (80–81). Understanding the lifeworld through an optical lens that contextualizes color and light within a world of things and beings, "nature is", in Sullivan's reading, "the processes of emergence of material reality of which we are a part" (90). In many ways, Sullivan's thoughts reverberate strongly with the volume *Prismatic Ecology: Ecotheory Beyond Green* (2013), edited by Cohen, which appeared slightly earlier than her essay in Iovino's and Opperman's *Material Ecocriticism* (2014). As its title indicates, *Prismatic Ecology* adjusts the focus of environmental philosophy, ecocriticism, and the new materialism to a more comprehensive take on colors and concerns, trading in an exclusive focus on 'green' matters for a wider spectrum of colors and environments. As Cohen makes clear in his introduction to the volume, "a preponderance of green prevents the eye from noticing that the aerial is as much part of an ecology as the arboreal" (2013: xi). However, this does not entail leaving behind the theoretical insights and committed practice of 'green criticism', but rather looking for new inter-relations between the experiencing I, the phenomena of the world, and the physical and material conditions that enable an exchange between the two.

Cohen takes the example of a rainbow to illustrate how this optical phenomenon involves nonhuman matter, the individual perceiver and their position, combining them in "a sudden relation that changes the quality of light itself" (xxxvi). In this context, "color is not some intangible quality that arrives belatedly to the composition but a material impress, an agency and partner, a thing made of other things through which worlds arrive" (xvi; cf. also Schliephake 2016). The phenomenological and posthuman outlook of *Prismatic Ecology* is, against this background, to be thought of as a seeing instrument which helps critics dissect the different natural, material, and sociocultural layers of our more-than-human worlds and how they interact to negotiate meaning. As Buell makes clear, this posthuman perspective also entails a new kind of ethics that does not abandon categories like human/nonhuman or culture/nature, but that realigns them and that, in Buell's phrasing, "insists that humanness including human 'culture' is embedded in ecological processes" (2013: xi).

I am not suggesting that ancient authors who wrote about colors – or purple for that matter – shared these thoughts or were remotely concerned with what we would refer to as environmental issues. However, in the way they reflected on the effects of purple (and other hues), in how they situated these sensations in a world where many elements (nonhuman animals and their shells, humans, sea, land and light) interacted, and how they rendered the embodied experience of wearing purple garments, they irrevocably present us with a holistic take on this world. The ancient texts that write about colors (and how they are made) thus present us with what we, with Yates, could refer to as a "'multispecies' archive or impression":

> Within such a modeling of the archive, color effects refer to a discontinuous, shifting world of sense impressions as perceived and hosted by perceptually competent actors. Color itself might best be modeled as a multispecies sensory process or network that generates biosemiotic-material effects that then take on a metaphorical life of their own as they are translated to different registers.
>
> (Yates 2013: 85)

The storied ecology of purple

In antiquity, myth was one such register to make sense of and/or attribute meaning to the world. As I have suggested elsewhere, the mythological stories of the ancient Greeks and Romans can be seen as a narrative mode of engaging with the nonhuman world, presenting us, in fact, with a "storied ecology" of it (Schliephake 2020: 24–31). I use the term in a metaphorical way of reflecting on how the interconnection between human systems and their natural surroundings found an imaginative echo chamber in the works of culture and meaning-making. It is a way of uncovering ecological implications in and between texts that are, to the naked eye, not environmentally related at all. It also means reconsidering the role of nonhuman nature and species in ancient texts, not solely as a backdrop to human actions and decisions but as a constitutive factor of vibrant environments. The copresence of nonhuman forms of life is an important aspect in this context because it has challenged conceptions of the human from the earliest written sources onward. This approach is based on the observation that myth "illustrates an intricate, integral relationship with its physical surroundings" (Hawes 2017: 1).

When it comes to the color purple, we can see this in the way the ancients imagined how it was discovered in the first place. According to a well-known version of the myth,

> One day, as Heracles strolled along the Levantine coast near the city of Tyre, his faithful hound began to worry some snails that lay on the beach. The dog's muzzle soon stained brilliant violet and, in this way, purple dye was discovered.
>
> (Poll. *Onomasticon* 1,45–49)

Heracles (associated with the Phoenician god Melqart), who had fallen in love with the nymph Tyre (also the name of a famous city on the Levantine coast), had made her a wonderful dress dyed with the pigment. The myth can be found in other versions during Roman Imperial times[2] and was most likely based on earlier narratives. What unites these versions is that they associate the story with a center of purple-dye production in antiquity, namely the Levantine Sea and especially the Phoenician settlements there. This is also the reason why, for a long time, scholars connected purple dye with the ancient Near Eastern world;[3] during the Roman Empire – and long before – especially Tyre and Sidon were famed for the quality of their purple (cf. Hom. *Il.* 6.288–292; Hor. *Ep.* 1.10.26–29; Strab. 16.2.23). While the Near East was certainly a center of production,

> the first archaeological evidence of murex snail exploitation for dyeing came from the Early Bronze Age levels at Aegina in the Saronic Gulf; the majority of sites in which crushed murex shells or purple-dye residues have been found are in Crete and belong to the Middle Bronze Age.[4]

Yet, what is also remarkable is how the stories ascribe a central role in the chance discovery to the dog of Heracles (or the fisherman). What is at the center of these narratives is not so much human ingenuity (the idea to dye clothes with the shell only comes later) but the agency of nonhuman animals: the dog looks for food and apparently bloodies his muzzle on the hard shells of the snail. It is only after closer inspection that it becomes clear that the snails ooze a liquid that stains skin (and other stuff) with the color of blood.

The ecology that the myths surrounding the discovery of purple tell about is thus tied both to a specific geography of the Eastern Mediterranean and to a highly relational account of lifeforms. It is a "multispecies" story of the encounter of different lifeforms in Haraway's sense – a story about entanglement and, although the structure may seem simple enough, of the complexity of multispecies relationships. It is also a story about the violence inherent in the chance encounter between human, dog, and sea snail. The semantic association between purple and blood that can be found in Greco-Roman literature is an implicit reminder of the fact the color and its brilliant effect can only be achieved by killing the snail. This is why Haraway posits that "multispecies storytelling is about recuperation in complex histories that are as full of dying as living" (Haraway 2016: 10). Purple is thereby not so much a synthetic process guided by human ingenuity, but an inbuilt property of nonhuman bodies, which needs to be extracted, but which, at the same time, changes the material composition or appearance of other matter it touches – so much so, that once a fabric had gotten in contact with the snail's liquid, it could never be removed again. As the Roman didactic poet Lucretius puts it in *De Rerum Natura*:

> The purple color of the sea-snail unites with the substance of the wool so that it can nowhere be separated, not if you do you your best to restore it with Neptune's flood, not if the whole sea would wash it out with all its colors.
>
> (6.1074–77)

In this elemental account of the dyeing process, the cultural practice of wool dyeing is re-imagined in nonhuman terms, where the agency of matter is brought center stage. I would claim that it is one of the first instances of "naturecultures", a synthesis of nature and culture that recognizes their inseparability in ecological relationships that are both biophysically and socially formed (Haraway 2003).

This passage, along with the mythological accounts of how purple was discovered, also resonates with Iovino's and Oppermann's observation in their introduction to *Material Ecocriticism* that

> the world's material phenomena are knots in a vast network of agencies, which can be 'read' and interpreted as forming narratives, stories. Developing in bodily forms and in discursive formulations, and arising in coevolutionary landscapes of natures and signs, the stories of matter are everywhere: in the air we breathe, the food we eat, in the things and beings of the world, within and beyond the human realm. All matter, in other words, is a 'storied matter'.
>
> (Iovino/Oppermann 2014: 1)

Building upon the posthumanist notion that nonhuman objects and things are not merely to be thought as passive entities or resources, waiting to be extracted or used by humans, but rather agents that stand in a dynamic interaction with human beings (2–3), Iovino and Oppermann invite a practice of interpretation that perceives of discourse and materiality as two sides of the same coin, entwined, as it were, in endless processes of signification. They argue that material configurations of the world, in fact, form stories that can be read and interpreted (and that, in turn, inform human relations to materiality). They also hold that literary texts possess a material dimension; not only because they are written down on a medium storing information, but because they constitute cognitive responses to a concrete, material environment that is translated into human language (2–10). While this act of symbolization is by no means to be confused with 'reality' as it presents itself at the respective moment in time, this idea builds on the biosemiotic notion that "the natural world is perfused with signs, meanings, and purposes which are material and which evolve. (…) What goes *inside* an organism, and *between* an organism and its environment always involves what (…) we must call interpretations" (Wheeler 2006: 279, emphasis original). As Iovino and Oppermann put it, this is "an embodied process of understanding" (2014: 4) that, in general, finds a concrete expression in literary world-making and that restructures our way of perceiving the world. Ancient takes on purple, its discovery, and the way it resonated in the sociopolitical realm can be seen as "material narratives" (6) in Iovino's and Opperman's sense. And it is from this understanding of purple and its material properties that it will be possible to analyze "the interlacements of matter and discourses" (6) as we can find them in the way the ancient cultures appropriated purple for the way they structured and understood society.

The sociocultural meaning of purple in Greco-Roman Antiquity

The story of purple is also one of the first stories of extraction. "As a material practice, to extract is to draw out", Insko writes by reference to the OED definition, "of a cavity or a body – by force, effort of contrivance". As he reminds us, "extraction always entails a degree of violence" (2021: 173), and this was no different in the case of the sea snails, from whose throats small veins would disgorge tiny drops of fluid when they were crushed by physical force. As the Roman polymath Pliny the Elder wrote in his natural history (authored in the first century CE), "when it (i.e. the snail) dies, it spits out this juice. From the larger ones it is extracted after taking off the shell; but the small fish are crushed alive, together with the shells, upon which they eject this secretion" (*NH* 9.60). It is no over-exaggeration to claim that there was a massive industry built around this extraction process in antiquity. From the middle of the second century BCE onward, the material finds of great heaps of murex shells become more and more numerous in places all around the Mediterranean Sea. While we have to infer that "marine animal products" encompass "both dietary and non-dietary products" and the snails could also serve as a food resource, "the exploitation of the sea as a means of subsistence" was also due to the "'luxurious' character of marine raw materials or finished items and products of marine origin" (Theodoropoulou 2017: 80).

The high symbolical value of purple had both to do with the special, almost magical, properties of the liquid, but also with the production process: as modern-day experiments show, about 10,000 murex snails would be needed to dye one kilogram of wool a deep purple, which had implications for the prize of the dye (Koren 2005: 146; Cardon 2007: 559–562).[5] The production of purple dye also involved other aspects that need to be taken into account, as recent environmental historical analyses show, that include "geo-environmental, scenic, and agro-pastoral" features (Giner/Nieto 2017: 145–146). Already in the Bronze Age, trade networks developed that illustrate that purple clothing was a sought-after and highly prized commodity. In the monarchies of the Near East, it seems to have turned into a status symbol of the court elite – and in the archaic period several Homeric passages show the symbolic value of purple, related to the display of luxury as well as the garments of mythical heroes (Ferrara 2017: 120–121; Reinhold 1970: 24). The stories attached to the material properties of purple as well as the extractive and economic processes connected to it partly explain why "certain substances could be endowed with specific values and highly esteemed as 'symbols of excellence', and therefore, credited with a kind of social 'agency'" (Grand-Clément 2016: 121). As Grand-Clément puts it,

> the chromatic experience is rooted in a sensitive, material reality. Color is corporeal, not abstract. So, before asking whether blue and red – which correspond to modern categories of thought – have this or that symbolic meaning, we should consider the value attributed to the chromogenic substances themselves. Some of them have an intrinsic value, linked to their extraction method.[6]

The high durability of the sea-purple dye also added to its value.

Moreover, we have to take into account social customs and regulations: in Rome, for instance, we can observe that there were attempts to regulate the wearing of purple garments from the late Republic onward. Even earlier, in Hellenistic times, a regulation of purple clothing seems to have been in place to create and symbolically mark a hierarchy at the royal court (Grand-Clément 2016: 136). Both in a diachronic and cross-cultural sense, purple thus indicated elite status and luxury. As Pliny remarked upon this:

> It is for this color that the fasces and the axes of Rome make way in the crowd; it is this that asserts the majesty of childhood; it is this that distinguishes the senator from the man of equestrian rank; by persons arrayed in this colour are prayers addressed to propitiate the gods; on every garment it sheds a lustre, and in the triumphal vestment it is to be seen mingled with gold.
>
> (*NH* 9.60)

Pliny's account equates the wearing of purple with the sociopolitical display of power. For while purple was not solely reserved for the imperial family, it was clearly a status symbol of the senatorial rank – a stark reminder of social hierarchy. Yet, as Bradley observes in his commentary of Pliny's account of purple, "Pliny playfully switches between using *purpura* to mean the dress-colour 'purple', and *purpura* to mean the snail: to really appreciate the former, one must understand the latter" (2009: 195). In this context, it is important to note that Pliny is not very appreciative of the elite's habitus, but rather questions the "this frantic passion for purple" on the grounds that "such a high value has been set upon the produce of this shell-fish" even though "that while in the dye the smell of it is offensive, and the colour itself is harsh" (Plin. *NH* 60). "One of Pliny's aims", Bradley writes, was "to demonstrate that traditional epistemology was being undermined by the Roman habit of taking bits out of the world – like *purpura* and colourful stones – entirely out of context, and deriving satisfaction from aesthetic appreciation alone" (2009: 195). Moreover, Pliny's account of purpura can be understood as "a typical *reductio ad absurdum*, a potent reminder of precisely where it was that all those senatorial purple stripes, triumphal garbs and emperors' clothes came from" (194).

That purple clothes, despite their high symbolic and social value, were also associated with particular olfactory and tactile properties that were not necessarily pleasant, but carried the smell of sea and shellfish, can be seen in other imperial texts (like Martial's writings, cf. 1.39; 4.4 and 9.63 with Bradley 2009: 196–197). "The dye's smell was a potent reminder of where it (i.e. purple)" originated from (196),[7] and the texts writing about it were material narratives in Iovino's and Oppermann's sense, substance stories that brought social habitus, status display, and the extractive practices upon which it was based in close conversation with one another. They reflected on the material fabrics upon which imperial self-representation was based. Implicitly, Pliny's *Natural History* presents us with an all-encompassing archive of the kinds of "extraction (that) might be described as a form of taking without giving in return" (Insko 2021: 173), reflecting "a nonreciprocal, dominance-based relationship with the earth" (Klein 2014: 169). This imperial Roman dominance found a concrete expression in the material practice of extraction, but also in its "underlying", "instrumentalizing logic" (Insko 2011: 173), which was predicated upon "the reduction of life into objects for the use of others" (Klein 2014: 173). Although it would lead too far to ascribe to Pliny (or other ancient authors that remarked upon the material practices of extraction for imperial self-representation and -fashioning) the role of a critic or even activist, who wanted to comment on Rome's politics that, for the sake of its elite, exploited human and nonhuman resources alike, his text nonetheless can be read from an environmentalist viewpoint in so far as he presents us with one of the most comprehensive accounts of human-nature relations in the early imperial period.[8] Bradley's comment that Pliny presents his readers with the ways in which Rome "takes bits out of the world" (2009: 195) finds a close parallel in the way Insko writes about the violence that is at play in forms of extraction:

> just as geophysical extraction literally removes minerals, metals, elements, and some humans from the geological strata or geographical locations in which they're embedded, so does extractivism (i.e. the logic behind it) remove them, conceptually from what the political ecologist Jason W. Moore calls the 'web of life': the interconnected geological histories and the social and ecological networks in which they are embedded. Extractivism alienates them from their contexts, transforming them instead into fungible commodities, converting 'nature' into 'natural resources'.
>
> (2021: 173–174)

By juxtaposing Roman elite status display and the material fabrics of the color that embodies their claim to superiority, Pliny invites an interpretation that focuses on the way sociocultural meanings and material extraction interrelate in order to shape environments and people. In a way, the story of the color purple and its many dimensions, particularly its material properties and practices, resonates with Tsing's discussion of the plantation model in *The Mushroom at the End of the World* in so far as the colonial plantation, in its total abstractedness from the contexts and subjects it involves, is a model based on alienation: it is a power structure that disconnects native species, exploits labor and transforms environments and ecological relationships (2015: 38–43). I would argue that the history of the color purple in antiquity, and particular in Roman imperial times, presents us with an ancient instant of such a, albeit violent, "history of encounter" (28), one that follows the "frictions" of natures and beings (28–29) and that calls for the need of "complex modes of analysis that travel through entangled territories" (282). While the multispecies stories surrounding the origin of purple discussed above are stark reminders of the entanglement of beings, the social uses of the snail's liquid and the symbolical value ascribed to it present a violent friction that takes – by force – one element out of nature and displaces it by transferring it to completely different settings. A narrative like Pliny's can, against this background, be read as a powerful reminder of and comment on the way the ancients dealt with and imposed meanings upon their lifeworld. His *Natural History* is an "epistemology of humanity's embeddedness in a nature of which we are, simultaneously, both part and apart" (Parham/Westling 2017: 10–11).

To conclude, I would claim that the multisensory experience of the color purple in antiquity, along with its material traces and the narratives it inspired, is a potent reflection of "how stories mediate different registers of power and how narratives have historically managed the manifold vitality of the cultures, bodies, and objects that shape and are shaped by them" (Foote/Cohen 2021: 2). Where the storied ecology of ancient myth perpetuated an understanding of humanity's immersion in nonhuman nature, the cultural appropriation of sea-shells as resources for human exploitation anticipates the extractive practices that have paved the way into the "Anthropocene".

Notes

1 I do not offer a comprehensive list, but a selection of important studies related to color and especially purple include Reinhold 1970; Stulz 1990; Blum 1998; Bradley 2009; Grand-Clément 2011; Jockey 2018; Thavapalan 2020; and Wharton 2021.

2 In Achilles Tatius' version, it is not Hercules' dog that bites into the shell, but the dog of a local fisherman, who believed his dog was wounded and bloodstained, but soon discovered the true cause of this color and then used the shell to dye a fleece (Ach. Tat. 2.11; on this passage see Bradley 2009: 191). Cf. also Soriga 2017: 80. The story of the discovery of purple finds an echo in how it was rediscovered: as Thavapalan points out, Henri de Lacaze-Duthiers (1821–1901), the zoologist credited with this rediscovery, "was apparently walking along a coastal village in Spain one day, when he observed a fisherman breaking open a sea-snail's shell and covering his shirt with its contents. To his fascination, the yellow smudge gradually changed to a brilliant purple color" (2020: 242).

3 Purple-dyed clothing is often mentioned in Old Testament; see *Exodus* 25.4–5; cf. Soriga 2017: 79.

4 See Soriga 2017: 80. Cf. Thavapalan 2020: 240–242 with a discussion of the trade networks that reached already in the Bronze Age reached all the way to Mesopotamia, possibly via Qatar.

5 In 1909, Paul Friedländer isolated 1.4g of the chemical compound from 12,000 murex brandaris. Cf. Cooksey 2017: 73.

6 "L'expérience chromatique s'ancre dans une réalité sensible et matérielle. La couleur possède une forme de corporéité et n'a rien d'une entité abstraite. Aussi, avant de se demander si le bleu et le rouge – qui correspondent à des categories de pensée modernes – revêtent telle ou telle signification symbolique, il faudrait s'interroger sur la valeur accordée aux substances chromogènes elles-mêmes. Certaines d'entre ells possèdent en effet une valeur intrinsèque, liée au mode d'extraction" (190; my translation).

7 On the stench and the chemical properties of the substance see Cooksey 2017: 73.

8 For a recent discussion of the discourses and practices surrounding nonhuman materials in Pliny's work see Angouissola and Grüner 2021.

Works cited

Angouissola, Anna and Andreas Grüner (eds.). 2021. *The Nature of Art: Pliny the Elder on Materials*. Turnhout: Brepols.

Becker, Hilary. 2021. "Technology and Trade". In *A Cultural History of Color in Antiquity*, edited by David Wharton, 35–48. London: Bloomsbury.

Blum, Hartmut. 1998. *Purpur als Statussymbol in der griechischen Welt*. Bonn: Habelt.

Boesken Kanold, Inge. 2017. "Dyeing wool and sea silk with purple pigment from Hexaplex trunculus". In *Treasures from the Sea: Sea Silk and Shellfish Purple Dye in Antiquity*, edited by Hedvig Landenius Enegren and Francesco Meo, 67–72. Oxford: Oxbow Books.

Bradley, Mark. 2009. *Colour and Meaning in Ancient Rome*. Cambridge: Cambridge UP.

Bradley, Mark. 2013. "Colour as synaesthetic experience in antiquity". In *Synaesthesia and the ancient senses*, edited by Shane Butler and Alex Purves, 127–140. London: Routledge.

Brøns, Cecilie. 2017. "Sacred colours: purple textiles in Greek sanctuaries in the second half of the 1st millennium BC". In *Treasures from the Sea: Sea Silk and Shellfish Purple Dye in Antiquity*, edited by Hedvig Landenius Enegren and Francesco Meo, 109–117. Oxford: Oxbow Books.
Buell, Lawrence. 2013. "Foreword". In *Prismatic Ecology: Ecotheory Beyond Green*, edited by Jeffrey J. Cohen, ix–xii. Minneapolis: University of Minnesota Press.
Cardon, Dominique. 2007. *Natural Dyes: Sources, Tradition, Technology, and Science*. London: Archetype.
Clarke, Michael. 2004. "The Semantics of Colour in Ancient Greek Word-Hoard". In *Colour in the Ancient Mediterranean World*, edited by Karen Stears and Glenys Davies, 131–140. Oxford: Archaeopress.
Cohen, Jeffrey J. 2013. "Introduction: Ecology's Rainbow". In *Prismatic Ecology: Ecotheory Beyond Green*, edited by Jeffrey J. Cohen, xv–xxxvi. Minneapolis: University of Minnesota Press.
Cooksey, Chris. 2017. "Recent advances in the understanding of the chemistry of Tyrian purple production from Mediterranean molluscs". In *Treasures from the Sea: Sea Silk and Shellfish Purple Dye in Antiquity*, edited by Hedvig Landenius Enegren and Francesco Meo, 73–78. Oxford: Oxbow Books.
Ferrara, Bianca. 2017. "'A Lydian chiton with a purple fringe …': The gift of the garment to the Hera of Samos and Hera of Sele". In *Treasures from the Sea: Sea Silk and Shellfish Purple Dye in Antiquity*, edited by Hedvig Landenius Enegren and Francesco Meo, 118–130. Oxford: Oxbow Books.
Foote, Stephanie and Jeffrey J. Cohen. 2021. "Introduction: Climate Change/Changing Climates". In *The Cambridge Companion to Environmental Humanities*, 1–10. Cambridge: Cambridge UP.
Gage, John. 1993. *Color and Culture: Practice and Meaning from Antiquity to Abstraction*. Boston: Little, Brown and Company.
Giner, Carmen Alfaro and Francisco Javier Férnandez Nieto. 2017. "'Purple Wars': fishing rights and political conflicts concerning the production of marine dyes in Hellenistic Greece". In *Treasures from the Sea: Sea Silk and Shellfish Purple Dye in Antiquity*, edited by Hedvig Landenius Enegren and Francesco Meo, 145–153. Oxford: Oxbow Books.
Gladstone, William Ewart. 1858. *Studies on Homer and the Homeric Age*. Oxford: At the University Press.
Gleba, Margarita, Ina Vanden Berghe and Luana Cenciaioli. 2017. "Purple for the masses? Shellfish purple-dyed textiles from the quarry workers' cemetery at Strozzacapponi (Perugia/Corciano), Italy". In *Treasures from the Sea: Sea Silk and Shellfish Purple Dye in Antiquity*, edited by Hedvig Landenius Enegren and Francesco Meo, 131–137. Oxford: Oxbow Books.
Goethe, Johann Wolfgang von. 2016 (1810). *Zur Farbenlehre*. Augsburg: Hofenberg.
Grand-Clément, Adeline. 2011. *La fabrique des couleurs: histoire du paysage sensible des Grecs anciens : VIIIe-début du Ve s. av. n. è*. Paris: De Boccard.
Grand-Clément, Adeline. 2016. "Gold and Purple: Brilliance, Materiality and Agency of Color in Ancient Greece". In *Essays in Global Color History*, edited by Rachel B. Goldman, 121–138. Piscataway, NJ: Giorgias Press.
Haraway, Donna J. 2003. *Companion Species Manifesto*. Chicago: Prickly Paradigm Press.
Haraway, Donna J. 2016. *Staying with the Trouble: Making Kin in the Chthulucene*. Durham: Durhman UP.
Hawes, Greta. 2017. "Of Myths and Maps". In *Myths on the Map: The Storied Landscapes of Ancient Greece*, 1–13. Oxford: Oxford UP.
Insko, Jeffrey. 2021. "Extraction". In *The Cambridge Companion to Environmental Humanities*, 185–198. Cambridge: Cambridge UP.

Iovino, Serenella and Serpil Oppermann. 2014. "Introduction: Stories Come to Matter". *Material Ecocriticism*, edited by Serenella Iovino and Serpil Oppermann, 1–17. Bloomington: Indiana UP.
Jockey, Philippe (ed.). 2018. *Les arts de la couleur in Grèce ancienne … et ailleurs*. Athens: École française d'Athènes.
Klein, Naomi. 2014. *This Changes Everything: Capitalism and the Climate*. New York: Simon and Schuster.
Koren, Zvi. 2005. "The First Optimal All-Murex All-Natural Purple Dyeing in the Eastern Mediterranean in a Millennium and a Half". *Dyes in History and Archaeology* 20, 136–149.
Llewellyn-Jones, Lloyd. 2021. "Body and clothing". In *A Cultural History of Color in Antiquity*, edited by David Wharton, 81–102. London: Bloomsbury.
Mohr, Christine and Domicele Jonauskaite. 2019. "Épilogue". In *Psychologie de la couleur dans le monde gréco-romain: Huit exposés suivis de discussions et d'un epilogue*, edited by Katerina Ierodiakonou, 347–368. Geneva: Fondation Hardt.
Olson, Kelly and David Wharton. 2021. "Power and identity". In *A Cultural History of Color in Antiquity*, edited by David Wharton, 49–62. London: Bloomsbury.
Parham, John and Louise Westling. 2017. "Introduction". In *A Global History of Literature and the Environment*, edited by John Parham and Louise Westling, 1–17.
Reinhold, Meyer. 1970. *History of Purple as a Status Symbol in Antiquity*. Brussels: Latomus.
Sassi, Maria M. 2015. "Perceiving Colors". In *A Companion to Ancient Aesthetics*, edited by Pierre Destrée and Penelope Murray, 262–273. Malden, MA: Blackwell.
Schliephake, Christopher. 2016. "More-than-green Ecologies". *Relations* 4(1), 117–119.
Schliephake, Christopher. 2020. *The Environmental Humanities and the Ancient World: Questions and Perspectives*. Cambridge: Cambridge UP.
Soriga, Elena. 2017. "Mari(ne) purple: western textile technology in Middle Bronze Age Syria". In *Treasures from the Sea: Sea Silk and Shellfish Purple Dye in Antiquity*, edited by Hedvig Landenius Enegren and Francesco Meo, 79–95. Oxford: Oxbow Books.
Stulz, Heinke. 1990. *Die Farbe Purpur im frühen Griechentum*. Stuttgart: Teubner.
Sullivan, Heather. 2014. "The Ecology of Colors: Goethe's Materialist Optics and Ecological Posthumanism". In *Material Ecocriticism*, edited by Serenella Iovino and Serpil Oppermann, 80–94. Bloomington: Indiana UP.
Thavapalan, Shiyanthi. 2020. *The Meaning of Color in Ancient Mesopotamia*. Leiden/ Boston: Brill.
Theodoropoulou, Tatiana. 2017. "A sea of luxury: luxury items and dyes of marine origin in the Aegean during the seventh century BC". In *Interpreting the Seventh Century BCE*, edited by Xenia Charalambidou and Catherine Morgan, 80–92. Oxford: Archaeopress.
Tsing, Anna Lowenhaupt. 2015. *The Mushroom at the End of the World: On the Possibility of Life in Capitalist Ruins*. Princeton, NJ: Princeton UP.
Wharton, David. 2021. "Introduction". In *A Cultural History of Color in Antiquity*, edited by David Wharton, 1–15. London: Bloomsbury.
Wheeler, Wendy. 2006. *The Whole Creature: Complexity, Biosemiotics and the Evolution of Culture*, London: Larence and Wishart.
Yates, Julian. 2013. "Orange". In *Prismatic Ecology: Ecotheory Beyond Green*, edited by Jeffrey J. Cohen, 83–105. Minneapolis: University of Minnesota Press.

7 The Byzantine Experience of the Natural World

Kirsty Stewart

Introduction

The Orthodox Christian faith present in Byzantium pervaded every aspect of life, whether consciously or unconsciously, influencing the practical engagement with the natural world, and even the ways in which pleasure could be derived from it. The relationship between humanity and wider Creation was not static, varying depending on time and context, but with certain core strands repeated throughout the history of the Empire. This article aims to provide an overview of some those ideas, before discussing the role of nature in day-to-day life, and some correlations between the religious ideology and reality.

Pious

In contrast to the relationship between Christianity and nature as defined by Lynn White, in which Christianity, in absolute contrast to ancient paganism and Asia's religions (except, perhaps, Zoroastrianism), not only established a dualism of man and nature but also insisted that it is God's will that man exploit nature for his proper ends, in the Byzantine world, there appears to have been a less clear sense of humans ruling nature, and thus able to manipulate it however they chose without fear of repercussions.[1] Instead we find positive and inclusive attitudes towards nature exemplified in the writings of the Cappadocian Fathers and Maximus the Confessor, all of whom remained greatly respected, known and copied throughout the Byzantine period. While mankind still held dominion, it was with a greater sense of the merits of other parts of Creation in their own right, and as a means of communication or message, the understanding of which would bring one closer to God. That is not to say that nature wasn't well utilised as a resource, and 'Byzantines found ways to put nature securely in its subordinate place'[2] at points, with nature appearing as 'corruptible, fleeting, transient, and false'[3] in at least some post-iconoclastic literature, but on balance, the religious understanding leaned towards a beneficial co-existence. Nature as Creation was valued as a reflection of God's glory and power. It was seen as an ordered system from which one could learn directly, particularly when discussing nature as described in the

DOI: 10.4324/9781003385202-8

Genesis narrative, though allegory definitely had its place as well. Nature is valuable as a contemplative bridge.

Perhaps the most influential of the early Byzantine religious texts addressing nature were the homilies on the *Hexaemeron* by Basil the Great, apparently preached around AD 378.[4] The homilies are deeply concerned with both the nature of God and with nature itself, expressing that the one can be known through the other:

> Let us glorify the supreme Artificer for all that was wisely and skilfully made; by the beauty of visible things let us raise ourselves to Him who is above all beauty; by the grandeur of bodies, sensible and limited in their nature, let us conceive of the infinite Being whose immensity and omnipotence surpass all the efforts of the imagination. Because, although we ignore the nature of created things, the objects which on all sides attract our notice are so marvellous, that the most penetrating mind cannot attain to the knowledge of the least of the phenomena of the world, either to give a suitable explanation of it or to render due praise to the Creator.[5]

Basil's wishes to depict the beauty of Creation and through it to inspire his listeners with a love of their Creator, their faith and to educate them regarding their duty and Christian life.[6] Animals feature prominently as exemplars for humans:

> Children love your parents, and you, "parents provoke not your children to wrath." Does not nature say the same? Paul teaches us nothing new; he only tightens the links of nature. If the lioness loves her cubs, if the she wolf fights to defend her little ones, what shall man say who is unfaithful to the precept and violates nature herself; or the son who insults the old age of his father; or the father whose second marriage has made him forget his first children?[7]

Basil even indicates specific psychological features such as the steadiness of the ox, the deceitful nature of the fox and the faithfulness of the dog.[8] Basil does not go as far as Origen who claimed that when God gave Adam dominion over the beasts of the sea, air and earth, 'these beasts were, in fact, interior dispositions or desires'.[9] Instead, Basil maintains a realistic, non-allegorical, ideology, and does not allow nature to exist solely to be subservient to mankind, though he still tends to an anthropocentric viewpoint. Like Origen, Basil believed the Bible contained different layers of truth and that these layers were all relevant, but that different audiences would find it easier to perceive different levels. Those present to hear his sermons would have included artisans, labourers, women and children, in short, the uneducated members of the population, as well as the more erudite.[10] His straightforward presentation of nature as Creation and a means of engaging with the Creator not only made the homilies accessible upon delivery, but are arguably a reason for their continued popularity.

Gregory of Nyssa, Basil's younger brother, did not write a *Hexaemeron* himself, but he did discuss his brother's work, and Creation overall. He takes a more scientific or technical approach to nature, man and the cosmos than his brother, but nevertheless examines the world through a spiritual lens. The connection between philosophy and the Bible, as well as the connection between man and animal, is key to his discussion of the natural world. Gregory is interested in the levels of creation, its order and form. He confirms the order of creation from vegetation to beasts to rational beings presented by the first guide to the world, Moses, who learnt of creation directly from God, and indicates that 'there are three orders of soul corresponding to each level: the nutritive, the sensitive and the intellectual'. In *On the Making of Man*, Gregory spends a considerable amount of time discussing the difference in form between man and beasts, relating it to the differences between their souls and the reason with which man was endowed. If man was endowed with protection as are the beasts not only would his dominion over them be questionable, but he would essentially be a beast himself. Man is not 'armed with prominent horns or sharp claws, nor with hoofs nor with teeth, nor possessing by nature any deadly venom in a sting'.[11] Instead man is 'slower than the beasts that are swift of foot, smaller than those that are of great bulk, more defenceless than those that are protected by natural arms'.[12] This lack of protection for one who is supposed to rule the other beasts is not seen as an unexplainable weakness by Gregory. Rather, if man had horns and hoofs and such-like things to protect himself he would be:

> a wild-looking and formidable creature, … and moreover he would have neglected his rule over the other creatures if he had no need of the co-operation of his subjects; whereas now, the needful services of our life are divided among the individual animals that are under our sway, for this reason—to make our dominion over them necessary.[13]

Thus creation relates to theology, the 'laws of nature reflect the laws of God. The study of nature leads to a greater appreciation for the Divine Nature'.[14]

The third of the Cappadocian Fathers, Gregory, bishop of Nazianzus (382–384) was highly popular, his homilies being transmitted in more than 1500 manuscripts dated before AD 1500.[15] He believed nature to be 'praiseworthy, surely, for the natural excellence of each of its parts, but still more praiseworthy for the proportion and harmony of all of them together', emphasising the variety of living things and their artistry, as interconnected and astonishing constructions of the Creator.[16] His *New Sunday Sermon* (Oration 44) uses the imagery of spring to describe the rebirth of man through Christ.[17] A highly sensory piece of literature, it paints a picture of bucolic harmony for the audience:

> For everything is conspiring together, rejoicing together, for the beauty of this feast. Look at all that meets your eyes! … Now heaven shines more brightly, the sun stands higher and glows more golden … Now the meadow is fragrant, the shoots burst forth, the grass is ready for mowing, and the lambs skip through rich green fields.[18]

Here it is the beauty of nature, and the peace and pleasure it creates, that are notable. This pleasure is however also the pleasure of faith. Though agricultural pursuit is part of the description, care for the world clearly sits alongside enjoyment of it. Gregory also warns his listeners not to be led astray by their senses as Eve was, but rather to understand the earth as reflecting Christian joy at the Day of Renewal, for it is 'the world's spring, the spiritual spring, spring for our souls, spring for our bodies, spring visible, spring invisible'.[19]

Maximus the Confessor was born around 580, and died in 662 following his mutilation and exile resulting from his belief that Christ had both a human and a divine will, but was venerated as a saint shortly after his death. Maximus agrees with the Cappadocian Fathers that the world could be read or understood as messages from God, though perhaps with the need for more interpretation as 'the whole spiritual world seems mystically imprinted on the whole sensible world in symbolic forms'.[20] Maximus saw God as the initial cause of beings, the final cause and the one *through* whom all things come to be.[21] At no point are any creatures, identical to God in essence, though all come from him and are thus connected to him. For Maximus humanity is a 'laboratory in which everything is concentrated and in itself naturally mediates between the extremities of each division'.[22] Maximus believes humans will be able to unite with God, but also able to bring the rest of Creation, whether rational or not, into union too.

> For this reason the human person was introduced last among beings, as a kind of natural bond mediating between the universal poles through their proper parts, and leading into unity in itself those things that are naturally set apart from one another by a great interval.[23]

Human impact is not wholly positive, the extremes within humans can lead to closeness with God, but are the reason for the rest of Creation's distance from God. Through the fall of humankind, all created beings fell. Human will contains vices, but nevertheless strives towards deification. Christ is the great unifier, and through his incarnation as man, death and resurrection all will be returned to exist as one. Arguably, Maximus' view is very anthropocentric, as man is at the centre of Creation, and acts as a facilitator in all things becoming closer to God. However, Maximus is very clear about the methods of reconciliation;

> Finally, beyond all these, the human person unites the created nature with the uncreated through love (O the wonder of God's love for us human beings!), showing them to be one and the same through the possession of grace, the whole [of creation] wholly interpenetrated [*perichoresis*] by God, and become completely whatever God is, save at the level of being, and receiving to itself the whole of God himself, and acquiring as a kind of prize for its ascent to God the most unique God himself, as the end of the movement of everything that is carried towards it.[24]

It is through love, not control, of the created world that all will return to God, and thus, despite the importance Maximus places on humanity, his theology was inclusive and required people to strive for a balanced interaction with nature.

None of these authors depict nature as existing only 'to serve man's purpose', an accusation levelled at Christian literature by Lynn White, who presents the notion of nature's servitude as ecologically destructive, though admittedly based primarily on Western theology.[25] Instead, there is a much more positive presentation, which continued throughout the Byzantine period, indicating that 'all creation reveals God', and was part of the Creator in the same way as humanity.[26] The New Sunday Sermon of Gregory of Nazianzus touches on the tension within the doctrinal view of the natural world, in which it is both God's creation and a temptation for man. The sermon would remain a model for the rhetorical presentation of nature throughout the Byzantine period, far more so than hexaemeral literature. There is little discussion of the hierarchy of nature, and the role of man as head of the terrestrial hierarchy under God. Instead, the relationship presented stresses the individual, divinely created traits of different aspects of nature, and their connection to mankind as products of the Creator. After all, 'we ought to show kindness and gentleness to animals for many reasons and chiefly because they are of the same origin as ourselves'.[27]

Practical

One of the most obvious ways in which any society engages with nature is in obtaining the resources required to survive. The Byzantine population, whether they lived in villages or cities, countryside or by the sea, would have encountered numerous animals and plants on a daily basis. The Byzantine diet included various grains and pulses, olive oil, wine, dairy products, fish and meat, fresh vegetables and fruit, and honey for sweetening, with some variations between regions. Crops and animals were farmed, and hunting and fishing were regularly undertaken for food as well as pleasure. Indeed we have a number of texts dedicated to food cultivation and to hunting, the *Geoponika* being perhaps the best known.[28] A manual compiled in the tenth century from a range of earlier texts, it includes various different ideas about the natural world, depending on the source material, showing both that trends in scientific thought, whether the belief in lunar planting, interpretations of weather, or the origins propagation methods of plants, changed over time, and that the Byzantines saw value in retaining these different elements, at least at certain times. It highlights key crops of interest, with books four through eight focusing on the care and preparation of grapes, book nine discussing the olive, book ten fruiting trees and book twelve, herbs and vegetables. There are also chapters on the pleasure garden, flowering plants and decorative trees, before the work moves on to discuss animals including birds, bees, horses, cattle, sheep, dogs, boar, deer and fish. Broadly speaking, the sections on animals are grouped into domesticated and wild animals, which could nevertheless be cultivated in large parks, such as the Philopation, which was just outside Constantinople, beyond the Blachernai walls.[29] Manuals on how to care for and train birds used in hunting were written throughout the period, and depictions of hunting were a key feature of early Byzantine art. Alongside birds we find discussions of the use of horses and dogs in hunting, as well as cheetahs, who would ride on the horse with the hunter then leap down to take the prey.[30] Interestingly, stories of the ancient Greek gods and their involvement with specific plants, including the rose, bay tree and myrtle, are easily found in the *Geoponika* but Orthodox Christianity has little to no presence in the collection, due to the choice of material.

Food was nevertheless closely tied to religious attitudes. There is some evidence, particular in censure, that ancient festivals relating to the planting and harvesting of foods continued in parts of Byzantium well into the seventh century if not later. Religious feast days and fast days were a core part of the Orthodox calendar, and observed by the wider population, not just monastic communities. The monastic diet in Byzantium excluded meat, though in some cases it would be served to guests, fish and seafood were present particularly for feasts, but monks were predominantly vegetarian. How well the rules defined by *Typika* were actually fulfilled is arguable, with a number of humorous texts directly mocking the gluttonous monk.[31] Even in these satirical texts, however, it is more the quantity and quality of food that is described, rather than any other infractions. Byzantines were well used to incorporating foods and terminology for foods into their world from across their spheres of influence, and acknowledged differences in diet in relation to religion. We have an account from a Syrian hostage of the tenth century, Harun Ibn Yahya, in which Muslim hostages are invited to the emperor's presence for a Christmas feast, where they are served hot and cold dishes and 'the imperial herald announces: "I swear on the Emperor's head that there is no pork at all in these dishes!"'[32] Pork itself does not appear to have been that well-considered a food in Byzantium, but the refusal to eat it was a mark of difference, for both the Jewish and Muslim populations. In at least some contexts, the Byzantines showed sensitivity to the religious prohibitions of others around food, though they also used them to demarcate and isolate.

Out of necessity, the Byzantines utilised as much of an animal as possible, obtaining clothes, tools and other products. A familiar example would be the tanners and furriers noted as using the hide and tail of foxes in the *Παιδιόφραστος διήγησις τῶν ζῴων τῶν τετραπόδων or Entertaining Tale of Quadrupeds*, a vernacular work of the Palaiologan period, which presents the process as dirty and violent, the oft denigrated practice of tanning was itself sometimes associated negatively with the Jewish population.[33] The same work notes less familiar examples such as the use of the hare's skin as headwear for noblemen, and for undergarments to ward off the cold, as well as noting that the animal's paws are used by goldsmiths to 'pick the gold and finely grated silver from his filings'.[34] It also repeats the use of deer horn to ward off snakes, a method noted in the *Geoponika*, as well as informing us of the role of bear fat in medicinal ointments, boar teeth for polishing books pages, and boar hair for thread used by saddlers, leather workers and cobblers, as well as for paintbrushes for icons and sprinklers used for spraying holy water, the latter two being recounted somewhat mockingly since the text also notes the boar's unclean habits.[35] Sheep's wool, goat fur and many other practical materials drawn from animals are described, some very specialist and, due to the nature of the text, all presented to express the inability of humans to cope without them, albeit as a means for the animals to boast about their value and status, which is broadly measured in terms of their impact, positive or negative, on humans. The religious elements of the text cover both Orthodox and Latin Christian practices, as well as referencing Jews and Muslims. No religious argument is overtly made in the text, though a political one appears to be, which may include the conflict the between the Latin and Orthodox churches. Comments instead reflect generally on the religious practices visible in cities like Constantinople, and on the prejudices held in relation to particular groups. Nevertheless, the 'hierarchy' or grouping and order of precedence indicated for the animals is of interest. Western depictions of Adam naming the animals are relatively easy to find, but this does not seem to be the case for Byzantium. In the text, the animals are arranged in 2, or possibly 4, groups, the clean, domesticated or useful animals, 'καθαρὰ καὶ εὔχρηστα', and the bloody and disgusting animals, 'αἱμόβορα καὶ βδελυκτὰ'. Paired as verbal opponents, they discuss their merits and denigrate each other, leading to a battle, whose ending is a possibly later addition, but which presents the work as a reflection of the internal conflicts of the period, whether civil war or potentially the conflict between the Latin and Orthodox Churches, the animals acting as stand ins for real or imagined personalities.[36] Regardless, the bloody and clean natures of the animals link to the Scriptures, and though the animals present their worth in terms of humanity, they seem to be aware of their own importance, with the apparent hierarchy between them being presented more as one of size than of inherent value or closeness to mankind, presenting both a practical and an ideological engagement with nature. The animals here, and in similar works, guide and critique, or even satirically mock, but from a position of safety, being somewhat indirect in the lesson or humour.

In another sense, texts like the *Tale* helped to establish and reinforce social hierarchies. As shown by the Cappadocian Fathers, the behaviour of animals towards each other could present appropriate behaviour to humanity, the example earlier in this article being the protective nature of parents and the filial role of children. Fable as a style of writing was one of the fourteen elements defined as being part of the *progymnasmata*, the rhetorical exercises that formed the basis of Byzantine education.[37] In fact fables, μῦθοι, are the first of the *progymnasmata* exercises. Easily recognisable, memorable, entertaining they taught young men, and possibly women, not only rhetoric but also how to behave in society. Fable is an enchanted world in which animals 'talk like us, and act like us, in distant and fictional worlds that help us reflect on ourselves'.[38] Because it is a fictional world, far removed from normal existence, animals in it can teach humans how to behave in a more direct fashion than in other genres. Thus, fables provided lessons to be followed, with a moral of some kind, usually stated in a *promythium* or an *epimythium*, a summary of the moral at the beginning or end of the narrative. However, the moral is not always overtly stated. Often multiple morals could be drawn from any one story and the audience could therefore be left to select their own, or the author could supply one. Sometimes fables were used more as a tool for criticism or a means to persuade, but there is always some kind of lesson. Animals, plants and weather are frequent features of these texts, and works like the *Tale* utilise the genre and its familiar images. In some ways similar to fable is the *Physiologus*, one of the most popular texts of the medieval period, in which an allegorical interpretation was provided to explain the messages divinely embedded in nature. There is considerable debate on the aim of this text, but at its most obvious, literary level, the presentation is clearly allegorical.[39] Like hexaemeral material, the *Physiologus* utilises Classical knowledge and provides apparently observable characteristics for each animal, plant or stone. Often the chapter will open with a quote from the Old Testament, and then provide information on the animals, plants and stones, which is followed by the Christian allegory and supported by a New Testament quotation.[40] Again, this is a text that presents a lesson on how to live, though in this case it is always a religious lesson, whereas in fable, the lesson may be purely social or even political.

A more visual reinforcement of hierarchy within society, at least economically, could also be seen through nature, and specifically animals. The horse in the *Entertaining Tale* describes its importance to mankind, highlighting the gold, silver, silks and bells used in adorning it and its accoutrements. Horses were used in hunting as already mentioned, and had a significant role to play in many different aspects of Byzantine life, alongside mules and donkeys. They provided a means of transport, locally and over extended distances, carrying people and goods throughout the empire. There is evidence that particular individuals were not allowed to ride horses, due to gender, status or religion, and donkeys were used often by women and monks, and in mock parades as mounts for denigrated figures. Horses did have a role in farming, but ownership was expensive, and the animals remained a mark of status, with those from Thessaly and 'Arab horses' seen as particularly valuable. Horses were also a key element of Byzantine military strategy at different periods, partly in response to the increase in mounted enemies. In the seventh century, cavalry were elite but numerous, and we find examples of heavy cavalry in the tenth and twelfth centuries. The image of a mounted knight was not uncommon for a Byzantine audience, and military saints such as St George and St Theodore were regularly depicted on horseback, and seen as pinnacles of male beauty.[41]

Nature could also have a negative impact on human lives. For the majority of the population, pests, that is, mice or rats, insects, as well as the ever-present impacts of weather and other natural phenomenon such as earthquakes were a continual concern. Hail and thunderbolts each receive dedicated sections in the *Geoponika*, and an entire book focuses on pests.[42] In addition to the flies, mice and locusts a modern audience would expect to find, this chapter also considers ways to keep snakes, weasels and cats away. Weasels and cats were disliked due to their tendency to attack birds and steal eggs, but cats at least were nevertheless also kept to reduce other pests. Insects and mice, damaged crops, spoiled food and were understood to be unclean. Many of the suggested preventions for pests in the *Geoponika* are not fatal, though plenty are, and numerous options are given as though at least some felt guilty about harming animals, possibly in relation to the teachings that they were part of Creation.

From a religious perspective, such prevalent and obviously negative aspects to nature clashed with the divine Creation behind it. The belief that nothing evil can come from God leads Basil, and other Byzantine authors, to state that all danger for mankind from plants and animals was simply caused by man's sin and that nothing was created without a reason.[43] Nature can only be good because God, who created it, is good, and it is therefore the perfect guide for the Christian life. On a day-to-day basis it would seem hard for the Byzantine population to truly accept this principle, but it is certainly evident in their responses to major natural disasters. John Chrysostom gave a homily following an earthquake in 400 blaming the sins of the rich for the disaster, and we have several accounts of similar responses to earthquakes, famines and other natural disasters.[44]

Individual sins could be connected with personal disasters, including diseases. Illnesses could be caused by elements of nature, such as food contaminated by mice, or insect bites, but nature, alongside prayer, may also provide the cure, and was a useful tool for explaining particular conditions. Snakes were feared for their poisonous qualities, and can be found depicted inside water vessels, the hope being that a real snake would see that another was already inside and leave the vessel alone. Though no Byzantine treatise on snakes has survived, ancient knowledge on these creatures continued to be copied. Snakes, and their relative the dragon, regularly appeared as foes for saints, in their guise as instruments or embodiments of the Devil. The image of the snake is also frequently found on amulets, 'magical' elements of popular religion, particularly in connection with the womb, and juxtaposed with images of saints on horseback, Christ or the Virgin.[45] Amulets as sympathetic magic used imagery and text drawn from nature and the Bible in combination as a preventative or as a cure. Not dissimilar to pilgrim's tokens in usage, they seem to have remained popular despite being denounced by the Church. Alongside prayer and miracles, plants and animal parts of various forms were used in medicine. The cures often mimicked an element of the disease, or had an allegorical connection. Referring back to the amulets, the uterus is referred to as dark and black, coiling like a snake, hence snake imagery to counteract it. In the same way, illnesses and even disabilities were likened to animals, aiding in the understanding and discussion of them.[46] An example would be lycanthropy and kynanthropy, wolf and dog-like madness, referenced in antique texts as well as in Byzantium.[47] Illnesses like this undermined the structures of society, causing sufferers to behave in unacceptable manners, and were sometimes seen as demonic possessions. Hagiography provides further examples, such as recounted in the *Life* of St Andrew the Fool, present at the death bed of a rich adulterer:

> The righteous man saw Satan going in and out of the mouth, appearing sometimes as a mouse, sometimes as a serpent or a viper, and crying through the organs of the hapless man, sounding sometimes like a cat (kata), sometimes like a dog or a pig.[48]

Again, nature provided a means of identifying and understanding in a practical sense, and arguments like those of Maximus on the extremes of humanity, a reason for such suffering.

While there are many other practical ways in which the Byzantines engaged with nature, even scientific approaches, the range above gives an idea of the directness of their experience, and the ways in which religious thought and practice intermixed with the necessary, and where there is some apparent distance.

Pleasurable

Like ourselves, the Byzantines understood the value of spending time in nature, and took opportunities to enjoy the world around them, not always to its benefit. The sensory pleasures of the natural world did not pass them by, even being highlighted in the religious works referenced at the start of this article, like the *New Sunday Sermon*. Although original hexaemeral literature was rarely written after the early Church fathers, we do find a twelfth-century example in the verse chronicle of Constantine Manasses, the *Historical Synopsis*. Manasses begins, in the tradition of chronicle writing, with the Creation narrative.[49] However, this hexaemeral passage is much more poetic than others found in chronicles and stresses the sensory beauty of nature as well as describing God as an artist and gardener who is visible in his creation.[50] The *Symbolic Garden*, a text of the tenth or eleventh century initially sets out the requirements for a well-cultivated garden/soul as including good soil, water, the warmth of the sun, and temperate breezes, which cause the plants to move melodically, all cared for by a diligent gardener, indicating the orthodox faith, Scripture, the Lord, the Holy Spirit and the intellect respectively.[51] The literal aspects of the garden in this work retain a sense of realism and a 'sense of closeness to and enjoyment of nature', intertwined with the importance of faith and the interpretation of nature, as well as its care.[52] The sounds of bird song, like that of the plants moving in the breeze, was pleasing to the ear, reminiscent of songs in praise of the Creator, and manufactured through the famous golden automata of Constantinople as well as featuring in a wide variety of literature.[53]

While it was arguably as a means of withdrawing from pleasure, the monastic habit of withdrawing to the remote countryside also allowed monks and nuns to contemplate Creation and live simply on the bounty of the earth produced through God, even to the extremes of subsisting on wild herbs and water. From our modern vantage point, we can appreciate this desire for seclusion in nature even from a non-religious angle, given the impact on human wellbeing. The sense of calm was probably augmented by the religious feeling of being closer to God the Creator without the distractions of city living. The sensory engagement with nature was also reminiscent of the sensory experience of a Byzantine church, with the brightly coloured mosaics of early churches showing religious figures with natural motifs, and the later opus sectile often described in terms of the earth, or meadows, as well as the flickering of candles and the smell of incense, all linking to the dappled light, smell and sights of summer. The senses were seen as aids in bringing the mind to the contemplation of higher things, not simply as distracting the faithful and tying them to their earthly presence. Danger lay nevertheless in the enjoyment of corruptible nature, which faded and died, or could cause harm, if one did not engage with it with a sense of faith, but rather one of indulgence or even of paganistic belief. It was this later fear that post-iconoclasm saw the reduction of artistic representations of nature in churches, though some were still present, and descriptions of churches after that period nevertheless utilise comparisons with Creation.[54] The impact of iconoclasm was also less in more personal spaces such as manuscripts, but the sense that nature could be used to mislead the faithful still lingered alongside the more positive interpretations.

Obviously, nature was a prime topic for poetry, and it would appear that *theatron*, literary groups in which works were read or performed aloud, could themselves also take place outside. It is possible to find an example of one such *theatron* taking place on a boat, with the sounds of the waves, birds and breeze providing a harmonious backdrop.[55] The connections between the potentially false aspects of nature were easily linked to the false tongues of rhetoric, whether poetic or otherwise.

Gardens as depicted in Byzantine texts are heavily sensory. In the romances, whether Komnenian or Palaiologan, they are directly tied with the heroines, and with sexuality. Though the sensual nature of gardens in the romances allows for the promiscuous behaviour of characters within them, a stress on the physical, and even a degree of female power, the sense is also counterbalanced by the enclosed nature of the garden indicating chastity, and the imagery of fertility with a religious angle. The concept of the garden as a representation of sexuality is present in earlier Christian material such as the *Song of Songs*, and repeated in works like Theodore Hyrtakenos' depiction of St Anna, whose barrenness he contrasts with her fruitful garden. Humanity is once again caught between extremes, with nature in its duality both existing in the glory of the Creator, and potentially leading one astray.

Descriptions of human beauty based in the natural world, are present throughout the Byzantine period and earlier, with images of male and female beauty in the Second Sophistic novels carry over to the later romances. The ideal of crystal-like skin coloured like petals and milk, alongside rose bud lips, and comparisons to cypress trees and saplings, and even hair as gold as the sun are frequently found in a wide variety of texts, including chronicles and hagiography. However, such beauty can mask evil natures, or even be entirely fake and created through make-up, and even when real is as transient and prone to decay as nature itself. True beauty was something more, and from the perspective of hagiography at least, the beauty of saints went far beyond the earthly and instead came from the divine light within the holy individual. Here then there is a contrast between the natural and the divine, but it is contradicted in other places, where holy men instruct women (and men) not to use cosmetics, for in doing so they hide their natural beauty bestowed by God.[56] In a late text entitled Τοῦ Νομοφύλακος [Ἰωάννου τοῦ Εὐγενικοῦ] πρὸς τάς καλλωπιζομένας γυναῖκας beauty is like the flowers in the field, pleasing to the eye and part of God's creation.[57] The users of cosmetics are remaking nature, and in doing so making it worse. Even for the romances, which are often very detailed and even sexual in their physical depictions of the heroines, there are examples of spiritual interpretation, A romance sometimes identified as *Kallimachos and Chrysorrhoe*, is described by the late Byzantine poet Manuel Philes, with the central characters being reinterpreted so that the hero's father symbolises God, the heroine as the beloved represents the hero's soul while the dragon killed by the hero is obviously an image of the Devil.[58] Though this may seem as much a need to excuse the author and popularity of such works as it may be an honest desire to use such stories and their *Song of Songs*-like imagery to engage with a literary audience, clearly such readings were possible, and the similarities between the nature in these texts and religious works, suggests that the natural imagery of gardens and wild spaces helped to signpost the religious interpretation.

The enjoyment of nature, though clearly present in at least some religious writings, and its connected sensual imagery, had the potential to be used to praise Creation to the fullest, and to risk further falling into sinfulness. The transience of the natural world, the falseness of rhetoric in describing it and the potential to worship the world rather than the Creator were areas of concern for the Byzantines, who nevertheless continued to use their experience of the world to describe their faith and hoped for heavenly end.

Conclusion

As a brief survey of the relationship between the Byzantine Empire and the non-human world in which it existed, it is obviously not possible to cover all experiences and beliefs. Key themes do emerge, however. The Byzantines had a strong sense of their reliance on nature, and of their human ability to manipulate it. However, such manipulation did not always sit comfortably with them. Nature was a reflection of, or even part of the Creator, could be learned from, and was to be treated with care. Humans as the centre point of the created world were prone to the extremes of it, and thus risked sinfully misusing, misunderstanding or harming the world, and in turn being harmed by it, due to their sins, rather than due to any inherent wickedness on the part of nature.

Man-made objects, and even people, were often described in terms of nature, the natural world being the standard for true beauty, but this rhetoric also risked going too far. Overall, though, it is possible to argue that the Byzantines, as much as they used nature for their survival, and overindulged either physically or in thought, held the natural world in a place of respect, valuing it for itself as part of Creation, and for its place in providing exemplars and support for the existence of mankind in striving to return to the Creator.

Notes

1 Lynn White Jr., "The Historical Roots of our Ecologic Crisis," *Science* 155 (1967): 1205.
2 Henry Maguire, *Nectar and Illusion: Nature in Byzantine Art and Literature* (Oxford: OUP, 2012), 77.
3 Mary Whitby, "The Devil in Disguise: The End of George of Pisidia's *Hexaemeron* Reconsidered," *The Journal of Hellenic Studies*, 115 (1995), 58–9.
4 Patrick Whitworth, *Three Wise Men from the East: The Cappadocian Fathers and the Struggle for Orthodoxy* (Durham: Sacristy Press, 2015), 61–63.
5 Saint Basil the Great, *The Treatise De Spiritu Sancto (The Nine Homilies of The Hexaemeron And His Letters)*. trans. Rev. Blomfield Jackson (New York: Christian Literature Company, 1895), 58; *PG* 29 *Hexaemeron* 1.11; Basil of Caesarea, *Homélies sur l'Hexaéméron*, ed. S. Giet (Paris: Éditions du Cerf, 1968), 134–137.
6 Basil, *Homélies sur l'Hexaéméron*, 18.
7 Basil, *Treatise de Spiritu Sancto*, Homily 9 section 4, 104; Basil, *Homélies sur l'Hexaéméron*, 498–499.
8 Basil, *Treatise de Spiritu Sancto*, Homily 9 section 3, 103; Basil, *Homélies sur l'Hexaéméron*, 488–489.
9 Blake Leyerle, "Monks and Other Animals" in *The Cultural Turn in Late Ancient Studies: Gender, Asceticism, and Historiography*, ed. Dale B. Martin and Patricia Cox Miller (North Carolina: Duke University Press, 2005), 150–171 citing Origen, *Homilia in Genesim* I.16 [SC 7bis.69].
10 Gregory of Nyssa, *On the Hexaemeron* (*PG* 44:66) cited in Peter C. Bouteneff, *Beginnings Ancient Christian Readings of the Biblical Creation Narratives* (Michigan: Baker Publishing, 2008), 130.
11 Gregory of Nyssa, "On the Making of Man," *Dogmatic treatises*: *Select Writings and Letters of Gregory, Bishop of Nyssa*, trans. William Moore and Henry Austin Wilson (Buffalo, NY: Christian Literature Publishing Co., 1892), 7.1, 392; Gregory of Nyssa, *La Création de l'homme*, trans. Jean Laplace (Paris: Les Editions du Cerf, 2002); Gregory of Nyssa *Dogmatic treatises,*; *La Création de l'homme*, 102.
12 Gregory, "On the Making of Man", 392; Gregory, *La Création de l'homme*, 103.
13 Gregory, "On the Making of Man", 7.2, 392; Gregory, *La Création de l'homme*, 103–104.
14 Doru Costache, "Approaching *An Apology An Apology for the Hexaemeron*: Its aims, method and discourse", *Phronema* 27:2 (2012) 53–81 citing D.F. Stramara, 'Surveying the Heavens: Early Christian Writers on Astronomy', *St Vladimir's Seminary Quarterly* 46: 2–3 (2002) 147–62 at 155, 68.
15 Gregory of Nazianzus, *Selected poems of Gregory of Nazianzus: I.2.17; II.1.10, 19, 32: a critical edition with introduction and commentary*, ed. Christos Simelidis (Göttingen: Vandenhoeck & Ruprecht, 2009), 59.
16 *Gregory of Nazianzus*, ed. Brian E. Daley (London: Routledge, 2006) Oration 38, section 10, 121; Grégoire de Nazianze, *Discours 38–41, Introduction, texte critique et notes* ed. Claudio Moreschini, trans. P. Gallay (Paris: Les Editions du Cerf, 1990),

122–123; Gregory of Nazianzus, *Five Theological Orations*, trans. Stephen Reynolds (2011). 13–44, esp. 33–44. Accessed 12 February 2024. https://tspace.library.utoronto.ca/bitstream/1807/36303/1/Gregory%20of%20Nazianzus%20Theological%20Orations.pdf.

17 Maguire, *Nectar and Illusion*, 64.

18 Daley, *Gregory of Nazianzus*, Oration 44, section 10, 160; *PG* 36.608–621.

19 Daley, *Gregory of Nazianzus*, Oration 44, section 12, 161; *PG* 36.620–621.

20 Maximus the Confessor, "Mystagogia," in *Maximus the Confessor, Selected Writings*, trans. George C. Berthold (New York: Paulist Press, 1985), 188.

21 Torstein Tollefsen, "Causality and Movement in St Maximus' *Ambiguum 7*," in *Studia Patristica XLVIII* ed. J. Baun, A. Cameron and M. Vinzent (Leuven: Peeters Publishers, 2010), 88.

22 Maximus the Confessor, "Ambiguum 41" in *On the Cosmic Mystery of Jesus Christ*, trans Paul M. Blowers & Robert Louis Wilken (New York: St Vladimir's Seminary Press, 2003), 1305A–B, 157.

23 Maximus "Ambiguum 41," 1305A–C, 157.

24 Maximus, "Ambiguum 41," 1308B–C, 158.

25 White, Jr. "Historical Roots," 1205.

26 Radu Bordeianu, "Maximus and Ecology: The Relevance of Maximus the Confessor's Theology of Creation for the Present Ecological Crisis," *The Downside Review* 127 (2009): 116.

27 Andrew Linzey, *Christianity and the Rights of Animals* (New York: Crossroad, 1987), 32 citing Donald Attwater *St John Chrysostom* (London: Catholic Book Club, 1960), 59–60.

28 *Geoponika: Farm work: A Modern Translation of the Roman and Byzantine Farming Book*, trans. Andrew Dalby (Totnes: Prospect Books, 2011).

29 Nancy P. Ševčenko, "Wild Animals in the Byzantine Park," in *Byzantine Garden Culture*, ed. Anthony Littlewood, Henry Maguire and Joachin Wolschke-Bulmahn (Washington, D.C.: Dumbarton Oaks, 2002), 69–75.

30 Ševčenko, 'Wild Animals in the Byzantine Park', 71.

31 See for example the third Prodromic poem discussed in Margaret Alexiou, "Ploys of Performance: Games and Play in the Ptochoprodromic Poems." *Dumbarton Oaks Papers* 53 (1999): 91–109 and the *Schede tou Myos* discussed in Marciniak, Przemyslaw "A Pious Mouse and a Deadly Cat: The *Schede tou Myos*, attributed to Theodore Prodromos," *Greek, Roman, and Byzantine Studies* 57 (2017): 507–527.

32 Ahmad Ibn Rustih, *Kitab al-a'lah al-nafisa* cited in Andrew Dalby *Tastes of Byzantium: The Cuisine of a Legendary Empire* (London: I.B. Tauris, 2010), 118.

33 *An Entertaining Tale of Quadrupeds: Translation and Commentary*, trans. Nick Nicholas and George Baloglou (New York: Columbia University Press, 2003), 170–1, lines 210–11.

34 *Entertaining Tale* lines 301–12, 176–7.

35 *Entertaining Tale* lines 422–427, 182–83.

36 See the discussion in *An Entertaining Tale* Appendix 2, 431–447.

37 Elizabeth Jeffreys, "Introduction", in *Rhetoric in Byzantium: Papers from the Thirty-fifth Spring Symposium of Byzantine Studies, Exeter College, University of Oxford, March 2001* ed. Elizabeth Jeffreys (Aldershot: Ashgate, 2003), 1–5.

38 John Heath *The Talking Greeks: Speech, Animals, and the Other in Homer, Aeschylus, and Plato* (Cambridge: Cambridge University Press, 2005), 15.

39 See Dora Faraci, 'Pour une etude plus large de la reception medievale bestiaries' in *Bestiaires médiévaux: nouvelles perspectives sur les manuscrits et les traditions textuelles: communications présentées au XVe Colloque de la Société Internationale Renardienne (Louvain-la-Neuve, 19–22.8.2003)* (Louvain-la-Neuve: Université Catholique de Louvain, 2005), 111–125, for a summary of scholarship on the topic.

40 *Der Physiologus nach den Handschriften G und M.* ed. Dieter Offermanns (Meisenheim am Glan: Hain, 1966); *Physiologus*, ed. F. Sbordone (Hildesheim: Georg Olms, 1991); *Physiologus: A Medieval Book of Nature Lore*, trans. Michael J. Curley (Chicago: University of Chicago Press, 2009).
41 See Myrto Hatzaki, *Beauty and the Male Body in Byzantium: Perceptions and Representations in Art and Text* (New York, NY, Palgrave Macmillan, 2009).
42 *Geoponika*, book 13, 268–80.
43 Frank Egleston Robbins, *The Hexaemeral Literature: A Study of the Greek and Latin Commentaries on Genesis* (Chicago: University of Chicago Press, 1912), 5.; Basil *Treatise de Spiritu Sancto*, Homily 5, sections 4 and 6, 77–78; Basil, *Homélies sur l'Hexaéméron*, 292–295 and 300–307.
44 John Chrysostom, *Homily after the Earthquake* (=*Terr. mot.*) (PG 50:716), trans. Bryson Sewell, https://www.tertullian.org/fathers/chrysostom_on_the_earthquake.htm. Further examples of connections between sin and earthquakes, particularly in relation to emperors, can be found in the Chronicle of the Logothete.
45 Jeffrey Spier, "Medieval Byzantine Magical Amulets and Their Tradition," *Journal of the Warburg and Courtauld Institutes* 56 (1993).
46 Disability in Byzantium is the current focus of a project by Dr Maroula Perisanidi and I am grateful for her recent presentations on the subject.
47 See for example the outbreak recorded by John of Ephuses in the sixth century discussed in Nadine Metzger "Kynanthropy: canine madness in Byzantine late antiquity," *History of Psychiatry* 26, no. 3 (2015): 318–331.
48 *The Life of St Andrew the Fool*, trans. Lennart Rydén (Uppsala: Acta Universitatis Upsaliensis, 1995), vol. 2, 238–9.
49 Ingela Nilsson, "Discovering Literariness in the Past: Literature vs. History in the *Synopsis Chronike* of Konstantinos Manasses," in *L'écriture de la mémoire: la littérarité de l'historiographie; actes du IIIe colloque international philologique EPMHNEIA, Nicosie, 6-7-8 mai 2004* ed. Paolo Odorico, Panagiotis. Agapitos and Martin Hinterberger (Paris: Diffusion de Boccard, 2006): 19.
50 Ingela Nilsson, « Narrating Images in Byzantine Literature: The *ekphraseis* of Konstantinos Manasses," *Jahrbuch der österreichischen Byzantinistik* 55 (2005): 134.
51 For an ecotheological discussion of the text see Kirsty Stewart, "The Symbolic Garden and the Spiritual Understanding of Nature in Byzantium" in *Natural Communions: Religion and Public Life*, Volume 40, ed. Gabriel Ricci (New York: Routledge, 2019), 92–105.
52 *The Symbolic Garden: Reflections Drawn from a Garden of Virtures, a XIIth Century Greek Manuscript*, trans. Margaret H. Thomson (Ontario: Captus University Publications, 1989), 9.
53 Gerard Brett, "The Automata in the Byzantine 'Throne of Solomon'," *Speculum*, 29, no. 3 (1954): 477–87.
54 See the discussion in Maguire, *Nectar and Illusion*, 106–34.
55 Marc D. Lauxtermann, *Byzantine Poetry From Pisides To Geometres Vol. I Texts and Contexts* (Wien: Austrian Academy of Sciences Press, 2003), 55. Full text in G. SOLA, 'Giambografi sconosciuti del secolo XI', *Roma e l' Oriente* 11 (1916), 20–21.
56 John Chrysostom has much to say on the topic, see for example John Chrysostom, *Homilies on the Gospel of St. Matthew*, 30.5, PG 57, 369–70.
57 See Phaidon Koukoules, "Περί κομμώσεως των Βυζαντινών," *Επετηρίς Εταιρείας Βυζαντινών Σπουδών*, 1930, Έτος Ζ', 3–37.
58 Borje Knös, "Qui est l'auteur du roman de Callimaque et Chrysorrhoé," *Hellenika* 17 (1962): 274–95.

8 An Eco-Spirituality of Wonder

An Aesthetic-Ethical Response to Myriad Nature

Carol Wayne White

> But I believe that the more clearly we can focus our attention on the wonders and realities of the universe about us, the less taste we shall have for destruction.
>
> – Rachel Carson, *Lost Woods*

Introduction

The concept "nature" is a historically evolving term that has also become a contested site among various thinkers. In the contemporary era, nature is primarily described in Anthropocene discourses as the environment, or an object of inquiry that helps underscore scientific consensus that humanity has become a global geophysical force affecting the planet's system. Lived realities associated with the Anthropocene are featured in the Great Acceleration graphs that indicate a series of distinct earth system changes are at play, including "atmospheric composition, stratospheric ozone, the climate system, the water and nitrogen cycles, marine ecosystems, land systems, tropical forests and terrestrial biosphere degradation."[1] Warnings from the Planetary Boundaries framework (at the Stockholm Resilience Center, Stockholm University) also delineate specific limits that should not be exceeded in order for the earth system to maintain its function as a life support system.[2] Accordingly, the Anthropocene epoch has gone hand in hand with sobering assessments of the adverse effects of human activity on "Nature" by earnest scientists, activists, and ecologists advancing diverse narratives or woven tales of "how we got here": the naturalist narrative, currently the mainstream one; (2) the post-nature narrative; (3) the eco-catastrophist narrative; and (4) the eco-Marxist narrative.[3]

Notwithstanding the varied models of environmental ethics these influential Anthropocene narratives have generated, they have also functioned—often with the best of intentions—to delimit what we can know about nature. Additionally, with an overarching concern for the human, or, more precisely, a concern about the activity of humans, these Anthropocene narratives have inadvertently (and ironically) perpetuated hegemonic culture's triumphalist view of humans as standing outside of nature and managing it to our own benefit.[4] Agostino Cera's description of an ethics of omni-responsibility is

DOI: 10.4324/9781003385202-9

helpful in grasping the fuller dangers lurking within this scenario. According to Cera, when technological capacity is aligned with moral obligation, ecological consciousness is geared toward an absolutization of the caretaking of the Earth System.[5] Whereas humans had formerly expressed *dis*-interest in and *ir*-responsibility toward the otherness of nature, the emerging *Aidosean Prometheanism* that Cera targets results in a paradoxical *hyper*-interest in and *omni*-responsibility toward the otherness of nature.[6]

I share Cera's sense of alarm in recognizing the nuanced paradox embedded in influential Anthropocene writings that imagine how we conduct our relationship with nature; the danger is heightened even more when noting that many religious traditions have enthusiastically embraced a stewardship model of religious environmentalism, or a variation of the managerial approach to nature.[7] In response, I introduce an eco-spirituality that promotes an expansive view of nature while attempting to address some of the ethical concerns expressed in Anthropocene discourses. Grounded in the tenets of religious naturalism, this eco-spirituality posits nature as all that there is, or what I describe as myriad nature—a term associated with the complex, endlessly evolving relations that constitute reality. This eco-spirituality also conceptualizes humans as natural processes in relationship with other forms of nature. In light of these claims, it contends that myriad nature cannot be fully grasped without also delineating the relational structure of everything that exists.

With its notion of myriad nature, this eco-spirituality identifies distinctive aesthetical-ethical responses to diverse forms of ecological injustice and degradation. I identify these aesthetical-ethical responses with the "turn to wonder," which will be described later. For now, suffice it to say that the category of wonder anticipates what possibilities and novel illuminations may occur when human organisms enact our evolutionary capacities as relational beings during a transitional time in which the old order strains and fractures. Throughout the essay, I trace the contours of this eco-spirituality in particular scientific, philosophical, and religious insights that feature the generative power of nature and its involvement in wonder and poesis. The result, I suggest, is an eco-spirituality animated by a humbling, appreciative, yet always tentative, grasp of evolving, myriad nature.

I: Introducing Eco-spirituality and Myriad Nature

Given the uniqueness of the term "eco-spirituality," I offer a few brief comments on my usage of it, beginning with "spirituality." I derive my meaning of spirituality from current trends in the United States, where individuals seeking deeper dimensions of their lived experiences, or those probing the fuller meaning of their existence, are not compelled by strict adherence to the tenets and rituals of the major wisdom traditions (or the historically evolving religions) that have shaped them.[8] As James Wiseman writes: Whereas the term "spiritual" was once often used in a negative sense to imply that a person was otherworldly or even irresponsible, today it is frequently used as a badge of honor, as when a person claims to be "very spiritual, though not religious."[9] This widespread interest in spirituality among those in Western cultures have led to more expansive views of the term and to cross-cultural approaches that are not just grounded in Euro-American conceptual frameworks, especially Christian ones.

My approach to spirituality emerges out of this critical awareness. An expression of spirituality is not necessarily centered in any specific tradition; rather, it is a mode of reflecting, experiencing, and envisioning one's relationality with all that is. Rightly understood, spirituality is a vital awareness that pervades all realms of our being alive in the manner in which we find ourselves. As Peter Van Ness observes, "the spiritual dimension of life is the embodied task of realizing one's truest self in the context of reality apprehended as a cosmic totality. It is the quest for attaining an optimal relationship between what one truly is and everything that is."[10] From the vantage point of religious naturalism that asserts the ultimacy of nature, this model of spirituality is crucial and fitting. As Donald Crosby asserts: "Nature requires no explanation beyond itself. It always has existed and always will exist in some shape or form. Its constituents, principles, laws, and relations are the sole reality. This reality takes on new traits and possibilities as it evolves inexorably through time."[11]

The cosmic context and sense of connectedness with all material reality (and particularly with every living thing on earth) that Van Ness describes also helps make sense of the "eco" qualifier in eco-spirituality. If an eco-spirituality is defined as a mode of ascertaining and embracing some type of relationality with everything that is, then it necessarily includes the general principles and aims of ecological studies. There is an investment in grasping the (often complex) forms of relationality between and among living organisms, including humans, and their wider contexts. Accordingly, this eco-spirituality embraces the notion of a biotic community in the fundamental sense that Aldo Leopold once described in *A Sand County Almanac* (1949). Leopold emphasized the radical interdependence of plants and animals in their natural environments, including the observation that human organisms are intimate participants in ecological relations and belong to a wider biotic community.[12]

Conceptualizing a biotic community in this sense is akin to affirming the intrinsic value of myriad nature (inclusive of both human and more-than-human existents). Additionally, in accepting basic ecological views of nature as forms and functions of matter imbued with a dynamic quality and self-organizing capacities, as well as with tendencies for ever more complex formations to appear, this eco-spirituality posits myriad nature as intrinsically dynamic and processual.[13] As suggested by Michael Hogue, certain epistemic claims are associated with the differentiated patterns and forms in material reality: that nature is all; that nature is differentiated, that nature is in process, and that the value of the things and relations of nature emerges out of their vulnerability.[14] Moreover, this eco-spirituality evokes Wesley Wildman's celebration of myriad nature's "beauty, terror, scale, stochasticity, emergent complexity, and evolutionary development."[15] Finally, advancing a holism that rejects mind-body distinctions, this eco-spirituality insists that all existents, or all things, living and non-living, are constituted of the same basic elements. As well, its approach to nature as matter is emergent or expansive in orientation, rejecting reductionist inclinations.[16]

Building on these theorizations, this eco-spirituality affirms several key steps in its approach to nature. First, it upholds constitutive relationality as a key element of the matrix of natural processes identified with myriad nature. Nothing exists outside of relationships. This crucial point is demonstrated wonderfully in Alfred N. Whitehead's process philosophy, which rejected the theory that reality is fundamentally constructed by bits of matter that exist independently of one another. Drawing upon insights from quantum physics, the theory of relativity, and evolution, Whitehead constructed a system of the natural order that went beyond the earlier atomistic views of both early Greek philosophers (Democritus) and later Western scientific thinkers (Isaac Newton). Whitehead noted that historically atoms were visualized as hard, lifeless matter; likewise, in being hard, atoms never penetrated each other and their relations to each other were seen as perpetually eternal.[17] Instead of focusing on atoms in this manner as fundamental units of reality, Whitehead began his inquiries with more fundamental "actual entities" or "actual occasions." These sub-atomic entities never exist in isolation; rather, they are intimately related to the whole field of life that throbs around them. Thus, for Whitehead, whether one is describing a grand phenomenon (for some, something like God) or "the most trivial puff of existence," the same principle of life is in all things because "actual entities are the final real things of which the world is made up."[18]

Whitehead's notion of prehension helps us understand how the elements of actual entities are related to each other and how these entities are further related to other entities. It suggests that every actual occasion absorbs, or is related to, the whole universe. All actual entities have a subjective nature (i.e., processes can experience, are partly self-determining, and can enter into relation with other processes); an objective nature (i.e., processes can be experienced by other subjects); and a temporal nature (i.e., processes happen through time or, perhaps, define time). As Whitehead observes: every prehension consists of three factors: first, the "subject" that is prehending; secondly, the "datum" which is prehended; and thirdly, the "subjective form," which is how the subject prehends the datum. Actual entities are brought together via processual creativity into sets, or societies, or nexus. In short, by way of processes of becoming, actual entities are formed through the concrescence of prehensions.[19] Grounded in this Whiteheadian context, the vision of myriad nature advanced here is fundamentally about change and temporality: becoming, not being, is the central metaphor for understanding reality, and contingency, emergence, and creativity are essential elements that take precedence over determinism and stasis. Within the context of ecological studies, this insight into the "nature" of nature is also illustrated by biologist Rachel Carson, who asserts, "There is nothing static about an ecosystem; something is always happening. Energy and materials are being received, transformed, given off. The living community maintains itself in a dynamic rather than a static balance."[20]

Second, this eco-spirituality brings us face to face with the Anthropocene paradox noted above. It does not merely point to particular shifts in ethical thinking toward more-than-human nature in current attempts to solve the urgent environmental problems that human activity has caused; rather, this eco-spirituality invites another level of inquiry into myriad nature that helps us grasp the category of the human. As Hogue has observed, this paradox consists in the determination that while the Anthropocene literally signifies the beginning of a human geological epoch for the Earth, it also signals the end of the idea of any ontological distinction of the human from the rest of nature. [21] This is a view of the human as intrinsically relational and material, compelling eco-spiritualists to reject human exceptionalism in its different forms, chief of which is the dominant Euro-American cultural fantasy that has anchored humans on one side of the Great Divide, away from all other species. Not only are humans becoming, evolving natural entities, we are also composite, material embodied beings. Consider, for example, Henry A. Regier's discussion of ecosystems as both part of the biosphere (from the top down) and an organism interacting with other organisms and non-living features of their shared habitat; in this context, the human also serves as a habitat for a variety of other creatures, together with some non-living material as in the gut. In short, the human can also be viewed as an ecosystem.[22] This insight is illustrated wonderfully in Donna Haraway observation:

> I love the fact that human genomes can be found in only about 10 percent of all the cells that occupy the mundane space I call my body; the other 90 percent of the cells are filled with the genomes of bacteria, fungi, protists, and such, some of which play in a symphony necessary to my being alive at all, and some of which are hitching a ride and doing the rest of me, of us, no harm. I am vastly outnumbered by my tiny companions; better put, I become an adult human being in company with these tiny messmates. To be one is always to *become with* many.[23]

Conceiving the human as embodied, composite materiality warrants new appreciation of ourselves as parts of myriad nature. The category of the human is creatively inscripted onto the tissues, bones, and liquids of which we are constituted. In this case, the eco-spirituality I am advancing vehemently rejects the earlier notion of humans outside of nature and protecting and controlling it. Rather, it shares Rachel Carson's passionate exposé of Western human hubris that has often been reflected in its purported knowledge and utilitarian approach to more-than human nature:

> The "control of nature" is a phrase conceived in arrogance, born of the Neanderthal age of biology and philosophy, when it was supposed that nature exists for the convenience of man ... It is our alarming misfortune that so primitive a science has armed itself with the most modern and terrible weapons, and that in turning them against the insects it has also turned them against the earth.[24]

Encouraged by such insights, this eco-spirituality accentuates a layer of nuance to the Anthropocene paradox: it does not lead so much to the demise of the human as it points to its transformation. Additionally, rather than insulating humanity from the risks and hazards of natural processes, this eco-spirituality brings critical awareness to the diverse and sometimes unrecognized levels and forms of vulnerabilities potentially experienced by myriad nature. In doing so, it introduces into Anthropocene discourses a host of entangled insecurities, vulnerabilities, and interconnected causes and effects that are radically unsettling our world.[25]

Third, this eco-spirituality focuses on the ethical and affective resonance of our claims when considering how humans might imagine and embrace myriad nature. If, as ecological studies uphold, all entities are interconnected and part of a greater whole, then any harm done to another entity is essentially harm done to ourselves. Such critical awareness shared among eco-spiritualists is quite clear: if we all influence each other, we have a responsibility to not only act in ways that promote the flourishing of all, but to also urge others to do the same. This worldview promotes kindness, empathy, and compassion (which in and of themselves are key aspects of the moral codes of most religions) for all relational existents.

Furthermore, experiencing the sublime grandeur of nature can often be awe inspiring on at least the same level as widely accepted forms of religious music and art. Consider the feelings that can come when gazing upon the stars at night and marveling at the vastness of the universe. A range of emotions can envelop you, illuminating the wondrous, felt sense that one is part of an amazing set of processes; of course, at the same time, a humbling sense of one's own individual insignificance can occur. Such moments (and even more systematic explorations of the natural world) can also provide humans with a feeling of community shared with the entire natural world, even while maintaining the mystery and ambiguity that is at the heart of religious experiences. In the next section, I continue to sketch the contours of an eco-spirituality that fosters ethical responses through the experiences of wonder—framed by the concept of myriad nature as structured by constitutive relationality.

II. Wonder and Myriad Nature

The concept of wonder is often framed as an aesthetic category in which humans value or appreciate more-than-human nature for the utility or pleasure it offers us. While there is inevitably a dimension of objectification in any expression of aesthetic valuing, one important insight of this eco-spirituality is that humans are an inescapable part of the very thing (myriad nature) we are valuing. Accordingly, in this context to wonder is not to express mere human intellectual curiosity or emotional awe at the facts of life out there; rather, to wonder is to exist in a particular mode of being in which humans embrace our constitutive, relational materiality with a sense of awe, humility, and responsibility. Wonder entails radically grasping and acknowledging ourselves as material, natural relational beings—it is embodied appreciation of just how entangled we are with other natural processes. Mark Nepo has suggested that enacting wonder is a certain kind of species recognition, noting that of all the endless species of plants and animals and minerals that make up the Earth, a very small portion of life has the wakefulness of spirit that we call "being a human."[26]

As may be true for many readers, I am charmed by Nepo's poetic language; at the same time, however, I think further elucidation of this species recognition is warranted for an eco-spirituality that aims to avoid any linkage with the Cartesian *cogito*, or with the rampant androcentrism in traditional humanistic thought that has often privileged the category of the human as an exceptional life force outside of nature. This model of eco-spirituality is clear: humans are natural, material beings through and through. Yet, as I have discussed elsewhere, becoming human, or actualizing ourselves as human beings, emerges out of an embodied awareness and aspirational goal to be more than a conglomeration of pulsating cells. Our humanity is not reducible to organizational patterns or processes dominated by brain systems; nor is it structured solely by DNA, diet, behavior, and the surrounding natural contexts. Rather, in positing fundamental questions of value, meaning, and purpose to our existence, human animals *become* human in those moments of recognizing our deeply structured emergence in materiality and ongoing constitutive relationality that marks evolving life. In other words, our coming to be human is structured by a crucial question: how do we come to terms with ourselves as distinct life forms and with life itself?[27] Or, better yet, how do we affirm life (myriad nature) in all its splendor and complexity, and amid vulnerabilities and excruciating forms of suffering?

This fundamental question requires our rapt attention, as it helps comprehend the turn to wonder as one measure (among others) of ascertaining myriad nature. In seeking responses to this question, the eco-spirituality advanced in this essay illuminates wonder as an aesthetic-ethical category that includes epistemic claims about the relational structuring of life. As such, it envisions an ethic that encourages humans' processes of transformative engagement with each other and with the more-than-human worlds that constitute our existence. In doing so, this ethic forsakes an either-or approach to addressing vulnerabilities expressed in and experienced by myriad nature. One example of this aesthetic-ethical approach is illustrated in an arresting passage by Patrisse Khan-Cullors, co-founder of the Black Lives Matter (BLM) movement. In her 2018 memoir (*When They Call you a Terrorist: Black Lives Matter Memoir*), Khan-Cullors shares a moving account of listening to a talk delivered by astrophysicist Neil deGrasse Tyson a few days after the September 2016 election. Khan-Cullors' friend, Asha Bandele (and co-author), had sent the link to DeGrasse Tyson's talk as a symbol of hope, and together they listened to the scientist expand upon an elegant truth that gave them pause. Khan-Cullors writes:

> We listen together, the very atoms and molecules in our bodies are traceable to the centers of stars that once upon a time exploded into gas clouds. And those gas clouds formed other stars and those stars possessed the divine-right mix of properties needed to create not only planets, including

> our own, but also people, including us, me and her. He is saying that not only are we in the planet, but that the universe is in us. He is saying that we, human beings, are literally made out of star dust. And I know when I hear Dr. deGrasse Tyson say this that he is telling the truth because I have seen it since I was a child, the magic, the stardust that we are, in the lives of the people I come from.[28]

In Khan-Cullors' personal rendering, the scientific truths from Tyson are inspirational, leading her to describe her activism as an act of love for her ancestors and for all African-descended peoples whose humanity has been stripped of its essential material value by white supremacy. Minimally, Khan-Cullors' wonder is an acknowledgment of the beauty of materiality and radical relationality, and the type of responsibility it elicits. Yet, on another level, Khan-Cullors' expression of wonder reinforces the cosmological affirmations articulated in Whiteheadian process thought: not only do humans belong to one another, we also are members of other natural processes. In a cosmological sense, to affirm we belong to one another is another way of asserting Whitehead's principle of internal relations associated with his description of the basic unit-events. According to this principle, the basic unit-events that constitute the world are not vacuous; rather, they possess a subjective nature with attributes that might be called feeling, memory, and creativity. In the final analysis, the dynamic cosmos that Whitehead describes is made up of is a host of entangling and ever-changing relationships, all of which leave their traces on life from its beginning to its end. At the same time, humans are also, within the context of those relationships, creative agents, making a difference, great or small, in the lives of other entities in the immediate present and in the long-range future.[29]

When conjoining Khan-Cullors' activism and Whiteheadian thought, a key aspect of wonder advanced by this eco-spirituality comes to the fore. As implied in those key questions raised earlier about becoming human—all of which are interchangeable—it raises crucial questions of how we come to terms with life: How do we affirm life in all its splendor, complexity, and excruciating forms of suffering? How do we affirm our wondrous, inexhaustible, endless constitutive relationality? One response of this eco-spirituality is clear: in becoming human we are affirming ourselves as emergent, interconnected life forms amid spectacular biotic diversity.[30] Understanding emergence is also crucial in this context and cell biologist Ursula Goodenough provides important insights. She speaks of emergent properties arising as a consequence of relationships—for example, the relationships between water molecules that generate a snowflake or the relationships between neurons that generate a memory. Emergent properties also give rise to yet more emergent properties, generating the vast complexity of our present-day cosmic, biological, ecological, and cultural contexts.[31] Reflecting meaningfully on the emergence of matter (and especially life) from the Big Bang forward promotes an understanding of myriad nature as complex processes of becoming. Acknowledging humans' deep, inextricable homology with each other and with the worlds around us draws our attention to an expansive view of our humanity as an emergent phenomenon, not an achievement. Embracing these naturalist perspectives, this eco-spirituality has the potential to counter a longstanding practice among humans that regards some aspects of myriad nature as more valuable and better than others. It is in this fuller sense that I believe the ethical force of Khan-Cullors' wonder is felt: affirming black lives as intrinsic, valuable life forms is not merely advocating for black humanity, but for all humanity in its variegated forms, and, by default, for the future of our planet.[32] In this wider, cosmological context, celebrating the bits of atoms and molecules that constitute human life's connection with the universe moves us beyond concern only about humans: celebrating "blackness" is one point of entry for thinking ethically about life, its continuance, and our planet. In short, the BLM movement's uncompromising demand that Black lives be viewed and treated as intrinsically valuable forms of life essentially promotes a relational ontology. Here, part of embodying justice in a materialist framework requires approaching materiality differently—we not only become daringly positive about our materiality; we also embrace materiality in its diverse manifestations. For eco-spiritualists, this means consciously resisting binary thinking that demarcates certain spheres of life as superior and others as inferior, justifying the exploitative practices of the former. This is an insight also promoted by the discourse of critical materialisms, inclusive of recent developments in feminist theories, animal studies, posthumanism, and vegetal studies, where older, static ontologies and impoverished, binary views of alterity have been challenged: human/more-than-human, white/black, divine/human, male/female, able-bodied/disabled, normal/queer, and so on.[33]

This insight moves us even further in grasping our constitutional relationality. To the extent that human life forms remain curious about our capacity to recognize and value our constitutive, embodied, material relationality, we learn anew (again and again) who and what we might become as we enact our relationality with *other* natural processes that constitute myriad nature. In other words, amid our struggles to institutionalize values that honor our expansive humanity, we continue to emphasize myriad nature as that reality that constitutes our own identities. As such, the concept of wonder offers social, affective, and ethical implications for our consideration, further evoking a provocative understanding of embodied, porous love. This sense of wonder says yes to abundant, wondrous life amid the complex, vexed, impoverished forms of relationality currently dominating U.S. cultural formations. It also mitigates against nihilistic tendencies and indifference amid alienating forms of relationality, pain, and suffering. In the final analysis, this conception of wonder affirms that gifting others emerges from the constitutive relationality structuring life, as Robin Kimmerer suggests:

> Even a wounded world is feeding us. Even a wounded world holds us, giving us moments of wonder and joy. I choose joy over despair. Not because I have my head in the sand, but because joy is what the earth gives me daily and I must return the gift.[34]

Kimmerer's expansive way of seeing can help shape the types of questions human life forms ask of ourselves and others in our aspirational goals of creating and sustaining responsive models of relational integrity and functioning. In the next section, I discuss further how the category of wonder underscores this type of appreciative ethic.

III. Wonder as an Aesthetic-Ethic

Rachel Carson has written poetically about natural processes in a manner that underscores a major theme we have been discussing, namely, that humans are a part of—and belong together—with myriad nature. Scientific truths for Carson are captivating, poetic forms of truth about myriad nature that alerts the reader to an ecology of being. In her introduction of Carson's *Lost Woods* (or collection of undiscovered and little-known writings), Linda Lear observes that Carson's "field notebooks are filled with incisive biological and ecological observation, but they also contain lyric vignettes that capture moments of dazzling humility."[35] Carson's observations are richly suggestive in evoking the power of wonder; when reading her, one is invited to pause, grasp, and be moved by the sheer facticity that we humans all belong to the flow of life itself—indeed, that each of us participates in life's grandeur. As she observes:

> I believe this affinity of the human spirit for the earth and its beauties is deeply and logically rooted. As human beings, we are part of the whole stream of life. We have been human beings for perhaps a million years. But life itself—passes on something of itself to other life—that mysterious entity that moves and is aware of itself and its surroundings, and so is distinguished from rocks or senseless clay—[from which] life arose many hundreds of millions of years ago. Since then, it has developed, struggled, adapted itself to its surroundings, evolved an infinite number of forms. But its living protoplasm is built of the same elements as air, water, and rock. To these the mysterious spark of life was added. Our origins are of the earth. And so there is in us a deeply seated response to the natural universe, which is part of our humanity.[36]

Her writings on the sea also invites readers reconsider to oceanic worlds as teeming with biotic communities of interacting agential beings and distinctive perspectives:

> Every marine animal, from the smallest to the sharks and whales, is ultimately dependent for its food upon these microscopic entities of the vegetable life of the ocean. Within their fragile walls, the sea performs a vital alchemy that utilizes the sterile chemical elements dissolved in the water and welds them with the torch of sunlight into the stuff of life.[37]

These perspectives from Carson are vitally important to an eco-spirituality invested in addressing difficult, thorny, and challenging issues that often diminish the wonder of life among us. Carson's naturalism can inspire eco-spiritualists to action when absorbing the symbolic and actual beauty embedded "in the migration of birds; in the ebb and flow of the tides; in the folded bud ready for the spring. There is something infinitely healing in these repeated refrains of nature."[38] In other words, Carson's work inspires eco-spiritualists to live in a state of wonder, pausing to absorb and take in the array of life's splendor that greets one in engaging every existent. Recognizing that materiality is interconnected in endless forms of relationality, eco-spiritualism endorses a key maxim: if any configuration of life is diminished, all forms of life lose illuminative power. Living in a state of wonder, you and I are determined to stay with the trouble, mindfully and constantly struggling against, or even resisting, detrimental, often violent forms of relationality supported by impoverished perceptions of differences among us. Living in wonder compels us to support efforts that provide right conditions in which all life can flourish and thrive. Moreover, we remain curious about our investment in upholding specific binaries that we love to construct in distancing ourselves from those perceived aspects of life that we have a harder time embracing. Why do we do this? What are we hoping to gain? Where did we learn to be so willfully anti-life, to war against the very fullness of life itself?

Existing in wonder is not easy, nor easily sustained. As part of myriad nature, humans are messy beings and, as we all know, in living through the years, each of us have accumulated pain, hurt, anger, and myriad misunderstandings that have occurred in our public, cultural spaces. Accordingly, this eco-spirituality reminds us to make living in wonder an intentional practice, cultivating habits of pausing and experiencing curiosity, compassion, and a sense of shared communion with our entangled, intimate partners. It reminds us that existing in a state of wonder can help transform us, sometimes bringing us back to our wholeness when we fall into momentary lapses of forgetfulness, or enshrouded in a haze or distorted perception of the organic wholeness of nature that constitutes our being here. Here again Carson's words are illuminating:

> A large part of my life has been concerned with some of the beauties and mysteries of this earth about us, and with the even greater mysteries of the life that inhabits it. No one can dwell long among such subjects without thinking rather deep thoughts, without asking himself searching and often unanswerable questions, and without achieving a certain philosophy.... Every mystery solved brings us to the threshold of a greater one.[39]

In certain moments of pausing, we may also be more receptive to experiencing epistemological humility, as Carl Sagan evokes with his narrative of apprehending Earth as a pale blue dot:

> Our posturing, our imagined self-importance, the delusion that we have some privileged position in the universe, are challenged by this point of

pale light. Our planet is a lonely speck in the great enveloping cosmic dark. In our obscurity—in all this vastness—there is no hint that help will come from elsewhere to save us from ourselves. The Earth is the only world known so far to harbor life. There is nowhere else, at least in the near future, to which our species could migrate. Visit, yes. Settle, not yet. Like it or not, for the moment, the Earth is where we make our stand. It has been said that astronomy is a humbling and character-building experience. There is perhaps no better demonstration of the folly of human conceits than this distant image of our tiny world. To me, it underscores our responsibility to be kinder with one another, and to preserve and cherish the pale blue dot, the only home we've ever known.[40]

Conclusion

In affirming there is no escape from the radical relationality and irrefutable materiality that structure the natural order, this eco-spirituality generates new avenues of thinking about myriad nature. To exist in wonder involves shifting attention away from alienating and problematic forms of human exceptionalism toward certain dreams and possibilities within a complex matrix of "becoming" and "experiencing" that are endemic to myriad nature's facticity. Roaming within this materialistic topography brings a sense of strange wonder to our human awareness of our constitutive relationality. It suggests the possibility of becoming other than what we are at any given moment when we are no longer captivated by an influential Euro-American narrative of the human. As intimated above, this modern narrative of human exceptionalism has denied humanity's inescapable relationality with other sentient life (and with each other) and it has been deadly and alarmingly violent.[41]

To the degree human life forms forget about our concreteness, our immediate matrix of interrelatedness, we continue to construct alienating forms of relationality. To what degree we grasp the communal matrix that constitutes our ultimate habitat, we can humbly create ethical principles that reject an either-or approach to myriad nature. The keener our appreciative sensitivity, the deeper our wonder and commitment to the cause of continuing creativity, a cause that far transcends what we alone can do. With other species, the human species is called to participate in life—and we can do that tentatively, humbly, and with much tenacity in our ongoing attempts to advance the flourishing of the entire community of life. The creative impulse that surges through the passage of all natural events in ways both great and small is the myriad reality, whose concern we acknowledge in the depths of our being. Indeed, we do belong to one another, to myriad nature, and human destinies are inextricably intertwined with others—most concretely and intimately. Here I evoke, once again, Whiteheadian imagery to help us think about the impoverished "worlds" some of us have constructed—particular forms of relationality in which we commit what Whitehead has termed the fallacy of misplaced concreteness. One commits this fallacy when one mistakes an abstract belief or theory about the way things are for a physical or "concrete" reality itself. As he writes: "There is an error; but it is merely the accidental error of mistaking the abstract for the concrete. It is an example of what might be called the 'Fallacy of Misplaced Concreteness' In this sense, one is rejecting the notion that a concrete physical object in the universe can be ascribed a simple spatial or temporal extension, that is, without reference to its relations to other spatial or temporal extensions."[42]

Honoring our radical relatedness and achieving our humanity is a task that can never be completed in an unfolding, mysterious universe. As noted above, such mystery destabilizes our sense of knowing in the pursuit of classifying and organizing the profound mystery of nature's profound alterity. Rather, a profound ecological perspectivism may emerge, which suggests our limited human perspective is part of a multiplicity of perspectives. In this illuminative discursive space, eco-spiritualists remain attentive to the alienating forms of relationality and actions that increase certain forms of injustice already operating in our various sociological and cultural settings. As a distinctive aspect of myriad nature, humans have been shaped by evolutionary constraints to understand and appreciate our constitutive relationality. As suggested elsewhere, as distinct life forms, humans can express love by creating alternative systems of interaction, and we can consistently question our values, behaviors, and resource uses. This wonder is seen in our local forms of activism when protesting mountaintop mining for coal, drilling for oil/petroleum, and fracking for gas and oil, or when deciding on the food we eat, determining how it is produced and transported, and considering ways of decreasing food waste in the United States and elsewhere. Certain possibilities may occur when human organisms begin to align our actions with the deeper mystery that we are not at the center of all that is.[43] Some (not all) humans appreciate the fact that we are part of an interacting, evolving, and genetically related community of beings bound together inseparably in space and time. In this context, whatever struggles, forms of resistance, and reimagining are possible, both collectively and internationally we are animated by musings, gentle manifestos, nudges and provocations of the sheer notion that we are all connected; and a some of us cannot ignore our inequalities, privileges, and vested interests in disrupting the flow of life in all its grandeur.

Notes

1 Will Steffen, Wendy Broadgate, Lisa Deutsch, Owen Gaffney, and Cornelia Ludwig, "The trajectory of the Anthropocene: The Great Acceleration," *The Anthropocene Review* (2015), 2 (1): 83.

2 https://www.stockholmresilience.org/research/planetary-boundaries.html. Accessed June 2023.

3 Christophe Bonneuil, "The Geological Turn: Narratives of the Anthropocene," in *The Anthropocene and the Global Environmental Crisis: Rethinking modernity in a new epoch*, eds. Clive Hamilton, Christophe Bonneuil, and Francois Gemenne (London: Routledge, 2015), 16.

4 Carol Wayne White, "A Poetics of Nature: Religious Naturalism, Multiplicities and Affinities," in *Religion, Materialism and Ecology*, ed. Sigurd Bergmann, Kate Rigby, Peter Manley Scott (New York: Routledge, 2023), 33–34.
5 Agostino Cera, *A Philosophical Journey into the Anthropocene: Discovering Terra Incognita* (Lanham, MD: Lexington Books, An Imprint of The Rowman & Littlefield Publishing Group, Inc., 2023), 5.
6 Cera, *A Philosophical Journey*, 5.
7 See, for example, Laudato Si' by Pope Francis (2015): http://www.vatican.va/content/francesco/en/encyclicals/documents/papa-francesco_20150524_enciclica-laudato-si.html; The UCC's Council for Climate Justice: "A Kairos Call to Action": https://www.ucc.org/a_kairos_call_to_action; https://faithfornature.org/news/; https://www.ccarnet.org/ccar-resolutions/ccar-resolution-climate-justice/; https://parliamentofreligions.org/programs/environmentalism-climate-and-the-rights-of-all-living-creatures/.
8 Linda Mercandante, *Belief without Borders: Inside the Minds of the Spiritual but not Religious* (New York: Oxford University Press, 2014), 28.
9 James Wiseman, *Spirituality and Mysticism: A Global View* (Maryknoll: Orbis Books, 2006), 1.
10 Peter H. Van Ness, "Introduction," in *Spirituality and the Secular Quest, World Spirituality 22* (New York: Crossroad, 1996), 5.
11 Donald Crosby, *Living with Ambiguity* (Albany: SUNY Press), ix–x.
12 Aldo Leopold, *A Sand County Almanac & Other Writings on Ecology and Conservation, ed. Kurt Meine* (New York: Library of America, 2013), 173.
13 Tim M. Lenton, "Gaia and Natural Selection," *Nature* 394: 439–47.
14 Michael Hogue, *American Immanence: Democracy for an Uncertain World* (New York: Columbia University Press, 2018), 5.
15 Wesley Wildman, "Religious Naturalism: What It Can Be, and What It Need Not Be," *Philosophy, Theology and the Sciences* 1, no. 1 (2014): 41.
16 Donald Crosby, "Matter, Mind, and Meaning," in Donald Crosby and Jerome Stone, eds, The Routledge Handbook of Religious Naturalism (New York: Routledge, 2018), 118.
17 Alfred N. Whitehead, *Process and Reality*, corrected ed., David Ray Griffin and Donald W. Sherburne, eds. (New York: The Free Press, 1978), 78.
18 Whitehead, *Process and Reality*, 18.
19 Alfred N. Whitehead, *Adventure of Ideas* (New York: The Free Press, 1967), 176ff.
20 Rachel Carson, *Lost Woods: The Discovered Writing of Rachel Carson*, ed. Linda J. Lear (Boston: Beacon Press, 1998), 231.
21 Michael Hogue, *American Immanence: Democracy for an Uncertain World (Critical Studies in Religion, Politics and Culture)* (New York: Columbia University Press, 2018), 3.
22 Henry A. Reigier, "The Notion of Natural and Cultural Integrity," in *Ecological Integrity and the Management of Ecosystems*, ed. Stephen Woodley, James Kay, and George Francis (Boca Raton, FL: CRC Press, Taylor & Francis Group), 3.
23 Donna Haraway, *When Species Meet* (Minneapolis: University of Minnesota Press, 2008), 3–4.
24 Rachel Carson, *Silent Spring and Other Writings on the Environment*, ed. Sandra Steingraber (New York: Library of America), xxxvi.
25 Sigurd Bergmann, "Is there a future in the eye of humans? A critical eye on the narrative of the Anthropocene," Forum: Consortium for the Study of Religion Ethics, and Society, 2016, 3. https://csres.iu.edu/. Accessed June 2022.
26 Mark Nepo, *The Book of Awakening: Having the Life you Want by Being Present to the Life You Have* (Newburyport, MA: Red Wheel/Weiser; Twentieth Anniversary edition, 2020), 1.
27 Carol Wayne White, *Black Lives and Sacred Humanity: Toward an African American Religious Naturalism*, (NY: Fordham University Press, 2016), 34–37.

28 Patrisse Khan-Cullors and Asha Bandele, *When They Call You a Terrorist: A Black Lives Matter Memoir*, with foreword by Angela Davis (New York: St. Martin's Press, 2018), 3. The crucial elements for life on Earth, often called the building blocks of life, can be abbreviated as CHNOPS: carbon, hydrogen, nitrogen, oxygen, phosphorus, and sulfur. Humans would have been the stardust generated by one or more stars that had exploded in a supernova explosion somewhere near the location where our solar system would have formed more than 4.6 billion years ago. The cloud that the solar system condensed out is thought to have been about 3 light-years across but the supernova explosion(s) could have come from significantly further away than that. A supernova creates an expanding cloud of gas and dust of the elements heavier than carbon and leaves behind a remnant—a black hole, a neutron star, or a white dwarf star.

29 Alfred North Whitehead, *Process and Reality*, 59ff; 65–67; 76; 160ff.

30 White, *Black Lives and Sacred Humanity*, 119.

31 Ursula Goodenough and Deacon Terrence, "From Biology to Consciousness to Morality," *Zygon Journal of Religion and Science*, 38 (4): 801–19.

32 Carol Wayne White, "Black Lives Matter and the New Materialism," in *Race, Religion, and Black Lives Matter: Essays on a Moment and a Movement*, ed. Christopher Cameron and Philip Luke Sinitiere (Nashville, TN: Vanderbilt UP, 2021), 109.

33 For sampling of new materialist readings, see Karen Barad, Meeting the Universe Halfway: Quantum Physics and the Entanglement of Matter and Meaning (Durham, N.C.: Duke University Press, 2007); Jane Bennett, Vibrant Matter: A Political Ecology of Things (Durham, N.C.: Duke University Press, 2010); Diana Code, New Materialisms: Ontology, Agency, and Politics (Durham, N.C.: Duke University Press, 2010); Rick Dolphijn and Iris van der Tuin, New Materialism: Interviews & Cartographies (Ann Arbor: MPublishing—University of Michigan Library, 2012).

34 Robin Wall Kimmerer, *Braiding Sweetgrass: Indigenous Wisdom, Scientific Knowledge, and the Teaching of Plants* (Minnesota: Milkweed Editions), 327.

35 Rachel Carson, *Lost Woods: The Discovered Writing of Rachel Carson*, ed. Linda J. Lear (Boston: Beacon Press, 1998), xiv.

36 Rachel Carson, *Lost Woods*, 160.

37 Enzo Ferara, *Rachel Carson – Undersea. Visions for Sustainability* (2014) 3:62–67: 64.

38 Rachel Carson, *Lost Woods*, 163.

39 Rachel Carson, *Lost Woods*, 159.

40 Carl Sagan, *Pale Blue Dot* (New York: Ballantine Book: Random House Publishing Group, 1997), 7.

41 Carol Wayne White, "Stubborn Materiality: African American Religious Naturalism and Becoming Our Humanity" in *Entangled Worlds: Religion, Science, and New Materialisms*, ed. Catherine Keller and Mary-Jane Rubenstein (NY: Fordham University Press, 2017), 256.

42 Alfred North Whitehead, *Science and the Modern World* (NY: Free Press, 1925), 52, 58.

43 Carol Wayne White, "Planetary Thinking, Agency, and Relationality: Religious Naturalism's Plea," in *Earthly Things: Immanence, New Materialisms, and Planetary Thinking*, ed. Karen Bray, Heather Eaton and Whitney Bauman (New York, NY: Fordham Press, 2023), 184.

9 The Sovereign Body of Country

Jennifer Evans

Introduction

This chapter seeks to ask the question 'can we know country as a sovereign body?'. As I work through this possibility, I deploy my nura mari ngarra (trust and listen to country) methodology, whereby I privilege Blak[1] voices, and make space for their powerful orations through their storying of sovereignty as Indigenous Peoples and their connections to their sovereign bodies of country. In my investigation I contrast their Blak voices to Sir David Attenborough's 'witness statement' narrative in seeking to understand the limitations of populist conceptions of nature and country. This chapter becomes a place for me to respond to the deep hurt of the rejection to the invitation of the Uluru Statement from the Heart (Uluru Statement from the Heart 2023, Chamberlain et al. 2024), from my Indigenous standpoint (Nakata 2007) by bringing attention to the silenced voice of country in the referendum experience in the place known as Australia (see Watego 2021).

In Aboriginal and Torres Strait Islander[2] cultures, country is a life force and being, one big living thing and is an entity within itself (see Bawaka Country et al. 2019; Evans 2023, Neidjie et al. 1985; Rose 1996). Country is sentient, a place of relationality, where Indigenous Peoples have a deep sense of belonging, and home and place are one with no incommensurable differences (see Brigg and Graham 2020, Moreton-Robinson 2020). First Law, the Dreaming Law and creation stories are alive in country and invite relationships of mutual obligation, where humans, non-humans and ancestral beings enjoy multifaceted deep time relations (see Burarrwaŋa et al. 2019; Perdrisat 2021; Poelina 2023). Country has agency and power, able to hold relations and shape, provide for and reawaken culture (see Perdrisat 2021; Darug Ngurra et al. 2020; Yunupingu & Muller 2009). All these phenomena are distinct from western concepts of wilderness, nature, bush, preservation and conservation (see Evans 2022; Evans 2019). Yet country is rarely narrated in terms of their[3] own sovereignty, within their own right. The notion of country as a sovereign body is hidden in current referendum, treaty, planetary sustainability and decolonisation debates.

DOI: 10.4324/9781003385202-10

Notions of sovereignty are contested, conflated and are 'far from being a concept of absolutes, can exist in many forms' (*tebrakunna* country et al. 2020, 19). Thus 'sovereignty means different things to different communities' and First Peoples (Davis 2012, 12). Some argue that current framing and promotion of sovereignty within constitutional theory renders it incapable of being transformed (Matthews 2021). Whilst others consider Indigenous sovereignty an oxymoron, 'given the colonial meaning of sovereignty (i.e. lord, ruler, master, highest, supreme or chief)' (Paradies 2020, 442). Mansell points out with irony the white possessive nature of the conceptualisation of sovereignty in Australia whereby 'the commonwealth of Australia was formed [because] … white people no longer wanted to be governed and controlled by someone else' (Mansell 2002, 84). Despite the broad views on the meanings and application of sovereignty in Australia, some warn that 'Australian nation-state sovereignty is a doxa composed of colonial foundations that are ill equipped for Indigenous-led truth-telling or social justice fixes' (*tebrakunna* country et al. 2020, 35).

Country has a right to sovereign life

Marlikka Perdrisat, a Nyikina Warrwa and Wangkumara Barkindji woman, digital storyteller and researcher, shares stories as a way to 'continue sharing of knowledge' and 'relegitimising ourselves' (Perdrisat 2021, 14:33). She talks about the Dreamtime as 'a way for us to communicate our science, our laws, and our philosophy' (Perdrisat 2021, 14:49). Collaborating with her brother, Perdrisat uses archive video footage to recreate a short one-minute story as a means to share messages of country:

> Rivers must have a right to life. And their contribution to all other life must be protected. Some are fearful that the humans lead their own extinction. I have hope in human beings. For living water lives within them too. It's their life that they must save if they are to continue living on mother earth. Mother earth covered in living waters cradled beneath the sky. I keep on singing as I believe a coalition of hope is coming. I sing this to you. Singing the river law song for people and country.
>
> (Perdrisat 2021, 13:21)

This one-minute story is impactful, moving and thought provoking. The footage starts with images of rock art depicting creatures, then rivers full of life with birds plenty, contrasted with water troughs being cleaned out for stock. It moves to people being dwarfed by an ancient boab tree, water lilies glistening in the sun, mountains and dirt roads leading off into the distance, and then men dancing in ceremony. All of this imagery is narrated by a voice that reverberates from the past into our present, with timelessness. The combination of sound and images creates a powerful message telling of the sovereign right that country has to full life, a life of respect and care. Perdrisat's work reflects Bill Neidgie's words: 'This earth for us. Just like mother, father, sister' (Neidjie 1985, 96). Perdrisat explains her intent of this work as 'the simplicity of this in comparison to something that goes for half an hour is really quite profound and shows the importance of what we can do in a minute in this modern society' (Perdrisat 2021, 13:00).

First Law

Perdrisat talks about First Law as the 'law of the land, not law of man. It is become from the land to teach us how to live with that ecosystem' (Perdrisat 2021, 11:04). She recalls

> as a little girl I used to think what part do we as human beings have in this ecosystem? And the old people told me that we keep country company, we sing to it, we dance, we look after it. And in any other form you have past this life you will not have the same privilege to look after the country in the same way.
>
> (Perdrisat 2021, 11:51)

Perdrisat explains that this deep harmonious relationship with country is understood through a removal of the hierarchy, where you are taught as young person that you are 'equal to and share respect with a non-human beings' and take responsibility for looking after everything with 'accountability, compassion, responsibility', as a 'lifelong project' (Perdrisat 2021, 9:33). Within this law of the land is its agency and power to hold memory and awaken us through memory, as 'memory is held in our bodies, it is known to us, it is something that we will always have and we will be able to share' along with the gift and commitment that 'we have to love country and country loves us' (Perdrisat 2021, 2:55, 8:06).

Country as ancestral beings

Aunty Professor Anne Poelina, a Nyikina Warrawa woman shares her knowledge and vision for understanding the Martuwarra River as an ancestral being (Poelina 2019, 2–23). When she asked the other Elders their thoughts about the proposition of granting personhood to the Martuwarra River (similar to that of the Whanganui River in Aotearoa), their response was 'the river is not a human being, it is an ancestral being' (Poelina 2023, 5:52). Country is an inseparable living entity (see Evans 2023), and rivers are lifeblood, more than a physical catchment or system, but a living ancestral being that has a right to life (Poelina et al. 2019). Integral to this life-force and right to life, the ancestral being of the river holds song, creation stories and agency to create values, virtues and ethics and teaches us how to self-regulate (Poelina 2023). In her storying of how the ancestral being of the Martuwarra River is present in country, Aunty Anne Poelina explains that the river unites people and country:

> Warloongarriy law is the First Law, law of the land, law of the river and so through Warloongarriy law along the Fitzroy River (the Martuwarra) 733 kilometres of it, that was the law that united all people from the bottom of the catchment into the top, from the top to the bottom, as well as out into the periphery of the desert. So Warloongarriy law was framed around one society, so all of us are seen as one society bound together under First Law, Warloongarriy law. So, the law was really a way, a song, a story of how to think about the river as such an amazing system. So in the song it says 'come in everybody, come in, come in, let the river replenish your soul your spirit, and then when you are full and re-nurtured go back to where you came from and tell your people why the river is so important why it needs to be here as the lifeblood of our people in this region, not just for indigenous people but for all
>
> (Poelina 2023, 30:31).

Original Sovereign Peoples

In Australia the phrase 'always was, always will be' and variations of this is a well-known Aboriginal and Torres Strait Peoples rights message (McBride 2021). It communicates that the Aboriginal and Torres Strait Islander Peoples are the First Peoples of the continent now known as Australia and all its islands, and that mobs[4] have been occupying, caring for country for over 65,000 years, and that sovereignty has never been ceded. This phrase was as important and visible as ever during 2023, when the Australian Indigenous Voice referendum was held. In 2017 an invitation was made to the Australian public to consider a proposal for substantive recognition of Aboriginal and Torres Strait Islander Peoples known as the Uluru Statement from the Heart (Uluru Statement from the Heart 2023), below is a portion of the statement that refers to maters of sovereignty:

> Our Aboriginal and Torres Strait Islander tribes were the first sovereign Nations of the Australian continent and its adjacent islands and possessed it under own laws and customs. This our ancestors did, according to the reckoning of our culture, from the Creation, according to the common law from the 'time immemorial', and according to science more than 60,000 years ago. This sovereignty is spiritual notion: the ancestral tie between the land, or 'mother nature', and the Aboriginal and Torres Strait Islander peoples who were born therefrom, remain attached thereto, and must one day return thither to be united with our ancestors. This link is the basis of the ownership of the soil, or better, of sovereignty. It has never been ceded or extinguished, and co-exists with the sovereignty of the Crown.
>
> (Uluru Statement from the Heart 2023)

It was predicted that the referendum would be the 'most consequential referendum in the history of the Federation of Australia' and 'will have a profound effect on the future relationship between Aboriginal and Torres Strait Islander peoples and other Australians' (Anderson et al. 2023, 2). The referendum was a question of morality, and the proposal offered a basic human right, the right to be heard (Fredericks and Bradfield 2023). Threats, abuse vilification, hate speech and racism against Aboriginal and Torres Strait Islander Peoples increased substantially during the lead up to the referendum and after the vote (Anderson et al. 2023, Carlson 2023, Oscar 2023). Indigenous leaders referred to the referendum result as shameful, gave license to vitriol and hatred that already existed and unleashed a tsunami of racism (Magann 2023). Decades of work and activism calling for Indigenous recognition and rights from Elders and community leaders culminated in the referendum. In the weeks leading up to the referendum some of these Elders and community leaders argued their final case for a yes vote at the National Press Club of Australia, including Senator Pat Dodson and Noel Pearson.

The love of country

Indigenous community leader and prominent referendum yes campaigner Noel Pearson, a Guugu Yimithirr man, addressed the National Press Club Australia weeks before the vote. He opened his address drawing on the mutual love that Aboriginal and Torres Strait Islander Peoples and other people have for the love of country and the land:

> My speech is titled for the love of country. I meant not love in a saccharine or sentimental way, nor romantic or patriotic love. I mean in the sense the late English philosopher Roger Scrutiny called oikophilia. Oikos the Greek for home and philia for love, the love of home … I said it's not the same patriotism because there is nothing political about this love of country. Rather it is about the wonderous land and seas. All of the nature and culture makes this country so beloved of us all and its effect upon our souls. No Australian can fail to feel this in the presence of Uluru and a million other places each of us call our special homes. I say today it is the love of country that is driving motivation for the recognition of Aboriginal and Torres Strait Islander Peoples in the Australian constitution. It is not love of each other that joins us it is our mutual love of country that unites us. We don't need mutual affection to succeed in this referendum. We need to recognise our mutually shared love for the land.
>
> (Pearson 2023, 4:12)

Not the Garden of Eden

Senator Pat Dodson is a Yawuru Elder and Indigenous rights activist. He addressed the National Press Club Australia in the last week leading up to the referendum in October 2023. Early in his address, Senator Dodson reflected on how he felt as a young person in the education system

> feeling the brunt of the dying race philosophy … I'd wonder whether we'd all disappear … as part of the passing race philosophy I didn't understand it, I didn't know where it come from but I had this fear that something mysterious was going to make us disappear and that was the philosophy underpinning the constitution.
>
> (Dodson 2023, 5:16)

When asked by a journalist at the event about his views on Senator Jacinta Nampijinpa Price's comments regarding her beliefs that there were no negative impacts from colonisation, Senator Dodson as part of his reply was:

> we're not in the Garden of Eden here, there are consequences of how we came to be colonised, and they have to be dealt with … There are legacy issues and responsibility and accountability issues for how you've taken someone else's country and subjugated them to the policies that you have. Assimilation, control, management, domination, determination of their futures, taking kids away, Stolen Generations, all of these things have consequences, from the first point of taking their lands and subjugating them to the policies of government to achieve the objective which benefits that society now enjoys.
>
> (Dodson 2023, 18:05)

Another question was posed to Senator Dodson regarding the transparency of Indigenous Land Councils. Senator Dodson responded with a masterclass on Indigenous rights, culminating with a clear message regarding the original sovereigns of this continent and its islands being the Aboriginal and Torres Strait Islander Peoples:

> The rights and interests that Aboriginal People have got, should be honoured … Mabo[5] showed clearly that Aboriginal Peoples and the Torres Strait Islander Peoples have got rights that preceded the colonisation of these lands … Our High Court discredited the legal fiction of terra nullius, it discredited our whole administrative land tenure system, and their public policy system has been based on a discredited lie. By voting yes in the referendum we are simply catching up in many ways to what the High Court has already said that the Aboriginal Peoples were the first peoples of this country and that they should be recognised as such and that they have certain rights that are not those rights given by the crown. They come from the thing called Native Title, so we as Australians are not dealing with the old colonial debate anymore. We're dealing with the new context where the rights and interests of the Aboriginal and Torres Strait Islander Peoples [have] got to be accommodated within our structure …

> You can't deny a people whose culture has been here for 60,000 years and if that's what happens with a no vote that's what you're doing. You're saying you people have no history here, you have no legitimacy here, you have no right to be here. Now that's an intolerable proposition.
>
> (Dodson 2023, 28:08)

Scars to our spirit, our bodies and our Land

Countering the yes vote campaign for the referendum was the Blak Sovereign Movement who opposed constitutional recognition and the Voice to Parliament (BSM 2024) and others referred to as the 'progressive No' movement (Farmilo 2023). Spearheaded by Senator Lidia Thorpe, their rights agenda focuses on their 'commitment to Truth-telling and the value we place in our Sovereignty over this land' (BSM 2024). Senator Thorpe asserts that 'Sovereignty never ceded. Our sovereignty does not co-exist with the Sovereignty of the Crown. We are the original and only Sovereigns of these lands. We are self-determined and self-governing nations' (Thorpe 2023). Other First Nations Peoples called for a Treaty before Voice, arguing that a Treaty being a legal agreement would commit the government to 'carrying out certain measures' rather than an advisory committee whose advice could be ignored (Andrewartha and Bolton, 2023, 9). Some claimed that the modest Voice proposal was purposely designed 'to attract support from more right-wing parliamentary parties' with the intention of 'demobilise[ing] the First Nations movement' creating space for the right-wing 'to whip up a racist consciousnesses among working class people' (Andrewartha and Bolton, 2023, 9).

A month before the referendum vote, the late Mr Wilson, Jabarda, jungai and proud Gudanji-Wambaya man, warrior and Chair of Nurrdalinji mob passed away. As a jungai (cultural lawman), he was a champion of country and the rights of Traditional Owners (Torre 2023; Morse 2023). Mr Wilson was a 'talented and powerful orator' and pushed back against gas mining in the Beetaloo Basin (Torre 2023, np).

> We do not want fracking on our country, full stop. Our cultural heritage and our sacred sites are so important to us – our songlines, our Dreaming and our water. Our water is our life. When that is contaminated, we are nothing. Where is our future?
>
> (Wilson quoted in Torre 2023, np)

Mr Wilson spoke powerful words of Blak resistance and stood strong against the brutal plunder the coloniser's sovereign white-saviour solutions (Chesterfield 2023):

> Scars to our spirit, our bodies and our Land, scars inflicted by settler-colonialism without let-up since they arrived, destruction and plunder being all they know what to do. Scars inflicted by a people that think they actually understand the stolen land on which they stand, arrogantly

> thinking they know better than us how to repair the Land and Sea Country they've destroyed. As far as the eye can see, these are the scars of surviving a Stray-Alien culture of rape and pillage. But our warriors resisted, survived and are reborn, guided by our Elders and those who came before, a resurgence of Strong Black Resistance – a seed that is constant in this ancient Land, and being watered by a new generation of mob finding new ways to resist and heal.
>
> (Wilson quoted in Chesterfield 2023, 8)

When asked about the referendum, Mr Wilson referred to it as process that imposed 'yet another layer' of brutal 'deep trauma' with 'waves of racism from both Yes and No' and that many (mob) felt 'too unsafe and unlistened to by the colony to contribute' (Wilson quoted in Chesterfield 2023, 8,9). At the heart of Mr Wilson's views of the utility of the referendum for Aboriginal People to protect country from destruction, he called for focus on the ability of the Voice to deliver 'practical allyship for Land and Sea Country protectors and decolonisation in the environment movement' (Wilson quoted in Chesterfield 2023, 9). Mr Wilson made it clear that Blak voices of resistance are demanding that it is time for settlers 'to listen up, and stand alongside us, or simply get out of the way' (Wilson quoted in Chesterfield 2023, 8).

Planetary universalism

Despite the requests from mobs to halt disrespect and destruction of country, populist conceptions of nature dominate social discourses. Current global climate change due to human activity during the Anthropocene has become a 'defining crisis of our time' (UN 2024, np). Movements such as the Extinction Rebellion, Indigenous youth led climate networks (see SEED 2023), and individual climate change activists like Greta Thunberg are calling for global action to address the climate and ecological emergency (XR 2024, BBC 2023), Meanwhile millions of viewers are watching environmental activist Sir David Attenborough's programme 'witness statement' illustrating the devastating impact humanity is having on our planet (Hayward 2020). Global responses to the Anthropocene have been interpreted as a 'new humanist and statist moment for universal politics' that generate totalising and naturalising rationales which 'reproduce colonial logics' (Jackson 2020, 698). Some call for push back against the 'hegemonic governance imaginaries' that feature in this new politic of planetary universalism invoked by the Anthropocene through seeking 'enduring ecological reciprocities' (Jackson 2020, 698). Others suggest that sovereignty has the potential to represent ecological forces and human social life in the negotiations required for the climate crisis of the Anthropocene (Matthews 2021). In this sense, the Anthropocene is allowing new totalising discourses which threaten sovereign movements and are not the way towards ecological flourishing.

The commencement dates of the Anthropocene and the modernity have been debated (see Saldanha 2020, Waters and Turner 2022). Regardless of their start date, some argue that the 'Anthropocene is inexorably racial as it flows out of a capitalist system which requires racializing populations and environments from early modernity to the present and into the future' (Saldanha 2020, 12). (Re)colonisation and decolonisation narratives represent small voices in current climate crisis debates, arguing that 'colonialism is inextricably intertwined with modernity' necessitating Indigenisation and 'radical land-based re-localisation, revitalised communalism and embedded kinship with all life' in order for flourishing of all (Paradies 2020, 447, 438). Even though the Anthropocene is used to designate a unit of geological time, and can become conflated with modernity and capitalism, others suggest that 'lurking just beneath the surface of the Anthropocene is a racialised narrative about white Earthly possession' (Baldwin and Erickson 2020, 3).

Witness statement

In his broadcast titled 'A life on our planet' Sir David Attenborough gives his 'witness statement' for the planet (Attenborough 2020; Hayward 2020). As I comment on this work, I recognise that not all may agree, however what I offer is an Indigenous Standpoint (Nakata 2007) as an individual Indigenous person. In his work, there is a subtle framing of a liveable and thriving world only possible as a result of the Holocene which started approximately 11,700 years ago (Walker et al. 2009). For example, 'The biodiversity of the Holocene helped to bring stability, and the entire living world settled into a gentle, reliable rhythm, the seasons … The Holocene was our Garden of Eden' (Attenborough 2020, 12:01, 13:08). This narrative silences the notion that Aboriginal and Torres Strait Islanders Peoples were thriving on this continent and its islands for approximately 65 millennia before the Holocene, living with country, with all that country had to offer. Sir David Attenborough also states that 'We learnt how to exploit the seasons to produce food crops … The history of civilisation followed … Each generation able to develop and progress only because the living world could be relied upon to deliver conditions we needed' (Attenborough 2020, 13:27). Similarly, this statement makes invisible that Aboriginal and Torres Strait Islanders Peoples are one of the oldest known continuous living cultures on Earth outside Africa with ancestors extending back approximately 75,000 years (Malaspinas et al. 2016, Rasmussen et al. 2011), not bound by western definitions of agriculture or civilisation (see Gammage 2011, Pascoe 2018).

Notions of country as an entity, personality, ancestral spirit, or a body with agency is also absent in Sir David Attenborough's 'witness statement'. Here he refers to the

> planet's wild places, its biodiversity … Billions of individuals and millions of kinds of plants and animals … Leading lives that interlock in such a way that they sustain each other … We rely entirely on this finely tuned life support machine.
>
> (Attenborough 2020, 2:40, 3:08, 3:19, 3:25)

There is large literature on how the western notion of 'wilderness' separates humans from nature (see Cronon 1996; Nash 2014). This storyline talks to an ecosystem from a distance, one that lacks a personal relationship between humans and non-humans, a desire and connection for closeness and care, one that mirrors wilderness tropes (Langton 1996) and dualisms of country without people (Evans 2023, Lee 2015) with mechanistic overtones.

Toward the end of his 'witness statement' Sir David Attenborough makes a gesture that people can have a relationship with 'nature' as an 'ally and inspiration' and that we need to move closer to nature: 'In this world a species can only thrive when everything around it thrives too … If we take care of nature, nature will take care of us' (Attenborough 2020, 1:10:57).

Methodology

Nura mari ngarra – trust and listen to country

When considering populist conceptions of nature and the silencing of country and deep time Indigenous occupation, I deploy a strategic Indigenous methodology to strategically 'situate Indigenous peoples within a web of power relationships' (Ray 2012, 89). As I explore the notion that country may be able to be storied (see Archibald, Lee-Morgan and De Santo 2019) as a sovereign body within their own right, having agency and power to invite and hold relations (see Evans 2022; Perdrisat 2021; Poelina 2023), I do with deep respect recognising that this may not be the way in which individual Aboriginal and Torres Strait Islander Nations and individuals see and know their country. I do not intend to cause harm nor disrespect to mobs and country themselves as I seek to learn if my enquiry might be considered. I come to this exploration as a Dharug[6] person, living on and in relation with *palawa*[7] country, knowing my place as a visitor and sharing my Indigenous standpoint (Nakata 2007). As I work through my notions of the sovereign body of country, I use my Indigenous methodology nura mari ngarra (trust and listen to country), as a way to listen to the legitimate Blak voices of our Elders and community leaders. I listen with trust; I listen with honouring them and country.

My nura mari ngarra methodology is counter to western academic approaches, as it makes space for the reader to listen to each single voice, to honour that voice and sit with each message as it comes. It is not a rapid fire of multiple voices and sources in exhaustive strings of references. The Blak voices in my nura mari ngarra methodology bring stories that are coming from country and the heart of First Peoples connected with country. I let their words speak for themselves with space and respect instead of western ways of justifying their words with the words of many others. I use block quotes of their words, standing on their own, as a way to give voice and fresh air to respected mob, since now that our voices have been ignored and silenced through the referendum rejection. Presenting their strong Blak voices this way is an important act, and my nura mari ngarra methodology reflects the sentiments of our times as we grieve collectively and individually of the rejection of the invitation of Uluru Statement from the Heart (see Chamberlain et al. 2024) and the prospect of an Indigenous Voice to parliament. My methodology is both a mark of respect and deep action of defiance and push back against colonising ways of research and how we represent our Blak voices. This is my nura mari ngarra substantive way of doing research, one that has cultural substantiveness through respect and giving space and sharing orations. This is my Indigenous and decolonising methodology in action.

Academic Elder and activist Distinguished Professor Aileen Moreton-Robinson, Goenpul woman of the Quandamoka people, eloquently describes the sovereign power of oration and its links to country:

> … orations [and other practices] enable the nourishing of sovereign well-being because they are places where our cultural discourses flourish sustained by our knowledges and our truths … sharing supports our sovereignty and discursive and material spaces for embodied cultural practices and knowledge production are where the capillary power effects of colonising by biopower becomes subdued or limited. As another form of power flows from the land and into bodies that are incommensurate in their ontological existence.
>
> (Moreton-Robinson, 2021, 45:07)

Country and their sovereign bodies

Conflation of sovereignty

Sovereignty as it applies to First Peoples can become conflated under white possession and restrict Indigenous embodied ontologies (see Moreton-Robinson 2015). This conflation is a convenience for colonising states where white male forms of sovereignty can become smokescreens in order to deny Indigenous sovereignty and reproduce (see Evans 2023; Moreton-Robinson 2007). This white possessive convenience disallows space for the possibility of country to be recognised, known or understood as a sovereign body, or multiple sovereign bodies as this would get in the way of settler occupation and their insatiable appetite for resource and life extraction from country. Under settler state occupation, the sovereign body of country is rendered invisible through their white possessive eyes. This denies space in settler state mentalities for conceptualising country as sovereign, as sovereign bodies with their own rights, life force, agency and power in relations with Indigenous Peoples.

Sovereignty can be understood as not a legal term but an emotive term regarding the power and authority to govern (see Falk and Martin 2020). Logics include terms such as 'supreme', 'absolute', 'external' and 'internal' sovereignty. Some argue that the concept of sovereignty is rooted in Christianity during the enslavement of pagans in the fourteenth century, giving supreme power within a state free of any restrictions (see Falk and Martin 2020; Daes

2004). By the nineteenth century the absolute sense of sovereignty waned, when under

> international law a 'sovereign' could be under the protection of another, greater sovereign without losing its 'sovereignty.' (Daes 2004, np). Sovereignty has also been described as the 'relationship of the ruler or state towards other states', thus 'the independence of a state from other states'.
>
> (Falk and Martin 2020, 35)

Further, contemporary understandings of sovereignty include 'external' sovereignty being delineated space/territorial and 'internal' sovereignty as political

> the bearer of supreme authority within a state (Falk and Martin 2020, 36). Regardless of the historical or contemporary understandings of sovereignty, in White Australia its laws and governments are ingrained in Christianity, devoid of Indigenous worldviews and sovereignty remains trapped in its historical framing (see Falk and Martin 2020). Despite the settler mentalities regarding sovereignty, 'Aboriginal people have never consented or ceded. Sovereignty did not pass from Aboriginal people to the settlers'.
>
> (Davis 2012, 14)

Distinguished Professor Aileen Moreton-Robinson explains the limitations of sovereignty and its white possession of country:

> Sovereignty remains contractually as a relationship between individuals and the state, between humans and the apparatuses they have created. This conceptualisation of sovereignty rests on enlightenment presuppositions that nature is servant to humanity and humanity is master of nature. Indigenous sovereignties are not limited in this way, instead our knowledge systems acknowledge the power of the earth, the limitations of human endeavour to control it and the need to live in respectful and proper relations with all living things.
>
> (Moreton-Robinson 2021, 49:25)

One true co-existing sovereignty

Although the Uluru Statement from the Heart states that First People's sovereignty 'has never been ceded or extinguished, and co-exists with the sovereignty of the Crown' (Uluru Statement from the Heart 2023) some argue that a co-existing sovereignty has not been consented to, or simply does not exist (see Evans 2023, Thorpe 2023). Regardless, I suggest that the one true co-existing sovereignty in Australia is that of one between the Aboriginal and Torres Strait Islander Peoples, their Nations and their ancestral country.

The body of country can be understood not in a 'supreme' or 'absolute' sense operating with hierarchy, but as a body that can love and hold relations with people, place, other non-humans, ancestors and ancestral beings (see Perdrisat 2021, Poelina 2023). This relationship is not about supreme rule; it's about First Law and the culture that is First Law (see Perdrisat 2021, Langton and Corn 2023). Although 'nature [country] makes no concessions for human needs and wants' (Langton and Corn 2023, 92), country can be loved, looked after and kept company in co-existing sovereign relationships. Aunty Professor Anne Poelina invites us to see country as ancestral beings (Poelina 2023):

> As Indigenous Peoples across the world we have not walked away from the garden of Eden, we have not walked away from nature, we see ourselves co-exiting with nature, and nature being very much a part of who we are as human beings.
>
> (Poelina 2023, 10:38)

Co-existing sovereignty between the sovereign body of country and their First Peoples can be understood from a rights-based approach, where Blak resistance and sovereignty are required to protect and care for country. Country needs protection from settler occupiers and their states as Mr Wilson describes; 'a theme runs through Blak resistance: we are still quarry – and a quarry – for empire, capitalism and colonialism' (Wilson quoted in Chesterfield, 2023, 9). Blak resistance and sovereignty gives voice to the sovereign body of country, together both inseparable sovereigns working together to heal and protect each other.

> It is an honour to amplify and weave together powerful voices of this Land … inspired by sixty millennia of care, community and custodianship of all beings across Land and Sea Country. Still standing strong against coloniser's brutal plunder of 250 years, and in living their sovereign solutions, adapting new ways and new allies to heal the damage caused by a terminally broken system of the invaders.
>
> (Wilson quoted in Chesterfield, 2023, 8).

Senior Nyikina Elders Lucy Marshall OAM[8] and Jeannie Wabi eloquently disentangle the difference between singular white possessive sovereignty (or what some may refer to as co-existing or absolute sovereignty) and the sovereign body of country: 'We are the Native Title holders of the Crown' (Marshall quoted in Poelina, Wabi and Marshall 2015, 35:58). 'Country belongs to us. Place, you know. Country. The river and the sea are for fishing, not for mining … We've got a strong culture. We still have our dances for this river. We are strong people' (Jeannie Wabi quoted in Poelina, Wabi and Marshall, 33:36, 34:17). 'You can't change anything. The white men's rules can change, but not the Blackfella's rules' (Lucy Marshall quoted in Poelina, Wabi and Marshall 2015, 9:53). The Nyikina women's description of country and sovereignty, and their relationships to country is clear; it is emotional, material, and relational. It is also specific and geographical; 'Our paddock, our boundary. From the hill country ... The river goes straight down to the sea' (Jeanie Wabi quoted in Poelina, Wabi and Marshall 2015, 36:01).

Similarly, the late Dr Galarrwuy Yunupingu OAM, Elder, senior lawman, land rights activist and member of the Gumatji clan from Yirrkala who 'spent a lifetime fighting for the rights of his people' and a strong supporter of the Voice (Allam 2023, np), clearly articulates his understanding of exclusive sovereignty between his people and his country:

> The clans of east Arnhem Land join me in acknowledging no king, no queen, no church and no state. Our allegiance is to each other, to our land and to the ceremonies that define us. It is through the ceremonies that our lives are created. These ceremonies record and pass on the laws that give us ownership of the land and of the seas, and the rules by which we live.
>
> (Yunupingu 2015, 4)

Dr Galarrwuy Yunupingu describes country as an entity, a body capable of being disrespected not only spiritually through interrupting Dreamings, but also through destroying parts of country such as rocks capable of holding and claiming sovereign powers within country. He makes reference to mining companies who were 'insulting the country', who 'damaged our country ... damaged a lot of Dreamings and Dreamings that where important' including 'a very important rock that claimed the country' (Yunupingu 2019, 3:31, 4:26,10:46).

Moral contract with country

Notions of country as a sovereign body is missing in sovereignty narratives and was absent in the referendum debate in Australia. However, Noel Pearson in his address to the Press Club, did bring to the debate, the notion of the love of country as a unifier, seeking people to see country as a common point and place of respect (Pearson 2023). His appeal chanced the prospect that if Aboriginal People can't be loved and respected through the referendum, then perhaps country could get people over the line and persuade them to vote yes? What seemed like an attempt to get settler colonisers to imagine what country is or might be, a place of love and strength was a worthy and a noble request by Pearson. His notion of country as home, as a place of love and a place that is loved, can be interpreted as an amplification of the sovereign body of country, as a stark counter to Westphalian notions of nature and wild places such as those of Sir David Attenborough (Attenborough 2020; Hayward 2020).

In Sir David Attenborough's witness statement, the notion of country is absent; instead, his manifesto is an exemplar of the universality of wilderness, as a place untouched by humans. This narrative is one of planetary universalism, a biophilic white possession of country, an othering of country, that reinforces the neoliberalist global north binary of planetary annihilation. It distils humans as a pest plague, as consumers without connection to country or place, and helps deny Indigenous sovereignty. Noticeably, this narrative is a noisy green wash, a complicated mess of unsustainability due to human progress, rather than a recognition of the ongoing colonisation of mobs who have cared for country for millennia.

Aunty Professor Anne Poelina offers an antidote to Sir David Attenborough's witness statement. Her vision is a way to move forward by making a moral contract with country as a sacred living ancestral being, to care for country in co-existence under

> strong contemporary governance based on First Law (Poelina et al. 2019, 248, Poelina 2023). In her description of a moral contract, Aunty Professor Anne Poelina explains that 'we bind ourselves through a moral obligation, a moral contract to protect all of the things that are with us because we live in a world that is reliant on that symbiotic relationship between nature and people. We learn that even the wind can have a name that it can bring a story but trees that are standing there as sentinel beings have watched how we've walked the landscape and managed the landscape'.
>
> (Poelina 2023, 16:49)

This is a storying (see Archibald, Lee-Morgan and De Santo 2019) of country as an interrelated body of many sovereign entities and lifeforces that make the sovereign whole, all under First Law and in relationship with their sovereign First Peoples. Her moral contract seeks action from the settler occupier state to know the sovereign body of country as more than 'nature' or a lifeless territory subsumed into a conglomerated planetary whole. Aunty Professor Anne Poelina's vision requests that

> we must get our nation state to realise we are all on one planet mother earth and that we all must have this moral obligation to right size the planet if we are to have a future as human beings.
>
> (Poelina 2023, 46:00)

Conclusion

At the outset of this chapter, I sought to ask the question 'can we know country as a sovereign body?' Through my exploration, I suggest that we as First Peoples share co-existing sovereignty with country, and that this co-existence is entirely distinct to the distorted notion of co-existing sovereignty between us and our colonising occupier state. This relationship between First Peoples and their country is sovereign to sovereign, equal, non-hierarchical, nourishing, loving, respectful and based in First Law. Thus, country does have a sovereign body, as it must if it is to be in relation with their sovereign First Peoples. So, if we conceive ourselves as sovereign First Peoples, our country must also be sovereign, as country is everything. If country is sovereign, then co-existence is with sovereign Nations and First Peoples, and as such not only have us as First Peoples been invaded and are under occupation, but also has/is the sovereign body of country themselves. Not only has no agreement been made with Aboriginal and Torres Strait Islander Peoples, but also not with the sovereign body of country.

All we can do in the face of the deeply hurtful rejection of the invitation of the Uluru Statement from the Heart via the referendum is to turn to each other (mobs) and to country for healing and guidance. To seek the true power of the sovereign body of country with our sovereign bodies in our combined co-existing sovereignty, the only true co-existence and co-existing sovereignty of this land, our Nation homes, our peoples. This is a deep process of recovery as we seek answers for our own self-determination away from the failures of our occupying colonisers and their racist democracies, and their attempts to thieve our sovereignty. This is our return to our heart. The wise words of Distinguished Professor Aileen Moreton-Robinson invite us to rethink our understandings of sovereignty and reflect on the power of the sovereign body of country:

> While western sovereign right can let live, make live and die, it does not possess the power to give life to everything. Perhaps it is a time for a new theory of sovereignty, one that is less concerned with western humanist endeavours and right than it is with the survival of the earth.
>
> (Moreton-Robinson 2021, 49:58)

Notes

1 I use the term Blak to refer to Aboriginal and Torres Strait Islander Peoples respectfully, following the work of Deacon in Baylis (2015, 16) and Evans (2023, 65) where Blak is 'defined by Ku Ku/Erb/Mer visual artist Destiny Deacon as a "vehicle to express identity and subvert the racist notion that Aboriginal People are 'black', or rather identifiable as having 'black' skin" (Baylis, 2015, p. 16; Evans, 2022)'.

2 I use the terms Aboriginal and Torres Strait Islander, Indigenous, First Peoples and mob interchangeably to refer to Aboriginal and Torres Strait Islander Peoples. I use the term Indigenous to also refer to Indigenous Peoples globally.

3 I use terms such as 'their', 'them' and 'they' as it applies to country, in the sense that country is a living entity and not an 'it' or an inanimate, subject or a thing, nor a fixed singular geography encompassing the continent known as Australia.

4 I use the term 'mob' respectfully referring to Aboriginal Peoples as a collective, acknowledging that the term is not accepted as a positively reclaimed term by all Aboriginal and/or Torres Strait Islander Peoples (see Carlson et al. 2014).

5 Mabo refers to the significant legal case in Australia whereby the High Court recognised that Aboriginal and Torres Strait Islander Peoples had a right to land according to their own laws and customs, under Native Title, and that "terra nullius" (land belonging to no one) did not exist at the time of colonisation (see Meyers and Mugambwa 1993; Mansell 1993).

6 I refer to Dharug as the Aboriginal Nation of the larger Sydney Basin, as Dharug Ngurra (country) and whilst doing so acknowledge that the geographical boundary

is contested as not confined to Western Syndey, to be respectful of varying Dharug views (see DCAC, 2022).
7 The term *palawa* is used by some to refer to Tasmanian Aboriginal People.
8 OAM refers to the Order of Australia.

References

Allam, Lorena. "Yunupingu, Yolŋu leader and campaigner for Indigenous rights, des aged 74." *The Guardian*. April 3, 2023. https://www.theguardian.com/australia-news/2023/apr/03/yunupingu-yolu-leader-and-campaigner-for-indigenous-rights-dies-aged-74 (accessed January 22, 2023).

Anderson, Ian, Yin Paradies, Marcia Langton, Ray Lovett, and Tom Calma. "Racism and the 2023 Australian constitutional referendum." *The Lancet* 402 no. 10411 (2023): 1400–1403.

Andrewartha, Jacob, and Sue Bolton. "Blak Sovereign Movement given a Boost by Senator Llidia Thorpe." *Green Left Weekly*, no. 1375 (2023): 9. https://search.informit.org/doi/10.3316/informit.900879859333368.

Archibald, Jo-Ann, Jenny Lee-Morgan, and Jason De Santolo, eds. 2019. *Decolonizing Research: Indigenous Storywork as Methodology*. London, UK: ZED Books Ltd.

Attenborough, David. "A life on our planet". Netflix, Los Gatos, California, United States, 2020.

Baldwin, Andrew, and Bruce Erickson. "Introduction: Whiteness, coloniality, and the Anthropocene." *Environment and Planning D: Society and Space* 38, no. 1 (2020): 3–11.

Baylis, Troy-Anthony. "Introduction: Looking into the mirror". In *Colouring the rainbow blak queer and trans perspectives: Life stories and essays by First Nations People of Australia*, edited by Dino Hodge, 1–17. Adelaide: Wakefield Press, 2015.

Bawaka Country, Sandie Suchet-Pearson, Sarah Wright, Kate Lloyd, Matalena Tofa, Jill Sweeney, Laklak Burarrwanga et al. "Goŋ Gurtha: Enacting response-abilities as situated co-becoming." *Environment and Planning D: Society and Space* 37, no. 4 (2019): 682–702.

BBC. "Greta Thunberg: Who is the climate activist and what has she achieved?". *BBC News*. November 15, 2023. https://www.bbc.com/news/world-europe-49918719 (accessed January 22, 2024).

Brigg, Morgan and Graham, Mary. "The relevance of Aboriginal political concepts (1): the current pandemic and the importance of ancient wisdom". ABC religion & Ethics. May 24, 2020, https://www.abc.net.au/religion/coronavirus-and-aboriginal-political-concepts-morgan-brigg-and/12281304.

BSM. "Blak Sovereign Movement". *Blak Sovereign Movement*. Accessed January 22 2024, https://blaksovereignmovement.com/.

Burarrwanŋa, Laklak, Ritjilili Ganambarr, Merrkiyawuy Ganambarr-Stubbs, Banbapuy Ganambarr, Djawundil Maymuru, Sarah Wright, Sandie Suchet-Pearson, and Kate Lloyd. *Songspirals: sharing women's wisdom of Country through songlines*. Allen & Unwin, 2019.

Carlson, Bronwyn. "There are two sides to the 'no' campaign on the Voice. Who are they and why are they opposed to it?". *The Conversation*. September 7, 2023. https://theconversation.com/there-are-two-sides-to-the-no-campaign-on-the-voice-who-are-they-and-why-are-they-opposed-to-it-212362.

Carlson, Bronwyn, Jeff Berglund, Michelle Harris, and Evan Te Ahu Poata-Smith. "Four scholars speak to navigating the complexities of naming in Indigenous studies." *The Australian Journal of Indigenous Education* 43, no. 1 (2014): 58–72.

Chamberlain, Catherine, Ian Anderson, Bronwyn Fredericks, Tom Calma, and Sandra Eades. "Indigenous peoples' health after Australia's No vote." *bmj* 384 (2024).

Chesterfield, Nick. "Blak Earth Rising-Strong Blak resistance to coloniser extractivism." *Chain Reaction* 146 (2023): 8–9.
Cronon, William. "The trouble with wilderness: or, getting back to the wrong nature." *Environmental History* 1, no. 1 (1996): 7–28.
Daes, Erica-Irene. "*Indigenous Peoples permanent sovereignty over natural resources*", June 3 2004. https://humanrights.gov.au/about/news/speeches/indigenous-peoples-permanent-sovereignty-over-natural-resources.
Davis, Megan. "Constitutional recognition does not foreclose on Aboriginal sovereignty." *Indigenous Law Bulletin* 8, no. 1 (2012): 12–14.
DCAC. "Darug boundaries". Darug Custodian Aboriginal Corporation. https://darugcorporation.com.au/about/. 2022
Dodson, Patrick. "In full: The Hon Patrick Dodson's address to the National Press Club of Australia". *Youtube*. 1:13:58. 11 October 11, 2023. https://www.youtube.com/watch?v=KfiuJmc_6-w.
Evans, J. "Can men weave baskets in Queer country? In *tebrakunna* country and E. Lee, & J. Evans." *Indigenous women's voices: 20 years on from Linda Tuhiwai Smith's decolonizing methodologies* (2022): 33–50.
Evans, Jennifer. "Decolonising the Sustainable Development Agenda: The Voice of Country and Treaty." In *Education and the UN Sustainable Development Goals: Praxis Within and Beyond the Classroom*, pp. 83–104. Singapore: Springer Nature Singapore, 2023.
Evans, Jennifer. "Giving voice to the sacred black female body in *takayna* country." In *Indigenous Perspectives on Sacred Natural Sites*, pp. 15–31. Routledge, 2019.
Farmilo, Kathleen. "These progressive No campaigners are looking beyond the vote. Here's what they want". *SBS News*. October 7, 2023. https://www.sbs.com.au/news/the-feed/article/these-progressive-no-campaigners-are-looking-beyond-the-vote-heres-what-they-want/tdyj2ilx6. (accessed January 22, 2024).
Fredericks, Bronwyn, and Abraham Bradfield. "Indigenous recognition and constitutional reform in Australia: The activism of 1967 and aspirations of 2023." *Journal of Australian Indigenous* Issues 26, no. 1-2 (2023): 31–36.
Gammage, Bill. 2011. *The biggest estate on earth*. Crows Nest. Allen & Unwin.
Hayward, Matt. "David Attenborough's 'witness statement' for the planet (commentary)". *MONGABY*. October, 27 2020. https://news.mongabay.com/2020/10/david-attenboroughs-witness-statement-for-the-planet-commentary/.
Falk, Phillip, and Gary Martin. "Misconstruing Indigenous sovereignty: Maintaining the fabric of Australian law." In *Sovereign subjects*, pp. 33–46. Routledge, 2020.
Jackson, Mark. "On decolonizing the Anthropocene: Disobedience via plural constitutions." *Annals of the American Association of Geographers* 111, no. 3 (2020): 698–708.
Langton, Marcia, and Aaron Corn. *First Knowledges Law: The Way of the Ancestors*. Thames & Hudson Australia, 2023.
Langton, Marcia. "What do we mean by wilderness?: Wilderness and terra nullius in Australian art [Address to The Sydney Institute on 12 October 1995]." *Sydney Papers, The* 8, no. 1 (1996): 10–31.
Lee, Emma. "Protected areas, country and value: The nature–culture tyranny of the IUCN's protected area guidelines for Indigenous Australians." *Antipode* 48, no. 2 (2015): 355–374.
Magann, Kathryn. "Indigenous voice campaigner lash referendum 'no' vote". *National Indigenous Times*. October 23, 2023. https://nit.com.au/23-10-2023/8268/indigenous-voice-campaigners-lash-referendum-no-vote (accessed January 22, 2024).
Malaspinas, Anna-Sapfo, Michael C. Westaway, Craig Muller, Vitor C. Sousa, Oscar Lao, Isabel Alves, Anders Bergström et al. "A genomic history of Aboriginal Australia." *Nature* 538, no. 7624 (2016): 207–214.
Mansell, Michael. "Australians and Aborigines and the Mabo Decision: Just Who Needs Whom the Most." *Sydney L. Rev.* 15 (1993): 168.

Mansell, Michael. "Finding the foundation for a treaty with the Indigenous peoples of Australia [Paper in: Treaty-Let's Get It Right.]." *Balayi: Culture, Law and Colonialism* 4, no. 2002 (2002): 83–89.

Matthews, Daniel. "Reframing sovereignty for the Anthropocene." *Transnational Legal Theory* 12, no. 1 (2021): 44–77.

McBride, Laura. "Always was, always will be, Aboriginal land". *Australian Museum*. July 9, 2021. https://australian.museum/learn/first-nations/always-will-be-aboriginal-land/.

Meyers, Gary D., and John Mugambwa. "The Mabo decision: Australian Aboriginal land rights in transition." *Envtl. L.* 23 (1993): 1203.

Moreton-Robinson, Aileen. "Dr. Aileen Moreton-Robinson "Aboriginal sovereignty, Foucault, and the limits of power"." *Youtube*. 1:34:10. January 14, 2021. https://www.youtube.com/watch?v=nN5zwy2Y8AY.

Moreton-Robinson, Aileen. "Our story is in the land: why the Indigenous sense of belonging unsettles white Australia." *ABC religion & Ethics*. November 9, 2020. https://www.abc.net.au/religion/our-story-is-in-the-land-indigenous-sense-of-belonging/11159992.

Moreton-Robinson, Aileen. "*Sovereign Subjects: Indigenous Sovereignty Matters.*" (2007).

Moreton-Robinson, Aileen. *The white possessive: Property, power, and indigenous sovereignty*. U of Minnesota Press, 2015.

Morse, Callan. "Beetaloo Traditional Owners at Odds with NT government as fracking moratorium is lifted". *National Indigenous Times*. May 3, 2023. https://nit.com.au/03-05-2023/5824/traditional-owners-of-the-beetaloo-basin-at-odds-with-NT-government-as-fracking-moratorium-is-lifted (accessed January 29, 2014).

Nakata, Martin. "An Indigenous standpoint theory." In *Disciplining the savages: Savaging the disciplines*, pp. 213–217. Canberra, ACT: Aboriginal Studies Press, 2007.

Nash, Roderick Frazier. *Wilderness and the American mind*. Yale University Press, 2014.

Neidjie, Bill, Stephen Davis, and Allan Fox. "Kakadu man… Bill Neidjie." *(No Title)* (1985).

Neidjie, Bill, and Keith Taylor. "Story about feeling." *(No Title)* (1989).

Ngurra, Darug, Uncle Lex Dadd, Paul Glass, Aunty Corina Norman-Dadd, Paul Hodge, Sandie Suchet-Pearson, Marnie Graham, Sara Judge, Rebecca Scott, and Jessica Lemire. "Yanama Budyari Gumada, walk with good spirit as method: co-creating local environmental stewards on/with/as Darug Ngurra." *Located Research: Regional places, transitions and challenges* (2020): 15–37.

Oscar June. "Social Justice Commissioner says Indigenous students are facing increased racism after referendum result". NITV. October 25, 2023. https://www.sbs.com.au/nitv/article/social-justice-commissioner-says-indigenous-students-are-facing-increased-racism-after-referendum-result/usmowsgwl.

Paradies, Yin. "Unsettling truths: Modernity, (de-) coloniality and Indigenous futures." *Postcolonial Studies* 23, no. 4 (2020): 438–456.

Pascoe, Bruce. *Dark Emu: Aboriginal Australia and the birth of agriculture*. Magabala Books, 2018.

Pearson, Noel. "In full: Noel Pearson address to the National Press Club of Australia ABC News". *Youtube*. 1:12:45. September 27, 2023. https://www.youtube.com/watch?v=5aDiWiDOU7g.

Perdrisat, Marlikka. "Keynote – Marlikka Perdrisat 2021 AIATSIS summit keynote day 5 plenary – country". 15:25. June 4, 2021. https://aiatsis.gov.au/presentation/keynote-marlikka-perdrisat.

Poelina, Anne. "Holding the fire: episode 2. Ancient wisdom with Anne Poelina". 56:19. October 9, 2023. https://www.resilience.org/stories/2023-10-09/holding-the-fire-episode-2-anne-poelina/.

Poelina, Anne, Kathrine S. Taylor, and Ian Perdrisat. "Martuwarra Fitzroy River Council: an Indigenous cultural approach to collaborative water governance." *Australasian Journal of Environmental Management* 26, no. 3 (2019): 236–254.

Poelina, Anne, Wabi, Jeannie and Marshall, Lucy. "Three Sisters Women of High Degree". *Vimeo*. 47:04. December 4, 2015. https://vimeo.com/147866161.

Rasmussen, Morten, Xiaosen Guo, Yong Wang, Kirk E. Lohmueller, Simon Rasmussen, Anders Albrechtsen, Line Skotte et al. "An Aboriginal Australian genome reveals separate human dispersals into Asia." *Science* 334, no. 6052 (2011): 94–98.

Ray, Lana. "Deciphering the "Indigenous" in Indigenous methodologies." *AlterNative: An International Journal of Indigenous Peoples* 8, no. 1 (2012): 85–98.

Rose, Deborah Bird. "Land rights and deep colonising: The erasure of women." *Aboriginal Law Bulletin* 3, no. 85 (1996): 6–13.

Saldanha, Arun. "A date with destiny: Racial capitalism and the beginnings of the Anthropocene." *Environment and Planning D: Society and Space* 38, no. 1 (2020): 12–34.

SEED. "Our story". SEED. Accessed February 31, 2023. https://www.seedmob.org.au/our_story.

tebrakunna country and Lee, Emma, Benjamin J. Richardson, and Helen Ross. "The 'Uluru statement from the heart': Investigating Indigenous Australian sovereignty." *Journal of Australian Indigenous Issues* 23, no. 1–2 (2020): 18–41.

Thorpe, Lidia. "Blak Sovereign Movement" Lidia Thorpe Senator for Vitoria. June 20, 2023. https://www.lidiathorpe.com/blak_sovereign_movement.

Torre, Giovanni. "Community mourns loss of Mr Wilson, champion of country and the rights of traditional owners". *National Indigenous Times*. September 13, 2023. https://nit.com.au/13-09-2023/7662/community-mourns-loss-of-mr-wilson-champion-of-country-and-the-rights-of-traditional-owners (accessed January 22, 2024).

Uluru Statement from the Heart. "The statement". *The Uluru Statement*. Accessed December 23, 2023. https://ulurustatement.org/the-statement/view-the-statement/.

Walker, Mike, Sigfus Johnsen, Sune Olander Rasmussen, Trevor Popp, Jørgen-Peder Steffensen, Phil Gibbard, Wim Hoek et al. "Formal definition and dating of the GSSP (Global Stratotype Section and Point) for the base of the Holocene using the Greenland NGRIP ice core, and selected auxiliary records." *Journal of Quaternary Science: Published for the Quaternary Research Association* 24, no. 1 (2009): 3–17.

Watego, Chelsea. *Another day in the colony*. Univ. of Queensland Press, 2021.

Waters, Colin N., and Simon D. Turner. "Defining the onset of the Anthropocene." *Science* 378, no. 6621 (2022): 706–708.

XR. "What is XR". Extinction Rebellion. Accessed January 22, 2024. https://rebellion.global/about-us/.

Yunupingu, Galarrwuy. *Short Black 12 Tradition, Truth and Tomorrow*. Vol. 12. Black Inc., 2015.

Yunupingu, Galarrwuy. "Welcome address YYF Chair Galarrwuy Yunupingu AM". *Youtube*. 14:16. August 3, 2019. https://www.youtube.com/watch?v=XU2NSP1cDS8.

Yunupingu, Djawa, and Samantha Muller. "Cross-cultural challenges for Indigenous Sea country management in Australia." *Australasian Journal of Environmental Management* 16, no. 3 (2009): 158–167.

10 When Coyote Stole Rabbit's Heart

O'odham Himdag, Environmental Sovereignty, and the End of the American Empire

David Martínez

The western discourse on ethics is a discourse on how to manage human behavior, which is based on theories of human nature, society, and morality. Plato's Idea of the Good, Kant's moral imperative, Mill's greatest good for the greatest number, and Nietzsche's transvaluation of values are highlights in the history of western ethical thinking. Through these works, and others like them, readers are challenged to think about the definition of key concepts (Good, Evil, Power), reflect on the meaning of moral principles (know thyself, act only if you will it to become universal law), as well as contemplate a world without meaning (God is dead). For many working in the western intellectual tradition, the critical thinking intrinsic to philosophical analysis sits in contradistinction to two extremes: the amoral universe of science, which is bereft of human agency, and the dogmatic cosmos of religion—namely, Christianity—in which human will is limited to orthodoxy. Philosophy defies the tenets of conventional thinking by asking questions that set the norms of society on edge, such as when Socrates disturbed the minds of Athens' youth, which led to his legendary sacrifice, as recounted in the *Phaedo*. Nonetheless, Socrates notwithstanding, philosophy is a literary genre, developed over time by white men who valued the idiosyncrasies, not of dialog and communitas, but of self-reflection, objectivity, and critique. Philosophy, moreover, as a written tradition, lends itself to the musings of the inner voice. In a sense, every philosopher is Descartes, exploring the limits of the cogito, searching for the boundaries between knowing and unknowing, in which even the existence of "God" can be doubted—at least hypothetically. But what is the cogito anyway but another social convention, a philosophical trope, spun from the Cartesian imagination? Descartes may be revered in the western philosophical tradition, but what did he know beyond the French Catholic world in which he lived? For that matter, what did Plato, Kant, Mill, and Nietzsche know about the world around them that was not European? What did they know that did not come from a book? Equally important, how do such lives and writings look to someone who does not identify as a European, or with any facet of western civilization, including the primacy of literacy, but rather as an Indigenous person whose radical

DOI: 10.4324/9781003385202-11

doubt puts the status of western civilization into dispute? What becomes of the discourse on ethics? Are we not now without a moral compass? Are we not mere savages without the Bible and Aristotle? My questions are meant to be ridiculous.

The reference here to the Akimel O'odham—whom the Spanish called "Los pimos" and their homeland "Pímeria alta," and which the Americans called "Pima Indians," once they occupied "Arizona"—is not arbitrary. Aside from the author's identity, Akimel O'odham, which translates as River People, and their jeved, or homeland, evokes a world beyond the reach of western theory. As Europe entered the Late Middle Ages, when the Continent was besotted with the Black Death, the Nahua, or Aztecs, in southern Mexico constructed their ancient capital, Tenochtitlan, complete with writing the codices that formed the basis of their civilization. In turn, ancestral O'odham, or Huhugam, entered what archeologists labeled the Classical Period,[1] during which sisivaní, chief priests, built great houses, ge' edaj kikih, along the akimel, the Gila River, and the onk akimel, the Salt River, such as the one at Casa Grande Ruins.[2] For the O'odham, Christianity and the Spanish Empire rode in with Eusebio Kino, a Jesuit priest of German descent, in 1694, to establish a mission system in the Pímeria alta, which was meant to parallel the one in Alta California. As anthropologist Eric R. Wolf might have said, the Akimel O'odham appeared in Kino's travel journals as a "people without history."[3] However, the racialized biases of anthropologists, historians, and missionaries need not be recounted here. What is important to note is that when the Spanish entrada arrived in Akimel O'odham land, the O'odham did not enter western history, but, rather, the Spanish entered the O'odham oral tradition.

According to Frank Russell, in *The Pima Indians*, when Kino met the Indigenous people living along the Gila River, he saw the land as terra incognita: "In 1694 [Kino] reached the Casa Grande [ruins] in company with native guides [likely Tohono O'odham][4] who had informed him of the existence of [this place]. Absolutely nothing is known about this expedition except that a mass was said within the walls of Casa Grande."[5] According to O'odham oral history, the name of "Pimas" was derived from the above encounter.

> The oldest reference to the name within the writer's knowledge is that by [Father Luís] Velarde: "The Pima nation, the name of which has been adopted by the Spaniards from the native idiom, call themselves Otama or in the plural Ohotoma; the word pima is repeated by them to express negation." This "negacion" is expressed by such words as pia, "none," piatc, "none remaining," pimate, "I do not know" or "I do not understand."[6]

Since Kino traveled northward from the Altar River, he inevitably encountered O'odham villages along the Santa Cruz River, which placed him in Tohono O'odham and Sobaipuri O'odham lands, entailing that Kino asked these O'odham their name before reaching Casa Grande.[7] Also, as a Jesuit, Kino obliged himself to learn O'odham ne'oki, though he was probably not fluent, hence the historic misunderstanding.[8] As for the O'odham name for the Spaniards, juhkam, it evokes a people arriving from the south, so is also applied to Mexicans. Milgahns, it should be noted, is O'odham for Americans, who also knew little about the O'odham jeved or the hemajkam, the people, who live here.[9]

With respect to the question, what becomes of ethics in the absence of western philosophy and religion, the implication of Spanish-O'odham history is that one must ask *Whose land is it?* before initiating a discourse on environmental ethics. In the case of the Sonoran Desert, one has to acknowledge it as ancestral O'odham homelands. The division of the O'odham Jeved between two settler colonial nations, the United States and Mexico, is the consequence of three centuries of colonization. Therefore, an environmental ethics must not only begin with an acknowledgement that the O'odham Jeved is occupied, but also the restoration of O'odham land relations entails decolonization. More exactly, O'odham communities must pursue—and non-O'odham must support, as a moral imperative—more environmental sovereignty over their homelands, especially sacred sites.

What is commonplace among Indigenous peoples is the notion that creation stories consist of teachings, and that these teachings are meant only for the people to whom these teachings are handed down from one generation to the next. Environmental ethics, as a nonindigenous intellectual field, be they applied or theoretical, tend to be premised on the assumption that what is articulated in principle is universal, meaning relevant to all humans, regardless of the differences that distinguish human communities. We all need drinkable water. We all need breathable air. We all need food security. Such prepositions lend credibility to the presumption that environmental ethics is global. After all, does not the global warming crisis necessitate a global response? An Indigenous environmental ethic, however, based on local land relations and cultural exclusivity sounds counter-intuitive.[10] Yet, insofar as global warming is the consequence of colonization,[11] which is also global,[12] then is it not sensible that the antidote is anticolonial, site-specific, and Indigenous?

When Kino visited Casa Grande in 1694, he joined others who pondered the origins of the great adobe houses.[13] Europe, of course, was home to a plethora of ruins. Some stood as monuments to well-documented civilizations, such as the Parthenon; while others, like Stonehenge, evoked an unknown people lost in the mist of history. Both known and unknown ruins, equally, marked the decline of an historical epoch and collapse of the civilization that once flourished where now only a few stones remain. So, what about Casa Grande Ruins?[14] An unfortunate consequence of the teleological theory of history is the presupposition that once a civilization declines it can never be reborn, especially if overtaken by a more powerful nation. According to Russell, O'odham queried about Casa Grande stated "However ready they may have been in the past to claim relationship with the [Huhugam] or relate tales of the supernatural origin of the pueblos, they now frankly admit that they do not know anything about the matter."[15] Curiously, Russell cites JR Bartlett, a member of the US Boundary Survey Commission, to corroborate his assertion:

> [Casa Grande], as well as the ruins above the Pimo villages, are known among the Indians as the "houses of Montezuma," an idea doubtless derived from the Mexicans.... We asked our Indian guide who Montezuma was. He answered, "Nobody knows who the devil he was; all we know is, that he built these houses."[16]

Was Montezuma a synonym for Siuuhuu, Elder Brother? While Russell noted that "Hohokam" referred to "the race that occupied the pueblos that are now rounded heaps of ruins in the Salt and Gila river valleys," he failed to comprehend the great houses as integral to the story of Siuuhuu, Elder Brother—also known as I'itoi—who taught the O'odham their himdag, their way of doing things. Indeed, generations of archeologists, from Jesse Walter Fewkes to Paul H Ezell disputed the relationship between the "Hohokam" and the "Pima," dismissing O'odham oral tradition as unscientific.[17]

Thin Leather, Komal Hok, was a mahkai, a healer or medicine maker, whom Russell respected as "intelligent, patient, dignified," who provided "the cosmogonical myth of the tribe, many speeches, songs, and much general information [about the O'odham himdag]."[18] Thin Leather also shared the creation story with J William Lloyd in *Aw-aw-tam Indian Nights*, in addition to serving as a consultant for Fewkes's archeological report on Casa Grande.[19] Although other O'odham mamakai, medicine people, and kekelimai, elders, have shared their knowledge of the creation story, Thin Leather is distinct for sharing the whole epic in two historically important sources.[20] Although Russell recorded the O'odham language in an obsolete, not to mention difficult to read, orthography, which he used to translate O'odham narratives into early-twentieth-century English,[21] as a record of Thin Leather's knowledge of the O'odham himdag, Russell's Bureau of American Ethnology report remains a significant source. The same is true for Lloyd, who, unfortunately, did not preserve a transcription of Thin Leather's O'odham neoki narrative. Instead, Lloyd concentrated on writing O'odham words phonetically (e.g., Ee-ee-toy for I'itoi), for the benefit of the general reader. Having said that, O'odham is a living language, so it is worth mentioning that the Akimel O'odham creation story persists today as a source of knowledge.[22] There are still mamakai availing their medicine to people suffering from illness. There are still kekelimai guiding the young and not so young. There are O'odham working in the fields of education, art and literature, music and dance, and other media that give expression to O'odham voices. And there is at least one O'odham, the author of this paper, who has a degree in philosophy. As hehemajkam, as ordinary people, O'odham are a part of the contemporary world. So too is the O'odham jeved. In which case, Thin Leather's legacy is more than historical; it evokes a himdag that endures today, even under layers of Milgahn colonization. In which case, the wia'i, ruins, that were "left behind" by the Huhugam, ancestral

O'odham, at places like Sivan Vahki, or Casa Grande, are not the vestiges of a fallen civilization—or "that which has perished," which was Haury's translation of "Hohokam"[23]—but the landmarks of a contemporary O'odham world.

Prominent among revered places defining the Akimel O'odham jeved is Muahudag Doag, Greasy Mountain, which the Americans christened "South Mountain." This is where I'itoi ki, Elder Brother's house, is located. I'itoi ki is commonly known as the Man-in-the-Maze symbol.[24] Each O'odham community regards a different mountain as where Elder Brother's house can be found. Tohono O'odham point to Waw Giwulk, Constricted Rock, or Baboquivari Peak. Hia-Ced O'odham regard Chuk Doag, Black Mountain, or Sierra Pinacante, as I'itoi's home. As for the Sobaipuri O'odham, it is unclear where they place I'itoi ki.[25] Our focus here is Greasy Mountain, in the heart of the Akimel O'odham jeved. When Thin Leather recounted the creation story, he lived in Ge'e Kih, Big House, or Sacaton Village, north of the Sacaton Mountains, along the south bank of the akimel, the Gila River. According to Thin Leather, after Jewed Mahkai cleansed the land of the first people, who turned on each other when they over populated their jeved,[26] they[27] made all things anew, including the first people, s*osanac.[28] However, not all was as before. Springing from tohav, white brittlebush, was Bán, Coyote, a trickster who embodies childish impulses and playfulness.[29] Also springing from the earth was I'itoi, later known as Siuuhuu, a powerful ma:kai whose powers challenged that of Jewed Mahkai I'itoi, in fact, declared to Jewed Mahkai that he would "destroy" the s*osanac.[30] In a sense, Coyote and Elder Brother mark a change from one world to the next.[31] As such, Elder Brother's appearance suggests that he is here to fulfill a destiny embedded in Jewed Mahkai's creation. More to the point, what the first people lacked when Jewed Mahkai made them, I'itoi would provide, namely their himdag.

Toward this end, Siuuhuu created an unnamed "handsome youth," who went to the Akimel O'odham and "wed" with whomever he pleased.[32] The unnamed youth impregnated women with children whose gestation was each shorter than the previous.[33] Eventually, a young man appeared at birth. "This was the child that caused the flood which destroyed the people and fulfilled the plans of Elder Brother."[34] However, the flood would not strike until Siuuhuu completed the "jar or olla"[35] he made for himself. Jewed Mahkai gathered their people together and described "the calamity that would befall them."

> [Jewed Ma:kai then] thrust his staff into the ground.... Some of the people went into the hole, while others appealed to Elder Brother. Their appeals were not heeded ... Coyote ... was [then] told to find a big log and sit upon it.... Elder Brother got into his olla and closed the opening by which he entered, singing in the meantime.[36]

After the flood, Jewed Mahkai saved themselves in their "reed staff,"[37] floating to a location "somewhere in the east."[38] Bán, Coyote, wound up at "Driftwood Mountain," whose location is also unknown. Siuuhuu, on the other hand, "came to rest beyond Sonoita, near the mouth of the Colorado river. The olla [nawaitakud], now called Black mountain [Chuk Doag], may be seen there to this day."[39] Chuk Doag, as noted above, is where the Hia-Ced O'odham say that I'itoi ki exists.[40] After locating the center of the world, Siuuhuu, Bán, and Jewed Mahkai sat down to make the creatures to inhabit the land, which was still "wet and muddy." After Bán and Jewed Mahkai cast their creations into the water and to the west, respectively, as Siuuhuu instructed, Jewed Mahkai suddenly sank into the earth. When Siuuhuu grabbed hold of Jewed Mahkai he was not strong enough to keep them from disappearing, leaving Siuuhuu's hands covered in "blood and dirt." When Siuuhuu shook his hands clean the blood and dirt scattered, spreading disease "over the land and in the water."[41] Siuuhuu then flung the Apache, the O'ob, to the other side of the mountains where they now live, which made them angry. In turn, Siuuhuu divided the people into four tribes. In addition to the Apache, he made the "Wä-akĭ Âp … the Maricopas, and, lastly, the Pimas."[42] Not long afterward, Siuuhuu decided mysteriously "to do mischief [among the O'odham], such as marrying the young women and then deserting them for others. The people began to be jealous of him and planned to destroy him."[43] At this juncture, the Akimel O'odham have learned how to build canals.[44] In fact, the last time Siuuhuu helped the people was when he assisted the Piipaash, the Maricopas, with building canals. What this means is that the people that wanted to destroy Siuuhuu were Huhugam, ancestral O'odham, dwelling in the big houses like Casa Grande.

Siuuhuu's mischief began with giving rattlesnake, ko'oi, their fangs, which they used to bite rabbit, chuhwi, which caused the first death. Moreover, it was during rabbit's cremation that Bán stole rabbit's heart, or ibdak, and ran away to Komadk Doag, Grassy Plains Mountain, which the juhkam named Sierra Estrella. When people went after him, Bán fled to Muahudag Doag:

> Again the people overtook Coyote, and he ran northward across the Gila, where he ate [rabbit's] heart, and as he did so the grease fell upon every stone of the mountain, which accounts for its appearance and the name it bears to this day—[Muahudag], Greasy mountain.

Finally, when he had his fill, Bán ran away "to live in the sea in the south."[45] However, Bán would always return when there was trouble to be made. As Thin Leather further recalls the downfall of the big houses, Siuuhuu alternates between playing a direct and indirect role. Suffice it to say, people endure difficult lessons, such as when they learn to hunt and fend for themselves, which was made all the more difficult when Bán's randy behavior caused the game to be chased away. Also, a sivani, Feather-breathing, who had strong powers but was childlike, built a vahki along the onk akimel, the Salt River, where a gambler, "Väntre," came to live. Väntre's name is not O'odham, implying that he is not human. Indeed, Väntre begins preying on the people. Because of the severe threat, Siuuhuu intervened to eliminate the monster from terrorizing O'odham villages. "Elder Brother," Thin Leather recalls, "then went home and told the people how to conduct themselves when they had killed an enemy." Once Siuuhuu taught the Akimel O'odham the enemy purification ceremony, "the people about Baboquivari wished to have Elder Brother come to them."[46]

At this point, Thin Leather focuses on Casa Grande. During a women's field hockey, or toka,[47] game, a "little green lizard"[48] appeared at the sivani's daughter's feet, under which "green stones"[49] were found. Women began wearing these as jewelry, which sparked envy in Tas Namkam, Meet-the-Sun, who sent a parrot[50] to eat the stones. The people had never seen such a creature,[51] so fed it more stones. When the sivani, Sial Tcu-utak, Morning Green, became angry and lashed out, Meet-the-Sun retaliated by sending a young man with a football, songiwul, which he gave to the daughter. When she hid the ball up her dress, playing a trick on the young man, it went into her womb and impregnated her. "The people wished to destroy the child, because it had long claws instead of fingers and toes; its teeth were long and sharp, like those of a dog."[52] Ho'ok, as this child was called, began preying upon everything, including children. Siuuhuu, once again, intervened to save the people. "After killing [Ho'ok] Elder Brother made his home at Baboquivari for some time."[53]

In time, Siuuhuu began his mischief again, and in earnest. "He sang the song of the menstrual period and accompanied it by reviling the family of the young girls. At last the people could endure his pranks no longer and drove him away."[54] Like Bán, when he stole rabbit's heart, Siuuhuu fled to Muahudag Doag. Four times, the people killed him. Four times, Siuuhuu revived. So, the people called a council, where they determined to seek Nuwi's help, the vulture ma:kai that lives in Komadk Doag.

> Vulture was a man who transformed himself into a bird with his own magic power and had gone through the openings in the sky and thus saved himself from destruction during the flood. After he came down from the sky he wandered about the country and finally built a va'-aki, magic house, the ruins of which yet remain, south of where Phoenix now stands, between the Gila and Salt rivers.[55]

Nuwi used his medicine to compel the sun to overheat Siuuhuu's house like a volcano. When the calamity was over, Siuuhuu was little more than a skeleton. However, yet again, Siuuhuu rose, alarming the people who wanted him slain. "Elder Brother sank through the earth and found the people that Earth Doctor [Jewed Mahkai] had assisted to reach that side in order to escape the flood. Elder Brother told the people there of his ill treatment and asked them to come through and fight with him and to take the land away from the Indians."[56] Thus, the decline and collapse of the big houses began.

Casa Grande was the first to fall. Then the big house at Santan.[57] From there, Siuuhuu's army went to Sweatwater.[58] At each village, Siuuhuu's allies sand a song. At Casa Blanca,[59] Siuuhuu's warriors sang of their concern at confronting such a powerful place: "It will be difficult, It will be difficult. To capture this pueblo. With its magic power." Each big house has a sivani, a chief priest. "[Chukchu Tadai Sivani at Casa Blanca] was the most powerful of all the chiefs who ventured to oppose [Siuuhuu's army]."[60] Thin Leather then describes an epic battle between Chukchu Tadai, Black Roadrunner, and Siuuhuu, in which each controlled natural phenomena like rain and wind. In the end, the Casa Blanca sivani was defeated. And in his heart, the victors found a small green stone, like the ones that Meet-the-Sun coveted. The next battle was against Nuwi. Siuuhuu headed to Gila Crossing. When Nuwi was captured, Siuuhuu scalped him, leaving the bald head that distinguishes him today. Afterward, big houses in Mesa and Tempe (Oidbad) also fell. Eventually, the fighting stops and Siuuhuu vanishes from the narrative.[61] Thin Leather concludes with an anecdote about a young man searching for food in the mountains near the Salt River valley for his pregnant wife, who "would eat nothing but green plants and game found in the mountains." While out there, far from home, the man was surrounded by Apaches and killed. When his remains were recovered, like rabbit before Bán stole his heart, "they [were] burned to ashes." One day, the young man's son, "Ka'kanyip," will avenge his father's death.[62]

In the aftermath of the Ge'edaj kikih, the Akimel O'odham emerge as a people of the olas kih, which is the small roundhouse that predominated across the O'odham jeved until the Americans' arrival in the 1850s.[63] The himdag that Siuuhuu embodied is adapted anew to the sacred places that Jewed Mahkai made at the time of creation, which were cleansed multiple times. Now, the hehemajkam, the people, without sisivani, sowed their fields, maintained their villages, and gathered for the wiigiida, the saguaro cactus harvest ceremony, which beckons the jegos, the summer monsoons to rejuvenate the O'odham jeved.[64] The cactus fruit, b'ahi:daj, which is turned into ceremonial wine, nawait, symbolizes this renewal. What Thin Leather told Russell and Lloyd teaches that the civilization the sisivani built collapsed because they rejected their himdag. Instead, they killed Siuuhuu. So, Siuuhuu sank into the earth and sought Jewed Ma:kai's people to assemble an army that would rid the land of the sisivani, which Siuuhuu's "mischief" exposed as corrupt. The people of the olas kih, of course, can still lose their way and stray from their himdag. Or, more exactly, they can violate the himdag of other beings, places, and events, causing these things to throw their power, their gewkdag, into them, becoming staying sickness, kacim mumkidag. When Siuuhuu scattered the blood and dirt from Jewed Ma:kai's arms, he infected the land with powers that oblige O'odham to treat their jeved, and everything in it, with respect, lest they contract something harmful. When illness occurs, mamakai, medicine people, can restore O'odham to health. Seeking a cure, then, when one is ill is also part of the himdag.

Today, the O'odham jeved is ill again,[65] and has been since American settlers stole the akimel from the hemajkam only a few years after they were placed on a reservation in 1859.[66] For decades, Indian agents pressured O'odham to abandon their himdag and take up the ways of Americans, who called themselves a Christian nation. When Charles H Cook—US Army veteran and, like Kino, of German descent—rode into Sacaton Village, which was home to the "Pima Agency," on December 23, 1870, he did so at the behest of the Board of Indian Commissioners,[67] which oversaw "Indian civilization," at a time when the Gila River community was undergoing a crisis due to severe water shortages caused by up-river settler damming.[68] The Presbyterian mission did not see itself as causing the water crisis, but rather the Army. Instead, the mission perceived itself as a humanitarian intervention motivated by the Gospel. Cook stated: "It was not until several months after I reached … Sacaton that I learned [of] others beside myself, who were anxious to have the gospel and christian [*sic*] civilization brought to a people, who are perishing for want of it."[69] In other words, Cook saw his work as charitable.

Nonetheless, the Presbyterian Church, as an agent of federal Indian policy, was quick to express disdain for the hehemajkam that clung too closely to their O'odham himdag, especially mamakai. Cook's colleague and contemporary, Reverend Isaac T Whittmore, in particular, did not hesitate to despise medicine men that he regarded as obstacles. "Ever since the missionary [Cook] began work here [at Gila River]," Whittmore recounted, "these … medicine men have been an annoyance and hindrance to his work."[70] Ultimately, *Among the Pima* is about how Cook's congregation grew at the expense of the O'odham himdag, complete with churches at Sacaton, Casa Blanca, Bapchule, Gila Crossing, and Maricopa.

> Many [Akimel O'odham] come regularly to church a distance of from two to twenty miles, and not a few twenty or thirty miles. In summer, when churches in town are closed from the intense heat, these overflow with a multitude who are attracted, not by the eloquence of the preacher or by the exquisite rendering of chants by a well trained [*sic*] choir, or soul-stirring peals of the organ, but from pure love to God and delight in the service of preaching, prayer and praise. This influence on a people just emerging from heathenism and breaking up old superstitions and vices, and instead of them, leading an industrious and virtuous life, must far exceed that of churches in town on a civilized people.[71]

While the Gila River Indian Community has gained political sovereignty, including water rights, over the generations,[72] their freedom is still limited by the same settler-colonial power—the United States—that drew the reservation boundary in 1859.[73] As stated above, one cannot articulate an Indigenous environmental ethics without first acknowledging that the land is stolen and occupied. Only the colonizers can afford to be color-blind, ahistorical, and transcendental. It is time, then, for Siuuhuu to assemble his army once again and cleanse the O'odham jeved.

Restoring the O'odham Himdag, however, is not something that the Akimel O'odham would will to become a universal law, but rather is limited to O'odham. Like other Indigenous groups, the O'odham do not proselytize nor expect other people to adopt our way of doing things. The O'odham Himdag is not a religion. In western civilization, religion is an institution governed by a set of beliefs oriented around a supreme being whose will for the people he created is expressed in a sacred text. As a text-based institution, religions—as opposed to being religious—are defined by the rules, beliefs, and missions that distinguish them from one another. In the case of the Presbyterian Home and Foreign Mission work, the goal of the Church in North America was to bring the gospel to the Indians, while organizing the frontier into presbyteries. Cook, in *Presbyterian Home Missions* by Sherman H Doyle, is recognized with the highest esteem:

> [Cook's] success has been marvelous. "Fourteen hundred Indians baptized is a fine record. Old men and young, mothers and grandmothers, warriors and medicine men, with children claimed in covenant, are written by name in his book of baptism." [citation omitted] Five church buildings have been erected under his care and through him thousands have heard the gospel, and hundreds been developed in Christian life and service.... What a record for one man. "Father" Cook is a great man, a patriarch among thousands of people, welcome in many villages.[74]

Even today, Presbyterian Mission is guided by influencing people's faith, beliefs, and values.[75] With respect to Cook's legacy, while the "apostle to the Pimas"[76] passed away long ago, in 1917, the role he played at disrupting the O'odham Himdag lives on in O'odham language declination and cultural loss. Consequently, an O'odham environmental ethic is not about encouraging non-Indigenous people to appropriate the O'odham Himdag, but about creating the cultural and political space in which the Himdag can be revitalized.

In addition to acknowledging that the O'odham Jeved is expropriated—which the United States seized from Mexico in its illegal 1846–1848 war of aggression—an environmental ethic appropriate to this jeved must be premised on a nonlinear, nonwestern concept of history, in which progress, manifest destiny, and the Anthropocene does not guide our thinking about the needs of this place. What American society, on the other hand, must will to be a universal law is the revitalization of Indigenous communities and cultures. Discourses on land management, conservation, and ecological restoration must open to an O'odham-centric worldview, in which O'odham can exercise their communal values. As such, America, as a colonizing nation, can do the greatest amount of good for the greatest number of Indigenous people. An anticolonial environmental ethic will enable Indigenous peoples to recover their original land relations. For the O'odham, this means reclaiming their relationship with Muahudag Doag, Waw Giwulk, and Chuk Doag. *The O'odham Himdag is not a religion, but higher than religion*. Whereas, in the western tradition, a religion is a man-made institution, and thereby prone to moral corruption, the O'odham Himdag springs from the jeved, the earth, just as Siuuhuu, Elder Brother, sprung from the earth during the time of creation. Only when the akimel and onk akimel are free from the "big houses" that the Americans have built on ancestral O'odham land, from their churches to their great cities, will our Himdag be back in balance.

Notes

1 See William W Wasley and David E Doyel, "Classic Period Hohokam," Kiva, Vol 45, No 4 (Summer, 1980):337–352.
2 See National Park Service, "Casa Grande Ruins National Monument": https://www.nps.gov/cagr/index.htm (accessed May 15, 2023).
3 For more about this concept, see Eric R Wolf, *Europe and the People Without History*, second edition (Berkeley: University of California Press, 2010).
4 The O'odham Jeved is divided into four discrete groups, namely Akimel O'odham, Tohono O'odham, Sobaipuri O'odham, and Hia-Ced O'odham. Each historically occupied a different area of the Sonoran Desert. See Native Land Digital, "O'odham Jewed": https://native-land.ca/maps/territories/tohono-oodham/ (accessed May 8, 2023).
5 Frank Russell, *The Pima Indians* (Tucson: The University of Arizona Press, 1975): 27. See also Anna Moore Shaw, *A Pima Past* (Tucson: The University of Arizona Press, 1974): 4, for an account of Kino's mass.
6 Russell, *The Pima Indians*, 19. Russell is referring to Padre Luis Velarde's *Relación* (1716). For more, see Rufus Kay Wyllys (editor), "Padre Luis Velarde's *Relación* of *Pimería Alta, 1716*," New Mexico Historical Review, Vol VI, No 2 (April, 1931): 111–157. For an O'odham synonymy, see Bernard L Fontana, "Pima and Papago: Introduction," *Handbook of North American Indians, Volume 10, Southwest*, Alfonso Ortiz (volume editor) (Washington, DC: Smithsonian Institution, 1983): 134. See also Ross Shaw, "Training an Indian Boy," *The Native American*, Vol 17, No 12 (June 10, 1916): 208; William A Aery, "Sixteenth Annual Founder's Day Services at Hampton Institute," The Southern Workman, Vol XXXIX, No 2 (February, 1910): 78, for anecdotes about how the O'odham phrase for "I do not know" became "Pima.".
7 Over the generations, because "Pima" became the historical name of the groups occupying the Gila and Salt River valleys, the story of Kino's confusion became a part of Akimel O'odham oral history. See Herbert Eugene Bolton, *Rim of Christendom: A Biography of Eusebio Francisco Kino, Pacific Coast Pioneer* (New York: Russell & Russell, 1960): 284–287; Eusebio Francisco Kino, *Kino's Historical Memoir of Pimería Alta*, edited by Herbert Eugene Bolton (Cleveland: The Arthur H Clark Company, 1919): 127–129.
8 Kino, *Pimería Alta*, 269–270.
9 See Edward H Spicer, *Cycles of Conquest: The Impact of Spain, Mexico, and the United States on the Indians of the Southwest, 1533–1960* (Tucson: The University of Arizona Press, 1962): 146–151.
10 Since the environmental movement began in the early 1970s, Indigenous people have had to respond to racially biased evolutionary theories that asserted that ancestral Indigenous people, ie "Paleo-Indians," were harmful to their environments, and that their "oneness with nature" is a modern myth. See Sam D Gill, *Mother Earth: An American Story* (Chicago: University of Chicago Press, 1991); Shepard Krech III, *The Ecological Indian: Myth and History* (New York: WW Norton & Company, 1999). For criticism, see Vine Deloria Jr, *Red Earth, White Lies: Native Americans and the Myth of Scientific Fact* (Golden, CO: Fulcrum Publishing, 1997).
11 See Sheila Watt-Cloutier, *The Right to Be Cold: One Woman's Fight to Protect the Arctic and Save the Planet from Climate Change* (Minneapolis: University of Minnesota Press, 2018).

12 See Marc Ferro, *Colonization: A Global History* (London: Taylor & Francis, 2005).
13 Among archeologists, the relation between "Hohokam" who built Casa Grande and the "Pimas" who now inhabit the area became known as the "Hohokam-O'odham Continuum," in which the Hohokam were presumed to be an extinct tribe. This persisted in the archeological literature despite O'odham assertions that the Huhugam were ancestral O'odham. See Chris Loendorf, *The Hohokam–Akimel O'odham Continuum: Sociocultural Dynamics and Projectile Point Design in the Phoenix Basin, Arizona* (Tucson: The University of Arizona Press, 2013); J Brett Hill, *From Huhugam to Hohokam: Heritage and Archaeology in the American Southwest* (New York: Rowman & Littlefield Books, 2018).
14 The ancestral ruins and artifacts of the Huhugam, or Ancestral O'odham, have been found across southern Arizona and Sonora, Mexico. See the Digital Archive of Huhugam Archaeology: https://daha.tdar.org/ (accessed May 8, 2023).
15 Russell, *The Pima Indians*, 23–24.
16 Russell, *The Pima Indians*, 24n*b*. For Barrett's full report, see John Russell Bartlett, *Personal Narrative of Explorations and incidents in Texas, New Mexico, California, Sonora, and Chihuahua* (New York: D Appleton & Company, 1854).
17 See Paul H Ezell, "Is There a Hohokam-Pima Culture Continuum?" American Antiquity, Vol 29, No 1 (July, 1963): 61–66.
18 Russell, *The Pima Indians*, 17n*a*. See also Bahr, *The Short Swift Time of the Gods on Earth*, 53–54n.
19 Jesse Walter Fewkes, "Casa Grande, Arizona," Bureau of American Ethnology, 28th Annual Report, 1906–1907, Smithsonian Institution (Washington, DC: Government Printing Office, 1912); J William Lloyd, *Aw-aw-tam Indians Nights* (Westfield, NJ: The Lloyd Group, 1911).
20 See David Martínez, "Pulling Down the Clouds: The O'odham Intellectual Tradition during the 'Time of Famine,'" The American Indian Quarterly, Vol 34, No 1 (2010): 1–32.
21 For example, Russell gendered words that are not gendered in O'odham (e.g., ma:kai). See also Bahr, *The Short Swift Time of the Gods on Earth*, 16–24.
22 I am deliberately citing Russell and Lloyd, not only to provide my reader resources that s/he can consult, but also to recognize Thin Leather as a historically important O'odham philosopher.
23 See Emil W Haury, *The Hohokam: Desert Farmers and Craftsmen, Excavations at Snaketown, 1964–1965* (Tucson: The University of Arizona Press, 1976): 5.
24 See Anna Moore Shaw, *Pima Indian Legends* (Tucson: The University of Arizona Press, 1968): 15–16.
25 Deni J Seymour is the leading authority on Sobaipuri O'odham historical culture. According to her, the Sobaipuri are a "subgroup" of Akimel O'odham, implying a relation with Muhuadag Doag. See "Unveiling Tucson's Namesake: The Sobaipuri O'odham Village of San Cosme del Tucsón," Journal of Arizona History, Vol 63, No 2 (Summer, 2022): 115–152. However, this author is skeptical of the categorization of subgroup. The Hia-Ced O'odham have, by comparison, been categorized as a subgroup of Tohono O'odham, which is misleading. Insofar as Seymour describes the Sobaipuri as a discrete community, it stands to reason that they believed that I'itoi ki could be found in a mountain, a doag, that was unique to their jeved, or homeland.
26 Thin Leather, in Russell's *The Pima Indians* (1908): "For a time they increased and over spread the earth until it became so populous that food became scarce and there was not sufficient water to supply their needs…Hungering, they began to kill one another and to eat human flesh. Earth Doctor pitied them in their extremity, but could devise no plan for relieving their distress, except to destroy all" (208).
27 Since mahkai, healer, is a gender-neutral term, I have chosen to refer to Jewed Ma:kai, Earth Medicine Maker, as they, as opposed to she/he.

28 The author wishes to thank linguistic anthropologist Mizuki Miyashita, University of Montana, for her assistance with deciphering Frank Russell's orthography. Email to author, dated July 3, 2017. Miyashita originally assisted with this perplexing question for my paper, "Earth Medicine Man Makes This Place: A Prolegomenon to an Akimel O'odham Environmental Ethics," which appeared in Natural Communions, Religion and Public Life, Volume 40, edited By Gabriel Ricci (New York: Routledge, 2019): 33–57.

29 See Amadeo M Rea, *Folk Mammalogy of the Northern Pimans* (Tucson: The University of Arizona Press, 1998): 193–199.

30 In his narrative for Lloyd, Thin Leather, according to Edward H Wood, who assisted with the translation, referred to the s*osanac as "new people." Moreover, in a differing account of the demise of original O'odham: "And now for a time the people increased till they filled the earth. For the first parents were perfect, and there was no sickness and no death. But when the earth was full, then there was nothing to eat, so they killed and ate each other," *Aw-aw-tam Indian Nights*, 29, 30.

31 Thin Leather makes it clear for Lloyd that Jewed Ma:kai created multiple worlds and first peoples prior to the flood, after which Siuuhuu made the Huhugam. See Lloyd, Aw-aw-tam Indian Nights, 27–35.

32 In the narrative for Lloyd, Thin Leather said: "And to this man whom he had made, Seeurhuh (whose other name was Ee-ee-toy) gave a bow & arrows, and guarded his arm against the bow string by a piece of wild-cat skin, and pierced his ears & made ear-rings for him, like turquoises to look at, from the leaves of the weed called quah-wool. And this man was the most beautiful man yet made," *Aw-aw-tam Indians Nights*, 36. For more about kuávul, wolfberry, see Amadeo M Rea, *At the Desert's Green Edge: An Ethnobotany of the Gila River Pima* (Tucson: The University of Arizona Press, 1997): 142–145.

33 According to Thin Leather, when Jewed Mahkai created another people after destroying the first for cannibalism, they too exhibited unusual traits: "And [Jewed Ma:kai], being now on the top of this fallen sky, again made a man and a woman, in the same way as before. But this man and woman became grey when old, and their children became grey still younger, and their children became grey younger still, and so on till the babies were gray in their cradles," *Aw-aw-tam Indian Nights*, 31.

34 Russell, *The Pima Indians*, 209.

35 The olla mentioned is a nawaitakud ha'a, the ceremonial vessel used to prepare the wine, nawait, for the Wiigiida, the harvest or rain ceremony. See Donald Bahr, *How Mockingbirds Are: O'odham Ritual Orations* (Albany: State University of New York Press, 2011): 57–67.

36 Russell, *The Pima Indians*, 210.

37 Common Reed or Carrizo that once grew in abundance along the banks of the Gila River, called "wahpk" (also vaapk). See Rea, *At the Desert's Green Edge*, 102–104. See also George Webb, *A Pima Remembers* (Tucson: The University of Arizona Press, 1959): 76–77.

38 Russell, *The Pima Indians*, 213. Given the Gila River flows west, from New Mexico to the Colorado River, Jewed Ma:kai's eastern destination remains clouded in earth's history.

39 Russell, *The Pima Indians*, 213.

40 See David Burckhalter and Paul Mirocha, "Hia-Ced O'odham Places," Journal of the Southwest, Vol 59, Nos 3–4 (Autumn-Winter, 2017): 515–519.

41 Russell, *The Pima Indians*, 214. What Siuuhuu spread were the powers or strengths, gegewkdag, that inhabit all beings, which, when something's himdag is violated, that being will throw its géwkdag into the violator. Thin Leather adds: "And he [I'itoi] washed himself in a pool or pond and the impurities remaining in the water are the source of the malarias and all the diseases of dampness," in *Aw-aw-tam Indian*

Nights, 49. See also Donald M Bahr, et al, *Piman Shamanism and Staying Sickness (Ká:cim Múmkidag)* (Tucson: The University of Arizona Press, 1974): 23–48.

42 Russell, *The Pima Indians*, 215. Thin Leather also describes the original tribes as playing together when trouble arose: "And they all took to playing together, and in their play they kicked each other as the Maricopas do in sport to this day; but the Apaches got angry and said: 'We will leave you and go into the mountains and eat what we can get, but we will dream good dreams and be just as happy as you with all your good things to eat.' And some of the people took up their residence on the Gila, and some went west to the Rio Colorado. And those who builded [*sic*] vahahkkees," in *Aw-aw-tam Indian Nights*, 49–50.

43 Russell, *The Pima Indians*, 215. Thin Leather adds to Elder Brother's tales of mischief: "[I'itoi] lived in the Salt River Mountain, which is called by the [O'odham] [Muhadagi], or the Brown Mountain, and whenever the girls had ceremonial dances because of their arrival at womanhood he would come and sing the appropriate songs. And it often happened that he would tempt these young girls away to his mountain, to be his wives, but after keeping them awhile he would grow tired of them and send them back." Moreover, "[I'itoi] would often shoot his hot arrows thru the [crop] fields, and wither up the growing things; and tho the people did not see him do this, they knew he was guilty, and they wanted to kill him, but they did not know how to do it," Aw-aw-tam Indian Nights, 125.

44 The Huhugam canal system is one of the wonders of the ancient world. See Arizona Museum of Natural History, "The Hohokam": https://www.arizonamuseumofnaturalhistory.org/plan-a-visit/mesa-grande/the-hohokam (accessed June 2, 2023).

45 Russell, *The Pima Indians*, 217. Thin Leather calls coyote "Toehahvs," or Tohavs, referring to the White Brittlebush, Bán's birthplace, *Aw-aw-tam Indians Nights*, 103-105. However, he does not tell the story of rabbit's death. See also Anna Moore Shaw, *Pima Indian Legends* (Tucson: The University of Arizona Press, 1968): 17–19. Similar to Thin Leather, Shaw omits Rabbit's death and cremation.

46 Russell, *The Pima Indians*, 221. In *Aw-aw-tam Indian Nights*, the gambler's name is rendered "Vaindah," in which no translation is provided. See Lloyd, *Aw-aw-tam Indians Nights*, 72–95. See also Bahr, *The Short Swift Time of the Gods on Earth*, 157–177.

47 See Webb, *A Pima Remembers*, 43–44.

48 If native to the O'odham Jeved, then either a Western Banded Gecko or Whiptails. See Arizona-Sonora Desert Museum: https://www.desertmuseum.org/books/nhsd_cloeonyx.php (accessed May 24, 2023).

49 May refer to turquoise. Chethagi means blue-green.

50 Parrots are also chehthagi. See Amadeo M Rea, *Wings in the Desert: A Folk Ornithologyof the Northern Pimans* (Tucson: The University of Arizona Press, 2007): 168–169.

51 Indicates the introduction of a nonindigenous species, likely Military macaw (*Ara militaris*). See Hohokam Archaeology Along the Salt-Gila Aqueduct, Central Arizona Project: Habitation sites on the Gila River (6 pts. in 3). United States: The Division, 1982. See also University of Michigan, Museum of Zoology, "Ara militaris, military macaw": https://animaldiversity.org/accounts/Ara_militaris/ (accessed May 25, 2023).

52 Russell, *The Pima Indians*, 222. Thin Leather tells an alternate version of Ho'ok's story in *Aw-aw-tam Indian Nights*, 99–102. See also Bahr, *The Short Swift Time of the Gods on Earth*, 141–150.

53 Russell, *The Pima Indians*, 224.

54 Russell, *The Pima Indians*, 224.

55 Russell, *The Pima Indians*, 225. Thin Leather tells an alternate version in *Aw-aw-tam Indian Nights*, 125–132. See also Bahr, *The Short Swift Time of the Gods on Earth*, 181–201; Rea, *Wings in the Desert*, 94–101.

56 Russell, *The Pima Indians*, 226.
57 See Gila River Indian Community, District 4: Santan: https://www.gilariver.org/index.php/districts/district-4-santan (accessed May 25, 2023).
58 See *infra* for "Sweet Water."
59 See Gila River Indian Community, District 5: Casa Blanca: https://www.gilariver.org/index.php/districts/district-5-casa-blanca (accessed May 25, 2023).
60 Russell, *The Pima Indians*, 227.
61 While not corroborated in Russell, I'itoi, who is often identified with Moctezuma, is believed to have left the O'odham Jeved, promising to return someday with a new era of peace and prosperity. See James S Griffith, *Beliefs and Holy Places: A Spiritual Geography of the Pimería Alta* (Tucson: The University of Arizona Press, 1992): 15–17. See also Wilton E Hoy. "Organ Pipe Cactus National Monument. Historical Research." Three Volumes: "Administrative, Early Period, and Frontier Period" (Ajo, Arizona: Organ Pipe Cactus National Monument, 1976): 136.
62 Russell, *The Pima Indians*, 229. See also Bahr, *O'odham Creation & Related Events*, 134–135.
63 George Webb observes: "Adobe houses were supposed to be more civilized than the old arrow-reed shelters. But the Pimas did not want to change. So the agency issued a wagon to any Pima family who would build and live in an adobe house," *A Pima Remembers*, 50.
64 See Russell, *The Pima Indians*, 347–352. See also Ruth Murray Underhill, *Singing for Power: The Song Magic of the Papago Indians of Southern* (Berkeley: University of California Press, 2021); Ruth M Underhill, et al, *Rainhouse and Ocean: Speeches for the Papago Year* (Tucson: The University of Arizona Press, 1997).
65 See Gila River Indian Community, Wildlife Program, Climate Change, Additional Information, "Gila River Climate profile": https://www.gricdeq.org/climate-change (accessed May 30, 2023).
66 President James Buchanan, White House, executive order, February 28, 1859 (11 Stat 401).
67 See Henry E Fritz, *The Movement for Indian Assimilation, 1860–1890* (Philadelphia: University of Pennsylvania Press, 2017).
68 See David H DeJong, *Stealing the Gila: The Pima Agricultural* Economy and Water Deprivation, 1848–1921 (Tucson: The University of Arizona Press, 2009).
69 Charles H Cook, et al, *Among the Pimas or, The mission to the Pima and Maricopa Indians* (Albany, NY: Ladies' Union Mission School Association, 1893): 31.
70 Cook, et al, *Among the Pimas*, 63.
71 Cook, et al, *Among the Pima*, 93.
72 Under the provisions of the 1934 Indian Reorganization Act (48 Stat 984; 25 USC 461 et seq), members of federally acknowledged tribes may organize into constitutional governments, which are empowered to manage the internal affairs of the community covered by its respective constitution.
73 See Paul H Ezell, "The History of the Pima," *Handbook of North American Indians*, Vol 10 Southwest, Alfonso Ortiz (volume editor) (Washington, DC: The Smithsonian Institution, 1983): 156–160.
74 Sherman H Doyle, *Presbyterian Home Missions* (Philadelphia: Presbyterian Board of Publication and Sabbath-School Work, 1902): 86–87.
75 See Presbyterian Church, USA, Presbyterian Mission, "Our Mission": https://www.presbyterianmission.org/who-we-are/our-mission/ (accessed May 13, 2023).
76 Both Cook and Kino have been accorded this sobriquet. See Minnie A Cook, Apostle to the Pima Indians: The Story of Charles H. Cook, the First Missionary to the Pimas (Jackson, MI: Omega Books, 1976) and Herbert Eugene Bolton, *The Padre on Horseback: A Sketch of Eusebio Francisco Kino, SJ, Apostle to the Pimas* (Chicago: Loyola University Press, 1986).

11 Ancient? Enduring? An Indigenous Lens on Time, Being, and Conversation

Phoebe Godfrey and Jacqline Wolf Tice

> There is no environment 'out there' that is separate from us.... Indigenous people are absolutely correct: we are born of the earth and constructed from the four sacred elements of earth, air, fire and water...
>
> —David Suzuki

Part I

As we begin our chapter, we, as co-authors and conveners of this atypical approach to academic participation, invite you to close your eyes and take a deep long relaxed breath. Breathe in and out, seated in your body, your lungs rising and falling like the waves of the ocean...

Conscious breathing is an awakening to our interconnectedness and interdependence with all life. Our breath is, in essence, the ultimate nonspecific 'land acknowledgement' in that it has the potential to continuously remind us that even if we who make up Western culture are no longer 'indigenous' in the ways that those enduring indigenous peoples/cultures are, meaning we are not from the places where we currently reside, nevertheless our bodies belong to the Earth. As such, our human experience of nature, like that of all living beings, is first and foremost, an experience of our own bodies. Bodies which Suzuki asserts, are 'born of the earth', of the 'sacred elements'; elements that have been here since the beginning and that will long endure beyond our own temporal use of them. As we discuss through narrative and conversation, our

DOI: 10.4324/9781003385202-12

understanding of the term 'ancient' – that the Earth, or rather Mother Earth (as understood by most global Indigenous peoples) (Caputi 2020) – has been on a 13+ billion-year journey and that we humans are merely one piece in Her vast cosmic tapestry that continues to unfold around and within us. This includes our vast inner landscape that is inseparable from the matter and materials, which breathe and breed our abilities to live, and in the case of humans, to create socially constructed worlds (Berger & Luckman, 1967) that are in essence social stories. As living beings we are in relation with Mother Earth, but more specifically as creatures that breathe oxygen, we depend on the prehistoric trees and plant *people*, whose miracle of photosynthesis have made, and continue to make, our existence possible. In referring to the trees and plants as 'people', we honor what Robin Wall Kimmerer (*Citizen Potawatomi*) (2013) refers to as the grammar or "language of animacy," which recognizes that in many Indigenous languages, unlike in English, the living world is made up of "a communion of subjects, not a collection of objects" (56; also see Little Bear 2000; also see Yunkaporta 2019). As such, part of our own perspective in this chapter is that we both, and those we converse with, practice the 'grammar of animacy' by asking to all we encounter in the world, "Who is that being?" as opposed to "What is it?" We have come to do this as a central part of our own life journeys, though not born into these teachings, by learning to 're-center the ancient-enduring Indigenous lens' from those who have always seen the world this way. Additionally, we seek to do so in a manner that acknowledges that this is a 'lens' that *animates* all life because it culturally recognizes and values that life *is animate* (Cajete 2000, 190). In other words, we recognize that Mother Earth has manifested life and that as Indigenous cultures recognize, albeit in different ways and terms unique to them, life has 'anima/us', as in soul/spirit. This perspective is very different from the norm in Western culture – a culture that sees the world as 'an inanimate collection of objects' – objects that ultimately only gain meaning as commodities – a "wilderness to be tamed, and as a non-living collection of resources to be exploited" (Tsosie 1996, 281).

Thus, our goal in this chapter is to begin by asking what is meant by 'ancient' from a human cultural perspective, thereby recognizing that not all human cultures are or historically have been the same and that the current largest cultural divide exists between Western culture and Indigenous cultures. Western thought asserts the only history worth telling is the so-called 'civilized' and 'civilizing' narrative; thus, the narrative of colonization and conquest of 'uncivilized' peoples and the Earth. Prior to the "discovered" colonies of the New World, there was Turtle Island, along with many other as yet unconquered/uncivilized lands, each with its sovereign civilizations operational with codes of conduct, governance structures, cultural mores, political dynamics, trade, and natural resource practices guided by a relational ethic with the Earth and all living things. In fact, as David Graeber and David Wengrow argue in *The Dawn of Everything: A New History of Humanity*

(2021), ancient, prehistoric humans were no different than ourselves in their abilities to create complex societies, despite dominant Western narratives framed by culturally supremacist language normalizing terms like 'primitive' and 'uncivilized.' To dispel the dominance of Western culture on our understanding of what it means to be human and consequently, as stated, on what is 'ancient', we are deliberately not focusing on all that is problematic with this colonial norm of civilization except to point out that Western culture perpetuates what can be called a 'cultural divorce' having progressively moved away from centuries-old embedded spiritual and relational interdependence with the living environment (seen and unseen) as still practiced among nature-relative civilizations. Instead, we seek to prioritize Indigenous ways of seeing and being in the world which, as Linda Tuhiwai Smith (*Māori from Ngāti Awa, Ngāti Porou and Tuhourangi*) states in *Decolonizing Methodologies: Research and Indigenous Peoples* (1999), represent "counter-stories ... powerful forms of resistance" to "...imperial and colonial practices" (2). She further affirms, "Research is one of the ways in which the underlying code of imperialism and colonialism is both regulated and realized" and as a result, "Much more is known about one side of those encounters than is known about the other side" (8). Consequently, in aiming to 'recenter the ancient-enduring Indigenous lens' after hundreds of years of being seen as not practicing "the 'arts' of civilization" (26), our choice of how to do so is 'by doing.' We are not just going to 'talk about' Indigenous ways of engaging with the world, we are also going to embody them to the best of our ability through conversations, including 'storytelling', which, as Tuhiwai Smith recognizes, "is a useful and culturally appropriate way of representing the 'diversities of truth' within which the storyteller rather than the researcher retains control" (146).

In our case, we listened to and engaged with the stories of two Indigenous elder women, as well as the stories of the living world around us. We have in fact been further inspired in this methodology by Aboriginal scholar, Tyson Yunkaporta's, *Sand Talk: How Indigenous Thinking Can Save the World* (2019), who, in speaking about his approach in his book, states that he "write[s] to provoke thought rather than to represent fact, in a kind of dialogical and reflective process with the reader" (20). Likewise, we aim to 'provoke thought' and to do so in 'dialogical and reflective' ways that include the reader, each other, and those 'beings' with whom we have shared emerging webs of conversation. Included among these beings are the stones. For Australian Aborigines, as Yunkaporta shares,

> Stones in the earth and the sky, all these stories and their connections can tell us more than the mere fact that they have existed for a certain number of millennia. They can tell us about how to deal with the complexities and frailties of human societies, how to limit destructive excesses in these systems.
>
> (24)

In fact, as Yunkaporta further recognizes and we agree, "Creation time isn't a 'long, long ago' event, because creation is still unfolding now and will continue to unfold if we know how to know it" (38). Thus, what is 'ancient' is also 'enduring', even as it is in, "a constant state of motion and we must move with it" (39).

In this spirit, the call of the wind, the rustling leaves, the eyes of a songbird, the sound of the stream, the bark of the tree, the face in a rock, the shape of the clouds – this is the stuff of ancient and informed knowledge – *beingness*. Western thought is but a young pebble in that landscape. Put your ear to the ground and look to the sky, feel the fire inside, and give thanks for water, the gift of life.

Part II

As we open Part II we are leaning into more Indigenous-traditional ways of sharing knowledge. Linda Tuhiwai Smith (1999) reminds us there are functional reasons to explore other contexts of research in Indigenous communities. There are

> the stories, values, practices and ways of knowing which continue to inform Indigenous pedagogies. In international meetings and networks of Indigenous peoples, oracy, debate, formal speech making, structured silences and other conventions which shape oral traditions remain a most important way of developing trust, sharing information, strategies, advice, contacts and ideas.
>
> (15)

What follows is a series of conversations, each leading with a question to serve as a guidepost for the reader. One could also imagine the four elements – Earth, Air, Fire, and Water – as instructive frameworks to apply to these narrative, inquisitive, and interactive dialogues. Overall we are reminded that this knowledge space is occupied by Indigenous thought and sharing of lifeways that is directly linked to relationships with and through these elements and all living things. We ask you, our readers, to accept for now that *all living things* encompasses what the Western-colonizing mind rarely examines as such, and even actively seeks to dominate and objectify.

Introducing Ourselves

SHIRLEY: Well, I'm Shirley Dean Khabbaz and I was born in the Midwest in Kansas, and we have amazing tornadoes there. In fact, when I was young and learned to really read the sky. My grandpa was Cherokee and I know that when I was about 7 years old, I was standing beside him and I knew he was going to tell me something important because I just felt it and he looked down at me and he said, "You're part Cherokee. Never forget it. Always be proud." So I never forgot it and I always was very proud to realize that that was part of my heritage.

And then I sought out Native Americans, whenever I could and I met this beautiful woman [*she reaches out her hand to Vera*] many, many years ago. We go back 40 years.

When we came here to Pennsylvania, I got my degree in education. I taught children until I retired and enjoyed every minute of it.

VERA: She:kon. My name is Vera Bauer Palmer, I'm Tuscarora of the Six Nations-Grand River Band, from the family of Hannah Groat and Albert Wright – We are the Skarò̇ rę̣ˀ – The Hemp Gatherers. I always told my students, "Don't judge me." [laughs]. I just retired from Dartmouth after teaching in the Native and Indigenous Studies Department for 23 years. It was a good run. There were challenging times, but it was a good run … I love sharing through teaching Indigenous perspectives and ideas.

My academic background is in western continental philosophy and English literature – both considered from an Indigenous perspective. I'm particularly interested in how western philosophers have written about Indigenous lives as savage, uncivilized, subhuman, etc. I taught Hegel, Rousseau, Mill, Foucault, among others, because I believe that the master's house can't be taken apart without knowing the master's tools. So I do part ways with Audre Lorde on that score.

So we need to understand how this schism between Indigenous thinking and ways of knowing – of understanding self and others and other-than-human animals, other than ourselves – how these terms differ from western thinking – and how western thought has gone off-track in claiming dominion over nature. It's crucial to be able to articulate [the meaning and consequences of] these differences.

PG: I'm Phoebe Godfrey. Both my parents were from England and moved to the North East in the early 1960s. As an environmental sociologist I was invited to submit a chapter to this book. I thought I would love to write a piece on indigenous perspectives but I'm not indigenous and wanted to work with those who are deeply rooted in those traditions. I do identify as Pagan and have a very earth-based perspective, which helps me to see how we were all once indigenous to somewhere and had a grounding to a place and thus to the sacred.

JWT: I'm Jacqline Wolf Tice. So this mom, Shirley, is my bonus Mom. [We say "hunka" mom – the word meaning 'relative' in Lakota.] I met her from age 26 and she became my mom. And that's part of an indigenous tradition – adoption, kinship. I lived in these ways since I was in my early 20s and claimed my indigeneity because there were three of my direct ancestors' names found on early Cherokee rolls, so I claimed it. And for over half my life I've walked the Red Road. I was adopted at Sundance, performed music as a

Cherokee descendent. And then my kids wanted me to get a DNA test with them and there's, like, nothing on my DNA that is Indian. And so I'm caught in this sort of sense of Indian identity without an Indian country.

I'm now an adjunct instructor for the Indigenous Peoples Law program at University of Oklahoma – Indian Country to 39 federally recognized tribal nations. Nearly 90% of our students are tribal citizens, so I work every day on issues in Indian Country.

What Is Ancient?

> And every stone, every little Stone that you pick up has a face, and if you look at it carefully you can see the face and if you hold that stone long enough and ask, 'will you talk to me?' it will just share some wisdom with you. It is beautiful.
>
> Shirley Khabbaz (Cherokee)[1]

Greg Cajete (*Santa Clara Pueblo*) says in *Reclaiming Indigenous Voices* (2000) that

> when you deal with Indigenous science, you begin to realize that everything is alive. The Lakota talk about, 'that abiding stone' (190). Everything, including the stone, can talk to you; Everything can teach you something. Everything is alive, related, and connected in the dynamic interactive and reciprocal relationships of nature. All events and energy unfold and enfold in themselves. That means that things come into being and things go out of being when they are needed. This principle, this idea, is a part of Indigenous ways of knowing.
>
> (190)

Conversation

JWT: When I wrote that original email reply to Phoebe about the chapter, I said, "what are they talking about … the ancient what? How do you express in Western terms what it is that the water is saying to you?"

VERA: Well, the lexicon in the academy that we're accustomed to using doesn't really even translate well [into Indigenous terms] … even the concepts of time, like the idea of being "ancient." What does that mean? If you speak of an Indigenous perspective or an *autochthonous* perspective, what does Ancient even mean, you know? It's not 400 years. It's not 500 years. It's not 1000 years. It's from time immemorial when Sky Woman came. Stories convey this differently than calculation. Time is a method of counting – time doesn't even exist, really. Time itself is a human construct. So, "ancient" … it's kind of a cute word, but it doesn't really describe anything, because we're just beings in a vast universe of other beings – we're so small [*holds up her hand to indicate something tiny between her fingers*].

I was just thinking how rich all these different ways of being are, Shirley. You come from the Tsalagi – Cherokee people – and yet your helpers were Lakota. I'm Haudenosaunee -Tuscarora – and our ways are totally different [from the Lakota]. Traditionally we don't have Sundance; we don't go into the Sweat lodge. But I've been a guest at these gatherings by generous Lakota folks.

Because so many colonizing events and institutions have thrown Native people together – like residential schools, forced removals, the reservation system, etc., we now can share tribal practices with each other in order to recuperate what has been lost in some cases. In reality, I don't believe that anything has been permanently lost, but it [traditional knowledge] has been sidelined and pushed out of access from so many of us. So sharing is the way we can come together to heal colonial losses and trauma.

You know, I've had a lot of wonderful help from Lakota people, and of course from my own [Huadenosaunee/Iroquois] people as well. Their stories, traditional practices are different. Our languages and cosmologies are different, but certain ideas are held in common among our tribes. For one, the knowledge that we humans are so very, very blessed and very significant, but also that we're not *all-important* – we're not central to all of creation. I think we share that perspective across tribes. That makes humility a necessity, not a weakness. You know, whether you consider the Maori of New Zealand, or Aboriginal tribes of Australia, or the Sami of Scandinavia – there are principles about our reciprocal relationship with the natural world, with the Earth, that are fundamental. We as Indigenous people share deeply on that basis, right?

SDK: And the earth is our mother to all. To all of the nations. It's the Mother Earth. It's always the Mother.

And it's understood differently, but still, yeah. I'll talk about the directions.

Yesterday, I started praying to the West in the morning. Because that's what I learned out in Lakota country. You have tobacco in your hand and you pray to the West and you pray that the West gives you the courage to pick up the lance as a warrior or a warrioress to pick up the lance for the battle, and then you have to face the North to give you the strength to fight. Through the battle to carry that lance into the year whatever cause it is you're fighting for, and then you have to turn to the east, which is wisdom.

Is this battle worth fighting for? And it, and also the East, has the ability to inspire others, gives you the ability to inspire others, and then you have to turn to the South, which is humility and love so that you don't become corrupt. And then the father, the father sky above that brings down rain and snow and gives our grandmother the moisture that she needs. And then our beautiful mother.

VERA: Then the seventh direction is within us.

SDK: You can't pray without giving a gift. So, you have to have enough tobacco in your hand for these directions to receive tobacco because you can't pray without giving a gift.

JWT: That's a really important concept.

PG: Yeah, it's about reciprocity.

VP: So, tobacco should be offered. And if you don't have tobacco, you give a strand of your hair. The sacrifice, the offering, is the reminder that you're participating in a relationship that's not primarily governed by ownership and land titles.

JWT: I like that you are clarifying about the directions and the commonality within the context of indigenous peoples around the world. I think part of what I'm wrestling with as a scholar are the kind of trivialized stereotypes. My understanding is there are such specific ceremonies, such specific belief systems, and specific ways of praying.

What Sustains Us?

> You can pray just about anywhere, and how you live is praying.
>
> Rosalie Little Thunder, Lakota

The Lakota have a word for what can be thought of simply as energy or life-force. That is *taku skanskan* – that which moves all things. This means from "subatomic particles to the blood moving in your veins into hurricanes crashing ashore out to galaxies colliding, everything is in motion and in this great motion we see beauty" (Tilson 2017). The ceremonial way of life is built into the identity of Indigenous communities and this lifeforce celebrates and sustains all that is. It is the ceremonies that provide "peace of mind" as the late former Cherokee Nation Chief, Wilma Mankiller (2004) reminds us, "for centuries outsiders have been trying to explain the … ineffable spirit of tradition-oriented indigenous people." Rather than focusing on analysis, categories and formalized logic systems, she continues, by interconnecting with the natural world "[it is] a more whole and complete [system] that considers the contradictions and unevenness of life" (42–43). The lens of knowledge systems passed through thousands of years of traditions and ceremonies based on functional relationships and reciprocity with the natural world in all its seen and unseen "contradictions" is not easily explicable in a "chaotic, materialistic world."

Westernized society exists in a framework of ordered production and consumption, managed by the clock of capitalism. For Indigenous communities, ceremonies are the framework marking time and existing as a "distinct community." These ceremonies in their communal and ancestral integrity sustain a meaningful life, create balance, and fulfill a "sense of identity, belonging, and understanding of … [our] place in the universe" (146).

While the unseen forces of nature characterize the ceremonial content, the process leading to the ceremony is filled with days and sometimes, months of preparation. From gathering supplies, to making prayer flags and ties, to drying and preparing specific foods, to sewing quilts and regalia. For those who directly lead or assist the intercessor of a ceremony, it takes years to learn and understand protocols and songs, to gather *medicine* and understand how to use it. Humility. Endurance. Gratitude. For those given this responsibility through vision, lineage, or necessity, this "divinely given power is an awesome and profoundly personal possession to be handled with the utmost care and concern" (Mails 1979, 5).

Conversation

PG: So, going to our question of 'what is ancient,' it's the Spirit World, right? And that's very ancient. And in that World time doesn't apply.

JWT: That's where I go when I think of what is ancient. I don't necessarily think of civilization. Like, who started farming and started agriculture? I don't think any of that. I think to myself, where are those places where there are ceremonies? How and why are those ceremonies?

SDK: A long, long time ago, my son was 18 at the time and he wanted to go out to a reservation and learn from elders, and he came to me and he said, "If you don't want to go with me, I'm going to go by myself. I'll hitchhike." And I said, "Well, I think I better go with you."

And so I told my husband and he said "well, goodbye." [*laughs*]

So we went together and we wanted to go to Pine Ridge because that's the place that we had heard of in South Dakota. So we were driving down the highway and we saw a herd of buffaloes.

They were fenced in, but it was a huge herd of buffaloes, so we stopped and went over to the buffaloes and they just, of course, never pay attention to you, but we just stood there for a long time.

I said, "Where does that road lead?"

And he said, "I think to Rosebud." And I said, "Well, maybe the Buffaloes are telling us something."

So we turned down that road and went to Rosebud and checked in at a hotel. And there was a native woman there and she said, "There's a ceremony tonight if you're interested."

I said "Well, yeah." [*laughs*]

So, we went to the ceremony of Uncle Robert Steed, who was an amazing healer. Uncle Robert's ceremony was something else because we were just sitting there, and the lights were all on and these two younger men brought in a Pendleton blanket.

They put it down and then they put something at the end of it. It looked like a shield of some sort with some kind of powder on. And then in came Uncle Robert Steed, who was the medicine man.

And they helped him sit down in front of everyone. And then the lights went out and so it was. Totally dark. They had plastic over the windows – garbage bags so that absolutely you can't have any light at all come in.

And then from the ceiling – I guess the best way to describe it would be like fireflies. There were just little lights and each one of those was a spirit and they came down and they were talking to Uncle Robert, and they were talking in old Lakota, ancient Lakota, which is not a language that anyone speaks anymore, except a medicine person and spirits. So, they were talking to him and then Uncle Robert said, "Stand up everyone, stand up." So, we all stood up. And of course, we couldn't see that anybody else was standing up.

So, you were there standing up and then these lights just came. Around all of us, they just all … all together.

They were in a bunch, and they went around the circle behind us and when they got to Dar [*Shirley's son*], who was right beside me, they stopped. And I thought, *why are they stopping at Dar*?

And they went down to his feet and up to where his head should be. And then they were about to the middle of his back. And then they shot back to Uncle Robert. And [the spirits] talked to him. And he said, "everyone sit down" and we all sat down and then he said, "Dar, stand up again."

So, Dar stood up again, and the lights again came and went up and down his back and stopped in the middle of his back and went around in a circle and then shot back then to Uncle Robert, and he said, "Dar, you're going to have to come back here for the next four years if you want to learn anything." Dar said, "OK." And that was the beginning.

Uncle Robert did a lot to heal Dar and help him. And so for four years we went back and then you learn how to have a pipe and how to offer it to a medicine person.

JWT: And that's the short version. [*laughs*]

SHIRLEY: That night I sat down beside this Indian elder and she introduced herself as Agnes White Mouse and she became my amazing teacher.

Agnes taught me all about ceremony and how to tie ties and how to give away, never keep something for yourself, but to give it. Let things pass from you to others, and so she was an amazing teacher.

Then I met another woman whose name was Rosemarie Mandan of the Mandan Hidatsa in North Dakota, and she taught me all about courtesy and etiquette, and how to treat an elder.

You never sit down in front of an elder. If you must get somebody's place and an elder is sitting there, you go behind them, never in front of them.

So, between those two teachers, I learned a lot and have tried to pass it on.

JWT: And we are always taught in the Sundance ceremony that we Sundance so that the people will live.

SHIRLEY: That's why we Sundance. And we started Sundance and vision quest. If you do a four-day vision quest, then you have to do four of them.

You have to come back for four years and do four of them and we both did that, and I had teachers and helpers and learned a terrific amount and finally, at one point, I just kept learning and learning and then they said, "Now you have to start teaching what you've learned. You can't just keep it to yourself. It has to be passed on."

So that's when I started teaching and helping women, and I was especially interested in women that have been sexually abused [and] was given an entire ceremony to help them get over that and to go on with their life – to reweave their web of life

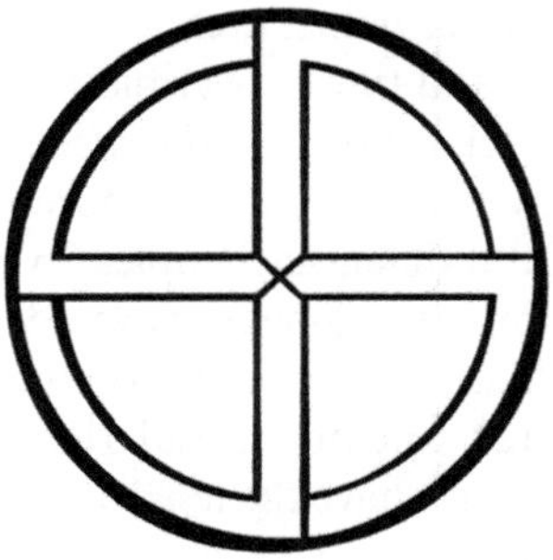

because if you see a spider web and you think of that as being your web … [*pause*]

… and then a hand comes in and pulls [*motions a downward pulling*] … that's what happens with sexual abuse.

And so I had to try and help these women reweave their web of life.

That was the special ceremony that was given to me by the spirits and there you have the spider right here. [*nods toward the living room table in the center of us*]

[*At this moment, we see there is a Spider crawling across the table*]

That's what they showed me.

Then, they showed me lightning coming down through. That was a vision that wasn't *from* me, but that came from the spirit world.

What Does Sovereignty Mean?

While there's a lot of activists who do have good intentions, non-native activists who have good intentions, I've seen it come from a completely ungrounded space, which hasn't worked in our favor or anybody's favor, and then it just puts this kerosene on a fire. You know, we're trying to get it to be a life-giving fire, not a consuming fire.

Chelsie Bonifer, Walla Walla/Cayuse/Metis[2]

Sovereignty is the ability to carry out your own direction.
If you think sovereign, you can be sovereign.

Audrey Shenandoah, Onandaga[3]

The dualism of nature and human nature is an ongoing theme in the development of modern societies and a primary barrier in reconciliation of the epistemologies of Indigenous and Western thought and lifeways. In recent decades while international law and policy has coincided and fostered the recognition of Indigenous sovereignty and its distinct rights while still under the umbrella of colonial governments, it remains greatly diminished by the dominant Western constructs of profit and power over nature-relative governance.

The United Nations Declaration on the Rights of Indigenous Peoples (UNDRIP) was adopted by the UN General Assembly in 2007, twenty years after the Working Group on Indigenous Peoples (WGIP) released its first draft of the proposed declaration. Its passage elevated the status of Indigenous Peoples as a distinct, identifiable constituency on the world stage with assigned individual, political, cultural, and territorial rights. While the UNDRIP is aspirational, meaning it does not have the legal authority of a treaty or convention in international parlance, its articles have the potential to be applied and codified as *customary law*. As such, although the United States, Canada, New Zealand, and Australia all opposed its adoption by the General Assembly in 2007, all four countries have since ratified the nonbinding document. Missing in the document, by choice of the WGIP, is a specific definition of Indigenous Peoples. Defining "indigenous peoples" as a collective introduces inherent inefficiencies. The grouping of any global classification of pre-colonized societies labels the recipient victims of colonization *en masse* and immediately creates categorical inconsistencies that elevate or dismiss the impacts of colonization. Jose Martinez Cobo (1983) reminds us,

> The fundamental assertion must be that indigenous populations must be recognized according to their perception and conception of themselves in relation to other groups; there must be no attempt to define them according to the perception of others through the values of foreign societies of the dominant sectors in such societies.
>
> (p. 49)

Conversation

VERA: All the way back to Sir Francis Bacon – the creator of the scientific method itself. He proposed that humans would [and should] conquer the natural world and create a *mundus alter* [another world] that ultimately would be for the glory of God. I mean, the idea that the natural world is nasty and imperfect and needs improvement is present in religion as well. [Bacon's] thinking that…humans can make it better, already dominated in the 16th and 17th centuries. Remember the incentives for Columbus and other colonizers: the three Gs, God, gold and glory.

PG: That the world as it is, in its 'natural state', is a mess.

VERA: Exactly - that it is 'naturally' wrong.

PG: And fallen, yeah.

VERA: It's [the earth, nature is] "fallen." And it's imperfect, evil, so [we humans] are obligated to bring it up to a standard that God would accept. So, with that idea, anything you do [to tame the wilderness] is an "improvement." I'm putting up air quotes here. You can't see my air quotes. [*laughs*]

JWT: I think that all ties into like this Judeo-Christian dominion-over construct that has really driven the development of the Earth and how we as humans have essentially dominated in so many ways, and that domination of the earth has extended onto the relations – the animals, the plants, all this – but then also, our fellow humans.

PG: I'm going to play devil's advocate for a minute, and I do this with my students who tend to use [the phrase] "those humans."

[My students might say] "We as humans have dominated" and I'll reply, "No, actually it has to do with different cultures – not all cultures have dominated the Earth. And so when you students say this you are implying that Indigenous cultures are not "human" [*uses air quotes*]. Further it implies that we are biologically destined to destroy the planet. But this is only what *our* culture would like us to believe."

JWT: So, you're delineating, saying, not all humans have done this, but that colonial nations have.

PG: Yes, not all humans have a story of domination.

JWT: And that ties back this Judeo-Christian ideology that has driven the dominion-over construct. And essentially, conquest of other humans.

PG: Yeah. My understanding of when Europeans first came to Turtle Island was [that their view of tribes was] you're not fully human because you're not conquering. You're not taming nature in the way that God has told you to do, and therefore you are savage, right? If you live in relationship with nature in a way that's more in balance with nature and/or more socially equitable, then that demotes you in the Judeo-Christian version of being human.

VERA: So in this logic, the "doctrine of discovery" is totally justified. Then conquest looks like an obvious necessity … and Indigenous societies are merely obstacles to "God's will."

PG: Yes, under that notion – exactly.

VERA: Well, you know, this starts to get a little sticky – especially when you begin to bring up the topic of sovereignty. Sovereignty [self-definition and self-determination] is important to every tribe that I know. Really, it's of a piece – we've talked about very spiritual experiences here, but the topic of sovereignty is where we go head-to-head, toe-to-toe, with western thought systems. When you bring up sovereignty in mainstream court systems, they don't give two hoots about what your [spiritual] experience is. It's a *thing* to own – property.

JWT: Right on.

VERA: I certainly do [care about spiritual practice.] But when we're talking about the *Realpolitik* of today, that's where it gets tough. That's what, for example, makes the Delgamuukw decision in British Columbia in 1997 so outstanding. Because for the first time Native oral tradition was accepted in the Canadian court system as validation of a land claim for the Gitxsan and Wet'suwet'en First Nations people – and there have been several other examples since then.

The Delgamuukw decision is important, but it's just one little, tiny toehold in an uphill battle. We're always faced with the objection: "but we own the [legal] title to this land. Native claims have no standing."

That brings up the ancient Aboriginal understanding of the term *autochthony* that I think anchors Indigenous views. The Greek prefix, *auto* refers to self, and *chthonos* refers to soil: self + soil. I remind my students that everyone born here is "Native American." But does that mean autochthonous? Indigenous? Is there a specific relationship to a specific place, a particular soil?

I think folks who [originate] from elsewhere … and this *is* a nation state of immigrants … folks yearn to belong. A relationship needs to be cultivated here, reciprocally, and in this place, this soil. Coming from a place other than Turtle Island means that that place was/is a *natural part* of the self + soil combo. The popularity of ancestry searches suggests the power of this desire … to know where we're from – what place – what soil.

Our nation state is made up mostly of immigrants, so a love relationship with *this* land needs nurture. It's really about love. So, you see, it's not *predominantly* a legal matter, or about flags or national boundaries. It's necessary to fall in love with this Earth, here, and care about it like a relative – honor her in particular ways, so self and soil means something. But that's where the spiritual significance meets hardball for tribal people. Hardball is sovereignty that insists on room for our land-love relationship to continue and thrive.

JWT: Absolutely, yeah.

VERA: You know the issues of tribal sovereignty [*looking at Jacqline*].

JWT: I think it is part of what has what has pushed me into this [legal] space in my life, which is why I have such strong commitment to the sovereignty and the identity of indigenous people in their soil and in their claims of identity and [to the] *reclaims* from what the colonial [constructionists] and posers have decided that they can interpret indigeneity as.

That's the thing that working in the legal realm has really created for me. It has narrowed my thinking, but it's broadened my understanding.

PG: It's interesting, I'm comparing sovereignty and indigenous claims to sovereignty through identity. Whereas from the perspective of ancestors from Africa, that was your exclusion from having any rights.

Remember the South, the one drop rule. That was your ticket to total social exclusion, and violent oppression. And so, on the one hand, for some as in the Native Americans their identity claim can link to sovereignty. Whereas for anyone of African descent, that was used to justify their social and economic exclusion. And this was part of the colonial project: using different mechanisms for entrapments and racist oppression for different human groups.

VERA: Yeah, that's right. A very specific way if you were going to use it this way – with one group and another way with a different group.

JWT: And when you can compartmentalize people, you can basically distance yourself from who they are. Distance yourself from their experience. Distance yourself from their pain and allow for the oppression because you are distancing yourself from their humanity.

PG: And the law becomes the mechanism, right?

VERA: And [the legal mechanisms] are totally arbitrary. I mean, they're set up for a reason – they create the right to disenfranchise. Africans became cogs in a slave economy … and laws [like the General Allotment Act[4]] created a "right" for the government to take away our Indigenous lands.

Where Do We Go from Here?

Water is life. It's a phrase made mainstream by the Standing Rock Sioux Tribe's resistance against the Dakota Access Pipeline. On its face and simply stated, all organic life springs from the interactions of hydrogen, nitrogen, oxygen, and other elements and molecules with the environment. In the cosmology of Indigenous cultures, though, "water is life" holds a deeper, ancient, implicit value beyond its scientific context.

Itazipco/Mnicoujou/Oglala Lakota writer Tiokasin Ghosthorse (2016) translates the spirit of water, called *Mni* in the Lakota language, as a *first consciousness*. In his piece *Living with Relativity*, he explains the meaning of water as a *Being*, adding, "One way of changing our perspective is to realize "Water as Life" and acknowledge the peoples of the world who still hold ceremonial respect for the "Being of Water." In Supai, Arizona, home to the Havasupai Tribe, an Indigenous village at the bottom of the Grand Canyon, children are taught from birth to swim, to respect the water, that this water spirit is their protector. The Yurok Tribe, in an effort to protect their inherent subsistence and spiritual life in relation to the Salmon in the Klamath River, has passed an ordinance giving the river the "rights of Personhood." In northern Minnesota, the White Earth Band of Ojibwe and others sued the Minnesota Department of Natural Resources (MDNR) on behalf of the "Rights of Nature, Rights of MANOOMIN (Wild Rice), and Rights of the Chippewa," which are protected and reserved by treaties. Their claim cites the threats of climate change and aims to nullify an MDNR permit for 5 billion gallons of public ground and surface water granted to Enbridge for the Line 3 pipeline. Winona LaDuke (*Anishinaabekwe/Ojibwe*) (2002) succinctly says, "To honor and protect the waters is our responsibility as people of the land" (55).

Critical to understanding this perception is the knowledge that water is not only a resource to be protected by statutes, regulations, and common law. It is a relative, which functions as a literal and symbolic source of nurturance, revitalization, guidance, and strength. Materially and spiritually, water teaches us. It needs to be carefully monitored, listened to, and observed for changes that signal necessary adjustments to practices that ensure its continuity as a creator and sustainer of life. Just as people have a purpose, so does this sacred water.

Water rights and the availability of water is the natural resource issue of our time. For Indigenous communities, who are on the front line of climate change, sovereignty is critical to the issue of climate change, according to legal scholar, Rebecca Tsosie (*Yaqui*) (2013) who asserts, "Developmental policies of global nation-states have created the current crisis, and these sovereigns have the capacity to mitigate the high level of greenhouse gas emissions or facilitate appropriate adaptation policies to deal with future climate events" (239). How do we get closer to bridging the gap from the mindsets of humans from Indigenous – *autochthonous* – cultures, as Vera Palmer noted, to those from consumptive societies?

Conversation

Vera: Going back to my own tradition, Haudenosaunee [Iroquois]: The Great Law, the *Kayanerehkow*, talks about The Good Mind at the center of our understanding of who we are as human beings. We say that we're "the real people." This doesn't mean other people aren't real. But rather, here are the principles required to be a real person – to be of the Good Mind.

Another way of saying it is the *thinking heart*. With western technology we think, like Bacon, that we can solve and improve everything. And as you said, Phoebe, all from here [pointing to her head]. Emphasis is on the human brain. But now even neuroscience states that human knowledge is created within the entire body. But mainstream thought still assumes that intelligence is produced here [*points to the head*].

In Haudenosaunee understanding, the *thinking heart* is the seat of knowledge and wisdom. The mind without the heart is nothing. The heart without the mind will accomplish nothing.

So, if only we can keep some version of the Good Mind as an ideal that brings balance to any innovation, any new initiative: "does it meet the standard of the *thinking heart*?"

That's what I want to share here from my own tradition. Our Haudenosaunee tradition. Thank you all for this conversation.

SHIRLEY: And I'd agree with that 100%, but it comes definitely not from just here [*points to her head*]. But it must come from the heart. Just put your hand on your heart. And you feel the connection, feel the beat of your heart. The connection between the two [*All placing hands on our hearts*].

It has to be one. It has to be and and if you forget any of it, just go out and walk on the earth.

VERA: And touch the tree.

SHIRLEY: And sit on the earth or sit on the earth with your back on the tree. And be thankful. Thank that tree for who it is and and the roots that are going down into our Grandmother and that tree is pulling sustenance up from the Grandmother ... and neither could be without it.

And we're just kind of here. We are walking around and trying our best to do a good job and to, you know, be true to ourselves, and to each other.

And it's wonderful that you two are trying to understand this and give writing to it. It's amazing and I appreciate that.

Thank you.

JWT: It's an honor always to listen to you and share your relationship here with Vera. I never wonder why I'm sitting here with you. I never wonder because every moment is a present one.

[*silence*]

VERA: So how do you see this evolving?

PG: There's so much reproduction of colonization within academia, right, where you're constantly taking somebody's idea and then claiming it, you know, so writing this *with* you is so important.

You know, and to our own capacity, allowing the trees [the water, and earth] to speak, you know.

VERA: And that is intelligence, that's the intelligence. We [humans] don't have a corner on it. We are not necessarily intelligent beings as a human race. The *intelligence* is all around us.

And much higher in many cases. I mean, the deep intelligence of trees and there's so much new science about all of the underground things that are happening chemically.

PG: Like the wood wide web or the mycelium?

VERA: Yeah, you know, I have so little botany understanding, but wow, it just blows your mind ... that you have a panther that can see 70 times better than a human, the intelligence and the richness ...

PG: I tell my students it blows our mind because we're so stupid. [laughs] But you know, we're so arrogant. They'll say wow – they did research and found out animals have consciousness. And I'm like – it's only a stupid culture would need, you know… [*laughs*]

JWT: It kind of brings that circle back to, like, how do you explain to somebody what the song of the stream is, what the song of the water is?

Because those who have come to listen and be and accept that, already *know* it is not some construct of the mind…

VERA: Or, that it's just poetry.

Notes

1 Quoted in Mankiller, 2004.
2 Chelsie Bonifer, *Walla Walla/Cayuse/Metis activist/scholar/musician* (Personal conversation with author Wolf Trice, June 2023).
3 Quoted in Mankiller, 2004.
4 General Allotment (Dawes) Act, 23 Stat. 388 (1887) imposed privatization of land, dividing reservations into individual allotment acreage and selling off "surplus" to white settlers, effectively breaking up Indian Territory guaranteed by treaties.

References

Berger, Peter and Luckman, Thomas. *The Social Construction of Reality: A Trestes in the Sociology of Knowledge*, New York, NY: Anchor Books, 1967.

Cajete, Gregory. "Indigenous Knowledge: The Pueblo Metaphor of Indigenous Education." In *Reclaiming Indigenous Voices and Vision*, edited by Marie Battiste, 181–192. Kelowna, British Columbia: UBC Press, 2000.

Caputi, Jane. *Call Your Mutha': A Deliberately Dirty-Minded Manifesto for the Earth Mother in the Anthropocene*. Oxford, UK: Oxford University Press, 2020.

Ghosthorse, Tiokasin. "Living With Relativity." *Humans and nature. Center for Humans and Nature*. August 26, 2016. https://humansandnature.org/living-with-relativity/

Graeber, David and Wengrow, David. *The Dawn of Everything: A New History of Humanity*. London, UK: Farrar, Straus and Giroux, 2021.

LaDuke, Winona. *The Winona LaDuke Reader: A Collection of Essential Writings*, Stillwater, MN: Voyageur Press, 2002.

Little, Bear Leroy. "Jagged Worldviews Colliding." In *Reclaiming Indigenous Voices and Vision*, edited by Marie Battiste, 77–86. Kelowna, British Columbia: UBC Press, 2000.

Mankiller, Wilma. *Every Day is a Good Day: Reflections of Contemporary Indigenous Women*. Golden, CO: Fulcrum Publishing, 2004.

Suzuki, David. *The Sacred Balance: Rediscovering Our Place in Nature*, Vancouver: Greystone Books, 2007

Tilsen, Mark. "There is No Lakota Word for god." *Last real Indians*. June 21, 2017. https://lastrealindians.com/news/2017/6/21/jun-21-2017-there-is-no-lakota-word-for-god-by-mark-tilsen

Tsosie, Rebecca. "Climate Change and Indigenous Peoples: Comparative Models of Sovereignty." *Tulane Environmental Law Journal* 26, no. 2: 239–257 (2013).

Tsosie, Rebecca. "Tribal Environmental Policy in an Era Of Self-Determination: The Role Of Ethics, Economics, and Traditional Ecological Knowledge." *Vermont Law Journal* 21: 225–333 (1996).

Tuhiwai, Smith Linda, *Decolonizing Methodologies: Research and Indigenous Peoples*, London, UK: Zed Books, 1999
Wall, Kimmerer Robin. *Braiding Sweetgrass: Indigenous Wisdom, Scientific Knowledge and the Teachings of Plants*, Minneapolis: Milkweed Editions, 2013.
Yunkaporta, Tyson. *Sand Talk: How Indigenous Thinking Can Save the World*, New York, NY: HarperOne, 2019.
United Nations Declaration on the Rights of Indigenous Peoples, Sept. 13, 2007, A/RES/61/295, http://undocs.org/A/RES/61/295, Images copied from https://www.vectorstock.com/royalty-free-vectors/traditional-native-american

12 Permaculture as a System for Designing Sustainable Human Settlements

Ahead of its Time or Impossible Dream?

Caroline Smith and Nick Towle

Introduction

The era we now call the Anthropocene is upon us. The window of opportunity to advance and achieve a more socially just and ecologically flourishing world is rapidly closing. Humanity and the more-than-human world in which we live are now in a state of polycrises (Tooze 2021), with anthropogenic climate change manifesting in more extreme and frequent weather events and related ecological and social disruptions. Human society finds itself in a state of overshoot, consuming more of Earth's resources and producing waste faster than can be replenished or assimilated. Humanity is using the equivalent of 1.7 'Earths' to meet current patterns of consumption, which cannot be sustained (Merz et al. 2023). Compounding this global ecological overshoot is the 'wealth pump' that has seen the extraordinary flow of wealth from poor to rich (Turchin 2023), leading to widening inequity and increasingly disruptive social upheaval.

Miller and Heinberg (2023) call this period of Earth's history 'The Great Unravelling'. Macy and Johnstone (2012) see it as one of three great cultural stories or competing narratives of our time. They name the other two stories "Business-As-Usual" and "The Great Turning". Business-As-Usual, powered by an abundance of fossil fuel, is currently shaped by a neoliberal worldview, where the values of individualism, competition, consumerism, and material accumulation are normalized and shape the way in which we – from individuals to governments and corporations – interact with the world around us. Business-As-Usual either largely ignores the developing polycrises, or naively assumes that our current technological trajectory and ingenuity will solve major threats. In stark contrast, The Great Unravelling directly faces and reflects the reality of the converging polycrises, but runs the risk of becoming stuck in a narrative of inevitable decline and inertia. Hope lies in The Great Turning – in lessons learned and opportunities to change course.

Macy and Johnstone (2012) go on to define three necessary dimensions of The Great Turning: holding actions (actions to slow and prevent the damage

DOI: 10.4324/9781003385202-13

being caused by those pursuing business as usual); life sustaining systems and practices (rethinking the way we operate in the world and creatively redesigning systems such as transport, agriculture, economics, and so on), and a shift in personal and collective consciousness and worldview (drawing on the insights of spiritual traditions and contemporary science to realize our sense of deep connection and interdependence with all life). Along with many others, Merz et al. (2023) identify human behavior, which they name the 'Human Behavioural Crisis', as the root cause of overshoot. Human behavior emerges from a complex of worldviews, values and ideologies and the institutions, structures, and technologies these favor (Beddoe et al. 2009). For at least the last 200 years, these have been increasingly shaped and enabled by psychosocial manipulation through colonialism and rampant capitalism, of which neoliberalism is its most recent form, by those who stand to gain most from exploitation of people and planet. These behaviors promote and privilege a consumption identity, where expanding material consumption has become a proxy for progress, success and quality of life, even when excessive consumption by the minority makes meeting basic material needs increasingly unattainable for the majority of the world's population. Science shows unequivocally that unlimited economic growth and related consumption is not possible on a finite planet; it is essentially an oxymoronic position (Carter and Smith 2023). Yet, for the majority of the affluent world, understanding the science is not sufficient to change this deeply entrenched consumption identity.

At the same time, while neoliberal Business-As-Usual is still the dominant worldview today, an increasing number of the world's population is being exposed to the damage wreaked by anthropogenic climate change. The convergence of ecological threats and social disruptions are provoking wide-ranging and critical debates in all spheres of human society that center on alternative ways of being in the world that envisage a reimagining of the relationship between humans and the more-than-human world, offering hope for The Great Turning.

We argue that permaculture contributes strongly to the dimension of designing and creating life sustaining systems and practices, and has the potential to contribute to the concomitant shift in personal and collective consciousness that underpin The Great Turning. For its many adherents, permaculture as a system for designing sustainable human settlements, offers a pathway to creating a flourishing society lives within Earth's ecological limits. It brings a strong ethical grounding along with principles and practices, for redesigning our relationships with each other and ways of living in the world. The question we discuss in this chapter asks whether the hope offered by permaculture, spawned in the counterculture of the 1970s, is still relevant and sufficient to enable its practitioners to engage with The Great Turning? Can permaculture provide a pathway that helps lead us out of the morass we find ourselves in toward a more desirable, prosperous, future? Or, as the title of this chapter asks, is it either still ahead of its time or just an impossible dream in a world in polycrises? As might be expected, answers are not straightforward.

What is Permaculture?

Originally a portmanteau of *perma*nent agri*culture*, then broadened to mean *perma*nent *culture*, the creative synthesis that became permaculture was conceived in Australia in the early 1970s by co-originators Bill Mollison and David Holmgren (Mollison and Holmgren 1978), and later added to with Mollison's substantial Designers' Manual (1988) and Holmgren's (2002) in-depth articulation of a systems view of permaculture. As permaculture evolved into permanent culture, it widened to embrace all elements of sustainable settlement. Among the many other significant contributing texts are Rosemary Morrow's 'Earth Restorers' Guide to Permaculture' (2022); David Holmgren's 'Retrosuburbia' (n.d.); The Transition Handbook (Hopkins 2008).[1]

With antecedents reaching back to the nineteenth century, such as Ebenezer Howard's social utopian Garden City movement in the United Kingdom (1898; 2003), permaculture developed within the broader countercultural landscape and environmental awareness that emerged in the 1960s and 1970s, aligning it ideologically within the postmodern eco-social movements of the political left. Then, a worldview was emerging that was beginning to challenge the legitimacy of the Western Enlightenment grand narrative of human progress through continued economic growth and material consumption, and recognizing its impact through the domination and degradation of nature. Beginning with its critique of modern agriculture, rejection of consumerism, globalization, and other dominant cultural forms, permaculture emerged as a novel system of design which was able to provide a practical pathway to addressing the polycrises of the Anthropocene. Closely allied to the concept of bioregionalism and to the energy descent peak oil concepts of the 2000s, politically, permaculture tended to align with those who propose rural self-sufficiency, economic degrowth, and political autarchy (Leahy 2021), though there has never been a serious attempt to include permaculture principles in mainstream political thought or action. For all that, permaculture practitioners have worked hard to establish intervention at the grassroots as an alternative, even if only implicit, to mainstream standard leftist politics (Leahy).

A key insight that sowed the seed for Mollison (1996) was recognizing that natural systems are self-sustaining, productive and flourishing and produced no waste, something that permaculture, through a process of design, sought to mimic. Mollison's deep experience in the ecology of the Australian bush had a profound influence on him, and on reflection he noted that:

> If we could create living systems, in which we were a part, then we could have a sort of Garden of Eden. It was as though I had unlocked a door opening on to a new world. I could see it all, for miles ahead and years away … all the rest was ways and means.
>
> (24–25)

Drawing its inspiration from Mollison's deep insights into ecology combined with the influence of Holmgren's background as a landscape designer, weaving in Indigenous approaches to land, sustainable polyculture farming practices and appropriate science and technologies, permaculture emerged as a conscious holistic design system (Mollison and Holmgren 1978). As he notes in his *Designer's Manual*, Mollison (1988) saw permaculture as:

> the conscious design and maintenance of agriculturally productive ecosystems which have the diversity, stability and resilience of natural ecosystems. It is the harmonious integration of landscape and people providing their food, energy, shelter and other material and non-material needs in a sustainable way.
>
> (ix)

Permaculture's intent was to enable the design of low input, highly diverse productive systems of plants and animals with no or minimal resort to chemical pesticides and fertilizers, to create a self-sustaining, dynamic, and regenerative system of human settlement.

Initially, the first applications of permaculture were to agricultural systems. From this perspective, permaculture can be seen as not dissimilar to other regenerative agricultural approaches such as agroecology. Both are based on the ecology of natural systems and seek to build agriculture systems that are regenerative rather than extractive, building soil and biodiversity in order to enable healthy plant (and hence animal and human) flourishing. Krebs and Bach (2018) note that the major difference between permaculture and agroecology is that design has been built in as conscious practice in permaculture from the beginning. Although agroecology has been a scientific discipline for a few decades, only recently have design principles for the reorganization of farming systems been formulated (Krebs and Bach).

Defining Permaculture

Despite all that has been written on it, because it offers philosophical, theoretical as well as practical foundations for transforming our world, permaculture evades simple definition. There is an apparent tension here. Resistance to a concise definition can invoke suspicions around the intent and purpose of permaculture, whereas a too narrowly defined definition may obscure its holistic intent and broad applicability. This was born out in a global survey in which permaculturists were asked to address the question "What is permaculture?" Ferguson (2014) rated their level of agreement with specific statements characterizing permaculture. The 698 respondents' views ranged widely from a framework for design and planning to a profession (Table 12.1).

Roux-Rosier et al. (2018) also point to the multiple, complementary, and systemic conceptualizations of permaculture; as a technical design practice; a holistic life philosophy, and as an intersectional social movement. So while

Table 12.1 What is permaculture?

Permaculture is:	*Level of agreement*
A framework for design and planning	99.4
A philosophy or worldview	95.9
A social movement	91.6
A set of farming and gardening practices	84.0
A profession	71.4

potentially confusing, the sheer complexity and diversity of definitions allow practitioners to imagine their roles in bringing about desired, sustainable futures in a range of ways even as they forge a strong, shared identity.

The Permaculture Design Framework

For the untrained observer or novice permaculturist, permaculture may appear as little more than an assembly of elements and techniques, such as swales, herb spiral and 'chicken tractors', useful for organic gardening. This fails to recognize the ethical grounding and principles-based design framework that give rise to its physical manifestations (Ferguson 2014), fueling the aforementioned definitional debates and confounding many who are new to its concepts. Adam Grubb (n.d.) has provided a useful diagram (Figure 12.1) that illustrates the relationship between the complex layers that form the elements of the permaculture framework.

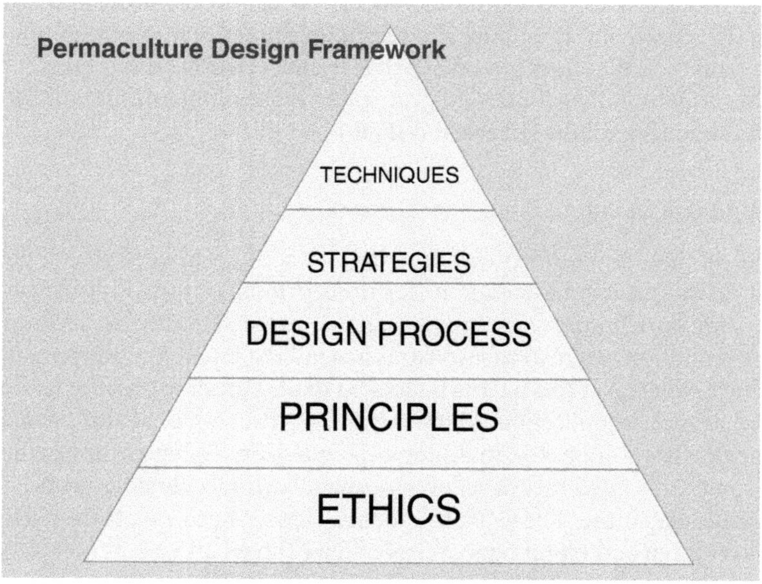

Figure 12.1 The permaculture design framework. Courtesy Adam Grubb, Very Edible Gardens.

Ethics

At its core are three ethics that underpin all design principles and decisions, and which articulate the values of a strong ecological worldview that also claims to understand and value the contribution and importance of Indigenous perspectives.[2] The three ethics are: Earth Care; People Care, and Fair Share. The third ethic, 'Fair Share', was originally expressed as "Setting Limits to Population and Consumption" (Mollison 1988, 2). This proved controversial, given the potential racist interpretations of population control, and was later reframed as redistribution of surplus or 'Fair Share'. Forms of these ethics are to be found in many traditional societies, representing culturally evolved mechanisms that regulate self-interest (https://permacultureprinciples.com/ethics/), in direct contrast to the competitive self-interest values of neoliberalism that have profoundly shaped and enabled current human behaviors. Mollison and Holmgren both advocate that Earth Care and People Care are of equal standing, and the practice of the Fair Share ethic is in service of both. This concern for all life on earth, beyond simply to advance the wellbeing of humans, conveys strongly ecocentric worldviews and values. The ecocentric worldview is characterized by a belief that humanity can find harmony, meaning, and purpose through affiliation and cooperation, and that nature has an intrinsic value rather than merely being a resource for human exploitation. In this worldview, all lifeforms are interconnected and interdependent. The related concepts of Planetary Health and One Health are a recent manifestation of this worldview (De Castañeda et al. 2023).

An apparent paradox exists between ecocentric worldviews and values clearly articulated and espoused within foundational texts and teachings (Holmgren 2002; Mollison 1988), and a design process heavily oriented to analyzing and assigning instrumental value to living and nonliving entities, pointing to an anthropocentric or human-centered bias. These competing worldviews and their underlying values are the subject of ongoing and dynamic debate within environmental education literature (Kopnina 2013), though rarely overtly referred to or examined as part of permaculture education. However, it is recognized that those with an ecocentric world view are more likely to act to protect the environment than those holding anthropocentric views (Kopnina).

Principles

Building on the foundation of the three ethics is a set of principles which are key to the process of creating a self-sustaining holistic system. They are regarded as thinking tools to inform the design process and provide a lens through which to evaluate the quality and robustness of strategies within the design. The principles are intended to be generalizable, enabling application to diverse contexts, from food production to intentional communities and at varying scales (Veteto and Lockyer 2008), from the design of a city unit, though to an entire bioregion. Permaculturist Geoff Lawton's work on large-scale permaculture projects, for example, greening the desert in Jordan, demonstrates that this is indeed possible (see https://www.actionecology.com/journal/?post=greening-the-desert). The framework has also been applied to less tangible domains such as developing innovative organizational management models (Vitari and David 2017), organizational resilience, and sustainability (Mannen et al. 2012), and even to a redesign of the school (Harney 1997). While historically the framework has typically been applied to the biophysical aspects of a design, the adoption and adaptation of permaculture to diverse geographical, climatic, and social contexts on every continent on earth (Ferguson and Lovell 2015) speak to the strength of an approach grounded in a shared framework.Two sets of principles exist, having been developed independently by Mollison and Holmgren. Mollison's (1988) principles are typically considered as very practical and, in some ways, more akin to design strategies, whereas Holmgren's (2002) seek to capture both practical and more philosophical ways of living. The existence of the two sets of principles has fueled heated debate within the permaculture community about which are 'best', and, along with other issues, culminated in a parting of the ways in the 2000s which was never really resolved between the charismatic Mollison and more reserved Holmgren. This even saw an unfortunate development of an increasingly hostile rivalry between followers of each of the originators (Dawborn and Smith 2011) and led some to label permaculture as a 'cult' with its own gurus and set of rules. Indeed Leahy (2021) highlights why permaculture may have been perceived as a cult, with features such as uncritically following the charismatic leadership of permaculture's founders by some adherents, and what might be seen as formal initiation through the Permaculture Design Certificate course (PDC), the first course of which was run in 1978. Fortunately, these historical rifts have largely subsided and not derailed the adoption of permaculture globally, with many new practitioners now active and with notable heterogeneity across different regions. The lack of an institutionalized permaculture 'movement', with interpretations supporting differentiated, localized solutions, shows how permaculture allows for heterogeneous ways of organizing (Mannen et al. 2012) and imagining sustainable earth habitation.

Today, Holmgren's (2002) 12 principles are the most widely applied. They are:

1 Observe and interact
2 Catch and store energy
3 Obtain a yield
4 Apply self-regulation and accept feedback
5 Produce no waste
6 Use renewable resources and services
7 Design from pattern to detail
8 Integrate rather than segregate
9 Use small and slow solutions

10 Use and value diversity
11 Use edges and value the marginal
12 Creatively use and respond to change

Design Process, Strategies and Techniques

The permaculture design process is an essential, deliberate, and conscious feature that draws on a philosophy of "protracted and thoughtful observation rather than protracted and thoughtless action" (Mollison 1988, ix). All PDCs include sessions devoted to deepening understanding of the biophysical and/or social context for which a design is being created, so that the choice of particular strategies and techniques is likely to be the most appropriate and beneficial for that context. This, however, is also the point where significant misunderstandings may occur, with problems arising when a set of appealing techniques are simply transposed by an enthusiastic individual to a location and culture for which they are not suited. For example, a strategy may be needed to slow down the flow and capture water in a particular landscape using the specific technique of creating swales (on-contour channels with planted earth berms). But the carving of swales in landscapes with high rainfall may exacerbate soil erosion. Hence, effective permaculture design will manifest quite differently across the world by virtue of the ethics, principles, and design process which seek to deeply understand the context in which they are being applied. To do otherwise would simply amount to another form of colonization.

Echoing this concern, there is little to prevent the principles, strategies, and techniques being applied through the lens of the survivalist 'prepper' cultures that tend to align with the radical right (Luke 2021). The critical difference is the articulation of the three ethics, without which the application of principles alone may enable extreme or exploitative practices.

Domains of Permaculture – The Permaculture Flower

As we have pointed out, permaculture is complex, even when the relationship between ethics, principles, and design approaches is understood. In an attempt to assist the designer to create an integrated system governed by the ethics and principles, Holmgren developed what he termed the 'permaculture flower'.

The flower is useful on a number of levels. As an organizational device, it promotes the application of the ethics and principles across seven domains of human endeavor considered necessary for holistic design and sustainable culture. The spiral radiating from the ethics and design principles (Figure 12.2) represents the evolutionary path that connects and integrates these domains. The flower has proved to be very useful within permaculture education. Exploration of its components enables participants to begin to understand the systemic and interconnected nature of sustainable systems, and, importantly, that permaculture is intended to go well beyond simply designing the vegetable garden. However, reach into some of the domains, such as finance and

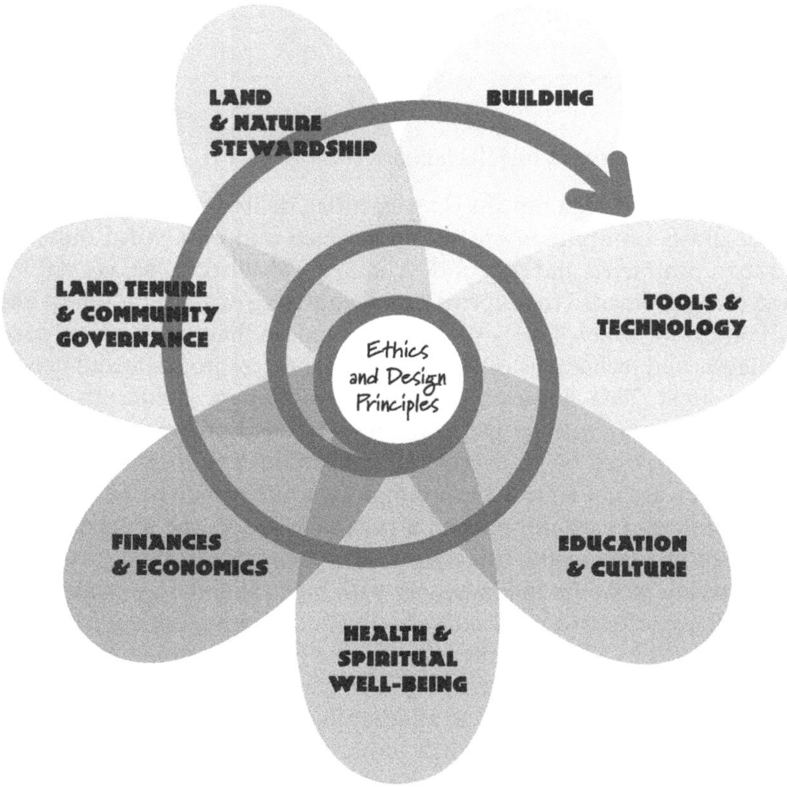

Figure 12.2 The Permaculture Flower.

Source: The Permaculture Flower was designed by Richard Telford (https://permacultureprinciples.com/), based on the ideas of David Holmgren (https://holmgren.com.au/).

economics, remains very limited, probably reflecting some educators' own lack of sufficient background in these areas.

The Transformative Potential of Permaculture

For many, permaculture seems exciting and empowering and offers a pathway to make an immediate positive change in their lives. Smith (2000) recounts comments made by participants after completing a two-week PDC in the 1990s:

- I feel wonderful. The feeling of power, of being able to achieve the plans I have in mind, is very strong.
- I feel inspired, more optimistic, hopeful, enthusiastic, enlightened, more direction.
- It has given a whole new perspective to the future.
- Because of what I've learned I have a responsibility to use the knowledge widely. I feel empowered by this experience and more hopeful for the future.

One comment in particular summed up the feelings expressed by many:

- The permaculture course was almost like a religious experience. I felt inspired with hope, excited that I could make a difference, touched by the realisation that I could be in control, empowered.

These responses suggest a strong transformational potential for permaculture. Why then, when it has the capacity to inspire and empower and provide deep insights and practical skills to support a more sustainable human presence on the earth, does permaculture appear to be languishing on the margins, even within sustainability movements themselves? Why does permaculture not interest or attract more people when it seems such an obvious way forward to its adherents, and that the principles are considered generalizable and can be taken up at different structural levels of social organization?

Dealing with Worldviews: The Spiral is Stuck

Superficial responses relating to difficulties in defining or communicating permaculture concepts were noted earlier. At a much deeper level, the genius of Mollison and Holmgren, in conceiving of a design system to achieve whole system transformations, actually set up a clash, a contest, an affront, to those with an identity deeply permeated by values shaped by a neoliberal worldview. Worldviews are tricky, yet vitally important. As renowned systems thinker Donella Meadows (2008) explains, our worldview forms the basis from which the values, goals, structures (physical and social), and rules of our human systems arise. They are the culturally formed and contextually based values and beliefs systems that arise from and are shaped by our need to make sense of the world. While potentially a place of deep leverage, where a small change can give rise to whole system change, worldviews are deep within us but often not examined or even recognized. Smith and McTiernan (2021) argue that most people attracted to permaculture tend to hold values constituted by an ecocentric worldview. However, this is considered the dominant view of only about 10% of people in the Global North, compared with some 50% who hold values more closely aligned to neoliberal or liberal values.

To emphasize this point, a critical reading of permaculture literature by Janzon (2018) comparing and contrasting permaculture and broader sustainable development discourses revealed profound differences that can be attributed to differing worldviews. Permaculture and sustainable development share commonalities around intended holistic systems thinking for sustainable futures, but Janzon argues that they diverge in the nature of the change required to actualize them. Permaculture is underpinned by strong assumptions around limits to growth, calling for a radical redesign in social systems, with economic degrowth and a contraction in material consumption (e.g., Jackson 2009). In contrast, sustainable development discourse as articulated in the UN Sustainable Development Goals (SDGs) (UN n.d.) is ambivalent about limits to growth, instead arguing that economic growth is a necessary precursor for sustainability within a broader position of low-level reform of existing structures. Indeed, Carter and Smith (2023) argue that the fingerprints of neoliberal worldviews and values are evident within the SDGs.

Drawing on Beck and Cowan's (2005) Spiral Dynamics (as opposed to Holmgren's flower spiral), Smith and McTiernan (2021) argue that people with a particular set of values tend to reject others, believing that they alone are right, and reacting in a closed-minded way to information that threatens their beliefs and values. The current polarization of socio-political culture in the United States is an extreme manifestation of this, and the permaculture community is not immune to it either. It has become too easy for permaculturists to live in tribal bubbles, communicating with like-minded others via social media that reinforces their view of the world and distances others.

Antagonism toward ecocentricity goes some way to explaining why more people are not attracted to permaculture (Smith and McTiernan 2021). Holmgren (2002) appears to recognize this in reflecting on a perception of permaculture as "theoretical, utopian and impractical because it was difficult to apply within the prevailing" system, with its "social, market and policy environment" (p. xxii). To advance Holmgren's spiral and the intention to create integrated, whole systems, it appears future teaching and practice of permaculture will need to develop processes that more overtly surface and recognize maladaptive worldviews, the social and physical structures these give rise to, and debate ways to move toward more ecocentric positions.

A Rising Global Social Movement

Having discussed the framework for design and the implications of competing worldviews, we now turn to the third conceptualization of permaculture as a social movement. If permaculture is unable to find a comfortable home within the tent of sustainable development, is it as a social movement that it will find its power to enact change? Increasing community awareness and experience of environmental threats such as floods and wildfires, combined with rises in the cost of energy, water, and food as well as the disruption of another pandemic could indeed be the catalyst for an increase in permaculture-inspired (if not named as such) orientation across cities, towns, and rural landscapes.

Building a social movement was always the intention of Mollison and Holmgren as from the outset they, along with a small cohort of like-minded others, set out to expand the philosophy and concepts of permaculture globally. Grassroots cultural niches grounded in permaculture ethics, principles, and design were seen as the central model for change (Dawborn and Smith 2011). Since its inception in the early 1970s, the number of people who have been involved with permaculture has grown considerably. Research data from the Next Best Step (NBS) project estimate that over 3,000,000 people worldwide have engaged with the ideas. Figure 12.3 shows the growth in the number of permaculture organizations globally from pre-1978 to beyond 2012, while the

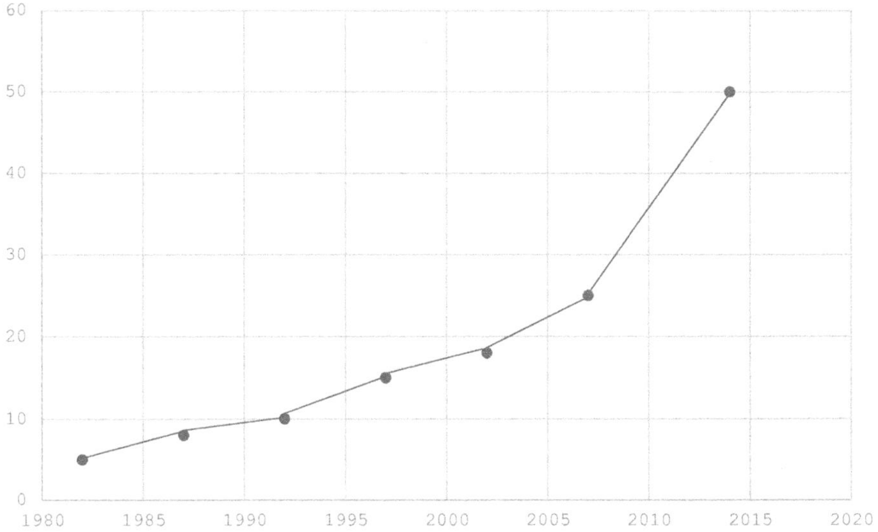

Figure 12.3 Growth in the number of permaculture organizations pre-1978 to beyond 2012.
Source: van der Velden 2017, The Next Best Step.

Oregon State University Permaculture Design Course provides a comprehensive view of the worldwide presence of permaculture training and permaculture organizations (https://workspace.oregonstate.edu/osu-permaculture-design).

Ferguson and Lovell (2015) provide the most detailed assessment to date of grassroots participation in the international permaculture movement. They define characteristics of grassroots movements as both enabling and hindering the adoption of permaculture. The capacity to mobilize resources in favor of locally relevant technical and social innovation has certainly helped to establish cultural niches where the teaching and practices of permaculture can be preserved, and resistant to commodification. Conversely, the limited extent of organization or institutionalization has left local innovation vulnerable to suppression or co-option by state or business interests. Further, there is the risk of becoming insular or exclusive, as noted in our consideration of worldviews, overlooking potential alliances with sister movements. Finally, the idea of scalar mismatch, where limited resourcing and narrow, technical solutions to challenges such as environmental degradation are insufficient to address issues that are more systemic in origin and beyond local technical applications.

As Leahy (2021) explains, a critical mass of bottom-up, grassroots participation has the potential to seed central aspects of a new society, such as via small ethical businesses, ethical cooperatives, with state- or bioregional-level governance, forming the basis for a cultural shift and radical transformation of present systems. Beddoe et al. (2009) describe the concept of cultural evolution, where cultural variants (akin to genetic variants) may emerge to form the dominant culture in times of crisis, of rapidly changing social and/or ecological conditions. Holmgren is convinced this is not a matter of if, but when:

> insofar as permaculture is an effective response to the limitations on use of energy and natural resources, it will move from its current status as 'alternative response to environmental crisis' to the social and economic mainstream of the postindustrial era. Whether it will be called permaculture or not is a secondary matter.
>
> (Holmgren, 2002, xvii)

In her detailed study of a PDC, Smith (2000) found that participants came from an eclectic range of backgrounds, not just from those relating to landscape or horticulture as might be expected (though the sample size of 22 was small). The majority were already attracted to the idea of sustainability long before they embarked on the course; they were already 'on the journey' as it were. They recognized the need to address the fragile state of earth's systems and expressed an ecocentric worldview consistent with the ethics of permaculture. While not clearly a catalyst for transforming worldviews, permaculture served to positively reinforce their existing worldviews and values. Most apparent was that participants emerged with an enhanced sense of agency, replacing potential fears of The Great Unravelling with a hope and determination to become active change agents working toward desired futures. However again, when asked where they might seek to apply their new-found insights and practical skills, the majority were keen to apply the principles to their own situation, mainly to their garden. This has implications for permaculture education, as we note below.

Critiques and Paradoxes

Previously we addressed some of the tensions and paradoxes that are contained within aspects of permaculture. We now turn our attention to two of the more significant critiques that have been leveled at permaculture. The first 30 years of permaculture were focused most intensely on establishing a grassroots movement, promoting participation and action, and developing an education model (the PDC) that was accessible and easily disseminated. As noted earlier, Mollison was vehemently opposed to permaculture being offered through mainstream academic institutions, and, along with the early pioneers, strongly resisted efforts to allow them to access permaculture theory and practice. This distancing from academia has led to consequences for both systematic enquiry and education in permaculture, in particular, investigation into claims to scientific rigor, and pedagogical processes.

Where is the Scientific Rigor in Permaculture?

A concern for the scientific rigor and evidence supporting permaculture processes is clearly recognized among some permaculturists themselves (van der

Velden, 2017). There are two prominent dimensions to this: first, the credibility afforded to permaculture by those within more mainstream institutions, particularly agriculture, if the evidence base is weak; and second, accountability, in that permaculturists need to be accountable for the techniques they promote. Apart from biennial country-based and international convergences, there are currently no systematic operational processes such as peer-reviewed journals for permaculture practitioners to collaborate globally (Van der Velden 2017), other than through a variety of internet forums of varying rigor and quality.

A grassroots movement with limited resources cannot afford to research and disseminate techniques that require resources for enquiry but may provide limited benefit. As noted earlier, the danger remains that permaculturists may unintentionally apply strategies that cause harm to ecosystems. As Ferguson (2014) puts it, there are concerns of overreach where extrapolations from ecological principles oversimplify the science, resulting in claims and prescriptions beyond what we really know how to do. For example, some permaculturists have a cavalier attitude to weeds such as weeping willows, seeing them as doing no harm as long as they fulfill a useful function within a permaculture design.[3] An example often quoted during PDCs is the claim by holistic management consultant and farmer Allan Savory in his widely viewed 2014 TED talk, that domesticated herbivores, principally cows, have the potential to sequester large quantities of carbon. Impressive photographs of lush landscapes are presented as evidence, when the kind of scientific evidence necessary to validate such claims is at best highly controversial and certainly not applicable to all situations (e.g., Briske et al. 2013).

Scientific foundations are beginning to emerge. Of the published literature within the past decade, Ferguson and Lovell (2015) identify several studies highlighting the potential for permaculture to enhance farm sustainability and ecological literacy and practices more broadly. Krebs and Bach (2018) examined each of Holmgren's 12 principles and concluded that scientific applications can be demonstrated for each. For example, the principle of 'Catch and Store Energy' can be understood through the example of organic mulch application, rainwater harvesting measures, or the retention of woody elements in agriculture.

However, permaculture design is oriented toward the creation of integrated (complex) adaptive systems, and researching its effectiveness requires a sophisticated methodology grounded in systems thinking and design. Mollison (1988) gives the example of 'yield', as in Holmgren's principle 3 – 'obtain a yield'. Yield may be simply measured in terms of tons per acre per year of a particular crop, or in terms of the wider impact on biodiversity, health, greenhouse gas emissions, and so on of the permaculture system. Those familiar with reductionist scientific methods which are able to generate discrete, measurable outputs will appreciate the complexity of such a task, and that the unique quality of emergence within complex systems further limits the extent to which a 'successful' demonstration in one context can be replicated elsewhere. Indeed Ferguson (2014) argues that discussions of practice consistently underplay the complexity, challenges, and risks that are faced in developing diversified and integrated systems, particularly for agricultural production. While he believes the permaculture principles and strategies are broadly well supported by science and complement and extend agroecology literature, they are not always backed by agricultural science.

In grappling with the complexity, Ferguson (2014) identifies "the critical unmet need is the social process that facilitates the identification, development, adoption, implementation, dissemination, and adaptation of techniques across sites and social contexts". With his understanding of a movement with limited access to resources and research expertise, he calls for permaculturists to embrace a model of "People's Science", involving practitioners in a culture of experimentation and critical thinking that 'mobilizes the resources of contemporary scientific research, and integrates them into our movement' (2014, 2). Ferguson argues that a 'People's Science' will strengthen permaculture, "by nudging us toward a healthier and more critical attitude toward science, and nudge us away from that part of our legacy that relies too heavily on charisma and plays too loose with evidence" (2014, 3)

Permaculture Education

There have been few efforts to improve permaculturists' lack of scientific literacy as part of a PDC, which limits capacity to critique. One example, in Australia, the introduction of knowledge and skills in experimental design within the permaculture curriculum in the Vocational Education and Training (VET) Level 5 course (see https://permacultureaustralia.org.au/apt/). With this in mind we now turn attention to the critiques of permaculture education, in particular, from the way permaculture education is delivered, largely still through the PDC. Having lectured in the university system, Mollison was strongly and legitimately opposed to what he saw as the institutionalization and fragmentation of knowledge. As a result, he refused to allow permaculture to be adopted and taught within the higher education sector, believing that its essential and intended holism would be lost. Further, he argued such a move would result in essential elements of permaculture such the ethics, being omitted or presented as 'optional' units. In response, Mollison developed the PDC curriculum based on his *Designer's Manual* (1988) and claimed that was sufficient to equip graduates to venture into the world to bestow their newly acquired know-how on to others. (It is perhaps ironic that the original PDC was designated a 72-hour course, based on the number of hours in a typical university unit.)

In the 1980s and 1990s, in Australia at least, Mollison's company was the sole issuer of certificates of attainment for a PDC, and only issued certificates for courses that demonstrated strict adherence to his curriculum. This tight control was initially an enabler for global distribution and access to teaching and buffered against commodification. Today, while offered in many parts of the world and a range of formats, the PDC remains in the informal education sector. Although there have been some attempts, nowhere is permaculture directly credentialed within Higher Education, though it has been modified to fit a Vocational Education and Training certificate course in Australia. However, there are now a number of postgraduate theses written with aspects of permaculture at their core, for example, Harney 1997; Janzon 2018; Krawczyk 2022; Loveday 2021; Palaniappan 2009; Smith 2000.

Further, there is an obvious tension and even contradiction between creating a defined curriculum, and enabling the integration of new forms of knowledge and thinking that reflect the dynamic and rapidly changing social and ecological contexts in which we live, and, indeed, of permaculture itself. While it was extremely useful for the early teachers of permaculture to have a curriculum as a resource on which to base their teaching, this has become somewhat of a barrier to innovation, as well as not being able to take account of knowledge of adult pedagogies other than a nod to the oversimplified and insufficient framework of 'visual, auditory, reading and kinaesthetic' (VARK) modes of learning.

Smith (2000) found that after completing a PDC, the participants still seemed content to operate at the level of their own garden or, at best, local community. While they were metacognitive learners in that they were highly intrinsically motivated, they brought their new understanding and application of permaculture into their existing situations such as applying it to the garden: 'that's my permaculture over there'. The hoped-for intent of education in permaculture toward transforming worldviews only manifested for the very few.

This reveals a notable deficiency within current teaching. Unless purposefully built into PDCs, other than through the creation of the design itself, deeper analytic, synthetic, and critical reflection practices as well as developing systems thinking competence may only emerge serendipitously through the nature and purpose of permaculture and the commitment of the participants. Permaculture is based on a systems approach, engaging with the domains in Holmgren's flower (Fig. 12.2). The flower is an attempt to unify the domains of permaculture and the tensions between this and the design exercise; however, systems thinking (e.g., Meadows 2008) is particularly poorly addressed at all levels in the Australian education system and PDC participants do not find it an easy task. What is still lacking is a framework for strengthening understanding of permaculture's holistic nature.

Foregrounding systems thinking techniques and tools in design, then, is critically important. It is the glue that helps people see the 'bigger picture', to see relationships and appreciate the likely scale, context, and nature of impact of their decisions as they create a permaculture design. As Vitari and David (2017) emphasize, it is the skills of systems thinking and design which enable permaculture to be translated to other domains of human endeavor, beyond agriculture or gardening. In our experience, PDC participants are often asked to develop a permaculture design for their own properties or situations, which, from a systems thinking perspective, creates a boundary that is too narrow and fails to identify important systems relationships that will shape the success of their endeavors. Neither is there typically any insistence that the design process reflect all the domains of the flower. This is the Achilles' heel of permaculture education; the emphasis on designing 'your own patch' plays strongly into the individualism of neoliberal ideology and the commodification of every facet of daily life, at least in the more economically affluent world of the Global North.

Discussion and Conclusion

In this chapter we reflected on the increasingly dire state of the world in the era of the Anthropocene, and the need for drastic action to reverse our currently unsustainable trajectory and conceptualize and build a flourishing future. We described how, in the societies of the Global North, the dominant worldview remains aligned to neoliberalism where expanding material wealth has become central to our identity and deemed a measure of societal success and progress. We also expressed the hope that the climate crisis may catalyze the search for more ecological values that align with a sustainable, flourishing future.

We then introduced and described the set of ethics, principles, and design strategies known as permaculture, as a holistic framework for creating sustainable human presence on Earth. We showed how permaculture's strength and attraction for its many adherents lies in its promise to provide both a theoretical framework and practical skills for designing systems to support sustainable human settlement that are fundamentally action-orientated and positive.

We also recognized that because of its very complexity both as an overtly ethical system and as a design framework, apart from the difficulty of even defining it, permaculture embeds a number of contradictions, paradoxes, and tensions. Key tensions are around whether permaculture has sufficient scientific rigor, the desire for permaculture to be part of transformational change in terms of working toward a lower energy sustainable future in the teeth of a dominant neoliberal culture, and the way in which permaculture is generally adopted without overtly challenging the dominant worldview. We further identified that education in permaculture, largely through the vehicle of the PDC, while free from the constraints of a formal system, may inadvertently exacerbate rather than resolve, these tensions.

While the influence of permaculture has expanded to fields such as agroecology, eco-design, urban planning, community development, and international aid and development (e.g., Ferguson 2017; Greenblott and Nordin n.d.; Holmgren n.d.), we believe that permaculture alone will not move us at a sufficient rate and scale to steer us toward a flourishing future. Within this is a call for permaculturists to confront their/our own biases, to make room for and form alliances with other 'cultural variants'.

Given the convergent polycrises, awareness of the necessity and urgency to address our collective overshoot will bring permaculture and its allied movements to the attention of our thought leaders – academics, educators, activists, planners, and policymakers, as well as even those in the corporate sector. So, we now come to consider the answer to the question posed by our title

"Permaculture: A System for Designing Sustainable Human Settlements. Ahead of its Time or Impossible Dream?" As the movie title says, it's complicated! Ahead of its time? Probably, as far as mainstream uptake goes. Beddoe et al. (2009) identify social and ecological crises as key catalysts for cultural evolution, yet continued growth in energy and material consumption currently feeds the illusion that human society is still in a period of ascendency. Coupled with the belief that technology will stave off any looming resource limits, there appears little incentive for mainstream neoliberal culture to see the need to explore alternative ways of being in the world that might mean an upheaval to their lives.

Impossible dream? No. Actually, it depends. Major transformations will occur, whether through design and engagement of a critical mass of global citizenry, or as a consequence of repeated crises and collapse. As a holistic system of design, permaculture clearly has an abundance to offer. It has traditionally been at its most successful at the small, the local, and the personal levels, and here lies both its strength and its weakness. A particular strength of permaculture is its ability to rapidly retrofit the suburbs (Holmgren n.d.) through cooperative redesign. It is also able, if necessary, to transform acres of high-energy input unproductive lawn and gardens to low-input food production as part of the creation of an urban agriculture, paralleling the 'Dig for Victory' garden movement in the United Kingdom during World War II where gardens and waste ground were turned over to food production. A more recent example where permaculture played a key role was to assist Cuba's need for rapid recovery after the fall of the Soviet Union in 1991, as told in the video 'The Power of Community: How Cuba Survived Peak Oil' (Morgan 2006). Today, urban and peri-urban agriculture are now estimated to produce between 15% and 20% of global food production (Peyan et al. 2022).

Application of permaculture at the local level is powerful in providing highly practical and inspiring contextually based solutions. However, identifying and pursuing the deeper leverage necessary for whole systems change will require permaculturists to engage more consciously in systems thinking and design which has always been a feature of permaculture.

Permaculture also offers a positive adaptive response. An intrinsic feature of the permaculture design framework is addressing risks such as flood and fire to withstand future extremes, while also enhancing system resilience through retaining water, building soil fertility, food storage capacity, and so on. With increasing prospects of extreme fire, flood, drought, or cyclone hit, a permaculture settlement may or may not survive any better than anything else, but what will survive is the knowledge, the systems understanding, the skills, and the attitudes to rebuild quickly, an important regenerative capacity in the face of adversity. In the face of such adversity there is no guarantee that individuals will seek out and engage in cooperative responses. At best, we may be doomed to see mere islands of permaculture in a climate-challenged, authoritarian world.

However, permaculture is now more visible as a global movement, offering an inspiring, ethically grounded, practical, and solutions-oriented framework for designing a sustainable human presence in the world. It offers the foundation for positive adaptation to rapidly changing circumstances.

We conclude with these lines from David Suzuki, quoted in HopeDance 2009, that capture our own hopes for a future enabled through permaculture:

> What permaculturists are doing is the most important activity that any group is doing on the planet. We don't know what details of a truly sustainable future are going to be like, but we need options, we need people experimenting in all kinds of ways and permaculturists are one of the critical gangs that are doing that.

Notes

1 For other prominent texts, see https://www.permaculture.org.uk/research/top-ten-permaculture-books.
2 This claim is contested, raising concerns of appropriation of cultural knowledge and practices. See Spangler, K., R.B. McCann, and R.S. Ferguson. 2021.
3 The issue of weeds on ecosystems versus their use in design remains a contested area within permaculture.

References

Beck, D.E., and C.C. Cowan. 2005. *Spiral Dynamics: Mastering Values, Leadership and Change*. Blackwell Publishing.

Beddoe, R., C. Costanza, and J. Farley et al. 2009. "Overcoming Systemic Roadblocks to Sustainability: The Evolutionary Redesign of Worldviews, Institutions, and Technologies." *PNAS 106*, no. 8 (February): 2483–2489. https://doi.org/10.1073/pnas.0812570106

Bendell, J. 2020. "Deep Adaptation: A Map for Navigating Climate Tragedy." *IFLAS* Occasional Paper 2 www.iflas.info. Revised 2nd Edition.

Briske, D., B.T. Bestelmeyer, J.R. Brown, S.D. Fuhlendorf and H.W. Polley. 2013. "The Savory Method Cannot Green Deserts or Reverse Climate Change. A response to the Allan Savory TED video." *Rangelands 35*(5): 72–74. https://doi.org/10.2111/RANGELANDS-D-13-00044.1

Carter, L., and C. Smith. 2023. The United Nations Sustainable Development Goals in a Neoliberal World. In: K. Beasy, C. Smith and J. Watson, eds. *Education and the UN Sustainable Development Goals. Education for Sustainability* vol 7. Springer, Singapore. https://doi.org/10.1007/978-981-99-3802-5_3

Dawborn, K. and C. Smith. 2011. *Permaculture Pioneers: Stories from the New Frontier*. Hepburn: Melliodora Publishing.

De Castañeda, R., J. Villers, C. Guzmán, T. Eslanloo, N. de Paula, C. Machalaba, J. Zinsstag, J. Utzinger, A. Flahault and I. Bolon. 2023. "One Health and Planetary Health Research: Leveraging Differences to Grow Together." *The Lancet Planetary Health*, Vol 7, Issue 2: 109–111. https://doi.org/10.1016/S2542-5196(23)00002-5

Evans, C. 2020. "Can Permaculture Play a Positive Role in International Development?" *Re-Alliance*. https://www.re-alliance.org/post/permaculture-development

Ferguson, R.S. 2014. "Toward 21st Century Permaculture: Peoples' Science or Pseudoscience?" *Permaculture Activist* No. 93.

Greenblott, K and K. Nordin n.d. *Permaculture as a 'Development Tool'*. https://neverendingfood.org/b-what-is-permaculture/permaculture-as-a-development-tool/#:~:text=Many%20Organizations%20are%20beginning%20to,are%20met%20by%20working%20seasonally%2C

Ferguson, R. S., and S.T. Lovell. 2015. "Grassroots engagement with transition to sustainability: diversity and modes of participation in the international permaculture movement." *Ecology and Society 20*(4):39. http://dx.doi.org/10.5751/ES-08048-200439

Ferguson, S. 2017. *Taking the Permaculture Path to Community Resilience. Smart Cities Dive*. https://www.smartcitiesdive.com/ex/sustainablecitiescollective/taking-permaculture-path-community-resilience/52646/

Harney, P.J. 1997. "Changing The Social System of a Catholic Secondary School: An Examination of Salient Design Features Pertinent to the Change Process from a Permacultural Perspective." PhD diss., Australian Catholic University.

Holmgren, D. 2002. *Permaculture: Principles and Pathways Beyond Sustainability*. Hepburn: Holmgren Design Services.

Holmgren, D. (n.d.) *Retrosuburbia: The Downshifter's Guide to a Resilient Future*. Melliodora.

Hopkins, R. 2008. *The Transition Handbook*. Green Books.

Howard, E. 1898, 2003. *To-Morrow: A Peaceful Path to Real Reform*. London: Routledge.

Jackson, T. 2009. *Prosperity Without Growth*. Routledge.

Janzon, T. 2018. "The Quest for Sustainability – A Critical Reading of Permaculture Literature." Masters diss., University of Uppsala.

Kopnina, H. 2013. "Evaluating Education for Sustainable Development (ESD): Using Ecocentric and Anthropocentric Attitudes Toward the Sustainable Development (EAATSD) Scale". *Environment, Development and Sustainability 15*: 607–623. https://doi.org/10.1007/s10668-012-9395-z

Krawczyk, M. 2022. "Staying with the Bioregional Trouble in Two Permaculture Sites: An Ethnography and Four Films." DOI https://doi.org/10.25904/1912/4703. PhD diss., Griffiths University.

Krebs, J. and S. Bach. 2018. "Permaculture—Scientific Evidence of Principles for the Agroecological Design of Farming Systems". *Sustainability 10*(9): 3218. https://doi.org/10.3390/su10093218

Leahy, T. 2021. *The Politics of Permaculture*. Pluto Press.

Loveday, M. E. 2021. "*The Use of Nostalgia at the Ideation Stage of Permaculture Design*." PhD diss., University of Leeds.

Luke, T. W. 2021. "Beyond Prepper Culture as Right-wing Extremism: Selling Preparedness to Everyday Consumers as How to Survive the End of the World on a Budget". *Fast Capitalism 18*(1): 50–62

Macy, J., and C. Johnstone. 2012. *Active Hope: How to Face the Mess We're in Without Going Crazy*. Canada: New World Library.

Mannen, D. S. Hinton and T. Porter. 2012. "Sustainable Organizing: A Multiparadigm Perspective of Organizational Development and Permaculture Gardening" *Journal of Leadership & Organizational Studies* 19(3): 355–368. https://doi.org/10.1177/1548051812442967

Meadows, D. 2008. *Thinking in Systems*. Chelsea Green.

Merz, J.J., P. Barnard, W.E. Rees et al. 2023. "World Scientists' Warning: The Behavioural Crisis Driving Ecological Overshoot." *Science Progress 106*(3). https://doi.org/10.1177/00368504231201372

Miller, A. and R. Heinberg. 2023. *Welcome to The Great Unraveling: Navigating the Polycrisis of Environmental and Social Breakdown*. Post Carbon Institute. https://www.postcarbon.org/publications/Welcome to the Great Unraveling: Navigating the Polycrisis of Environmental and Social Breakdown.

Mollison, B. 1988. *Permaculture: A Designers' Manual*. Australia: Tagari.

Mollison, B. 1996. *Travels in Dreams*. Australia: Tagari.

Mollison, B. and D. Holmgren. 1978. *Permaculture One*. Australia: Tagari.

Morgan, F. (2006). (Director). *The Power of Community: How Cuba Survived Peak Oil* (video). AlchemyHouse Productions Inc. https://intercontinentalcry.org/the-power-of-community-how-cuba-survived-peak-oil/

Morrow, R. 2022. *Earth Restorers' Guide to Permaculture*. Hepburn: Melliodora.

Palaniappan, G. 2009. "*Learning to Practice Transitional Agriculture: An Action Research Thesis*." PhD diss., The University of Queensland.

Payen, F. T., D.L. Evans, N. Falagán, C.A. Hardman, S. Kourmpetli, L. Liu et al. 2022. "How much food can we grow in urban areas? Food production and crop yields of urban agriculture: A meta-analysis." *Earth's Future*, 10: e2022EF002748.

Roux-Rosier, A., R. Azambuja and I. Gazi. 2018. "Alternative Visions: Permaculture as Imaginaries of the Anthropocene." *Organization 25*(4): 551–572. https://doi.org/10.1177/1350508418778647

Savory, A. 2013. *How_to_Fight_Desertification_and_Reverse_Climate_Change*. Video TED talk. https://www.ted.com/talks/allan_savory_how_to_fight_desertification_and_reverse_climate_change

Smith, C. 2000. "The Getting of Hope: Personal Empowerment through Learning Permaculture." PhD diss., The University of Melbourne.

Smith, C. and D. McTiernan. 2021. "Integral permaculture: Making sense in a complex world." *Permaculture Works*: 54–62.

Spangler, K., R.B. McCann and R.S. Ferguson. 2021. "(Re-)Defining Permaculture: Perspectives of Permaculture Teachers and Practitioners across the United States." *Sustainability 13*(10): 5413. https://doi.org/10.3390/su13105413

Suzuki, D. 2009. In HopeDance. *Quotes about Permaculture. HopeDance: Celebrating Transition, Opportunity & Resilience*. http://www.hopedance.org/index.php/home/soul-news/996

Tooze, A. 2021. *Shutdown: How COVID shook the world's economy*. Viking.

Turchin, P. 2023. *End Times: Elites, Counter-Elites, and the Path of Political Disintegration*. Allen Lane.

Van der Velden, N. K. 2017. *Next Big Step Report*. The Permaculture Association of Britain https://www.permaculture.org.uk/sites/default/files/materials/5664/permaculture_next_big_step_report_on_international_survey_responses.pdf

Veteto, J.R. and J. Lockyer. 2008. "Environmental Anthropology Engaging Permaculture: Moving Theory and Practice Toward Sustainability". *Culture & Agriculture*. *30*(1-2): 47–58. https://doi.org/10.1111/j.1556-486X.2008.00007.x

Vitari, C. and C. David. 2017. "Sustainable Management Models: Innovating Through Permaculture." *Journal of Management Development 36*(1): 14–36. https://doi.org/10.1108/JMD-10-2014-0121

United Nations. (n.d.). *Sustainable Development Goals*. https://www.un.org/sustainable development/sustainable-development-goals/

13 A Paradox of the Anthropocene

The Radicalization of Techno-Scientific Modernity and the Future of Solar Geoengineering

Jean-Daniel Collomb

The concept of Anthropocene stems from a recognition that our Modern vision of nature has involuntarily led to the end of nature, i.e. a situation in which the human species has become a geological force in its own right, making the Modern idea of a clear-cut separation between nature and culture increasingly obsolete. This has come to pass because of a profoundly instrumental approach to the natural world – first in a few Western countries, then in several regions of the world – which rests on the notion that nature is above all a pool of resources which can be bent to human desires through the use of science and technology. This perspective on the relationship between humans and nature set the stage for industrialization from the early nineteenth century onward, and continues to shape many of our intellectual habits to this day (although it does not go unchallenged). In practice, the techno-scientific project has produced both tremendous human progress (and no longer just in a few Western countries) and terrifying environmental challenges, in the form of biodiversity loss, water scarcity, ocean acidification, and acid rain, to name only a few.

No wonder, then, that the increasing salience of climate change in national and international politics and the rising awareness of the Age of the Anthropocene have prompted a lot of soul searching about what the proper relationship between humans and their natural environment ought to be – even though it should be noted that ecological reflexivity has a rich history that long predates the emergence of the term Anthropocene.[1] Many activists, academics, and public officials have set out to advocate profound cultural change, which they present as necessary to change our ways and drastically lower our footprint on nature. Most of these reflections lead to the conclusion that the techno-scientific vision of nature is the root of the problem, and that it is therefore urgent to move away from a starkly instrumental conception of nature in order to embrace a new vision founded on a clear recognition of our dependence on natural cycles and of inescapable physical limits to economic development.

And yet, for all the efforts made by those activists and intellectuals, the world still remains very far from a paradigm shift away from techno-scientific Modernity. The large discrepancy between global commitments to decarbonization and the reality of global emissions (the so-called "emissions gap"[2]) is a

DOI: 10.4324/9781003385202-14

case in point. Furthermore, many influential proposals to grapple with our environmental challenges continue to bear the hallmark of the techno-scientific *ethos* and suggest that many twenty-first-century humans are merely striving for an updated and enlightened continuation of the Modern project of human mastery of nature.

Thus, humankind finds itself in a disorienting and somewhat paradoxical situation in the early twenty-first century: although a series of daunting and potentially devastating environmental challenges, whose root cause has been attributed to the Modern techno-scientific *ethos*, have been duly identified, alternative visions, though attractive to some, have turned out to be unable to replace it – at least for now. Proponents of a technologically driven green transition assert that humans do have the capacity to secure an absolute decoupling of our environmental footprint and economic growth, while radical critics of economic growth, such as Richard Heinberg and William Rees, make apocalyptic predictions about the collapse of human societies and civilizations. This chapter will explore another potential scenario, in which some governments or private organizations turn to more desperate and extreme attempts at ameliorating and adapting to our environmental challenges by doubling down on highly complex and speculative technological solutions, in the form of solar geoengineering schemes.

Solar geoengineering is portrayed by its supporters as a contingency plan that humankind may have to activate in the event that it fails to avert catastrophic climate change in the twenty-first century. Put simply, solar geoengineering would consist of manipulating the Earth's temperature in order to limit warming, at least temporarily. Paradoxically enough, this means that, in the context of the Anthropocene, a group of political, scientific, and corporate leaders may be tempted to take the techno-scientific Modern project a step further. In this chapter, I will argue that the growing allure of solar geoengineering in some elite circles may mean that the Modern vision of nature, which was instrumental in precipitating human-made climate change, may be given a new lease of life as a consequence of the predicament it helped create in the first place.

What is the Modern vision of nature?

The Modern vision of nature is said to have emerged in Europe in the course of the scientific revolution in the seventeenth century, and is often portrayed as a techno-scientific quest to dominate nature and exploit it in order to cater to human needs.[3] According to the anthropologist Philippe Descola, the Modern vision of nature rests on a clear-cut division between nature and culture, so that Modern dualism prescribes that society, humankind, and human-made artifacts be perceived as fundamentally different and separate from the natural world.[4]

Because of its mixed record – an unparalleled rise in the material living standards of hundreds of millions of people combined with major negative environmental externalities – the Modern vision of nature came in for much criticism in the last decades of the twentieth century.[5] As early as the 1980s, Deep ecologists, such as Bill Devall and George Sessions, faulted the "Dominant, Modern Worldview"[6] for fostering a destructive brand of anthropocentrism and promoting a delusional belief in unlimited economic growth and resource consumption, which led them to call for "a shift in worldview based upon a metaphysics consistent with ecological interrelatedness."[7] A couple of decades later, the theory of the Anthropocene, singling out humankind as a distinct geological force, further undermined Modern dualism.

As early as 1990, Bill McKibben's book, *The End of Nature*, had foreshadowed the Anthropocene thesis. McKibben held that human activities had transformed the natural world so deeply that nature could no longer be said to exist as a distinct entity, separate from human civilization:

> The idea of nature will not survive the new global pollution – the carbon dioxide and the CFCs and the like. This new rupture with nature is different not only in scope but also in kind from salmon tins in an English stream. We have changed the atmosphere, and thus we are changing the weather. By changing the weather, we make every spot on earth man-made and artificial. We have deprived nature of its independence, and that is fatal to its meaning. Nature's independence *is* its meaning; without it there is nothing but us.[8]

More to the point, the ecocentrist philosopher J. Baird Callicott argued that it was actually "the modern idea of nature" that was no longer relevant, since it boiled down to seeing humans as external to nature, despite mounting evidence to the contrary.[9] This questioning of the Modern vision of nature culminated in the widely discussed 2000 article by Paul Crutzen and Eugene Stoermer, in which they posited that humankind should now be regarded as a geological force in its own right. In the Age of the Anthropocene, they warned, nature could no longer exist beyond the reach of human influence.[10]

The implications of the advent of the Anthropocene not only for our common understanding of the meaning of nature, but also for the future of environmental advocacy, are enormous. Building on the notion that nature is "a social construct, and thus its meaning changes across time and space" and that "one cannot consider (let alone defend) nature independent of human experience," Paul Wapner urges twenty-first-century environmentalists to turn their backs on Modern dualism, celebrate the hybrid nature of the world we live in, and replace the Modern project of human domination of nature with a sober and humble acceptance of our role in shaping our environment.[11] From this perspective, the Anthropocene may be interpreted as both the by-product of the Modern vision of nature and the reason why we should give up on it.

Unsurprisingly, the growing prominence of the Anthropocene has led many scholars and activists to articulate new visions of what the proper relationship between humans and their natural environment ought to be. They call for profound cultural change, as a means to both redefine the good society and foster a more sustainable way of inhabiting the earth.

In the field of economics, for example, Herman Daly and Joshua Farley have called for "neoclassical economics" to be replaced with "ecological economics," which, they claim, "envisions the macroeconomy as part of a larger enveloping and sustaining Whole – namely, the Earth, its atmosphere, and its ecosystems."[12] From the perspective of ecological economics, our dependence on nature's services ultimately implies that the search for human mastery over nature is nothing short of a dangerous fool's errand. In the end, this means that we ought to pursue justice and equity within the physical limits imposed by the Earth and its multiple ecosystems, and stop regarding nature as a pool of unlimited resources at our disposal. This worldview echoes the spirit of the 1972 bestseller *Limits to Growth* and dovetails with the warnings issued by the likes of the economist Tim Jackson, drawing inspiration from Amartya Sen's work, who recommends that our understanding of sustainability be predicated, not on "a set of disembodied freedoms," but on "a range of 'bounded capabilities' to live well – within certain inevitable limits."[13]

In like manner, several philosophers and ethicists see the advent of the Anthropocene as a golden opportunity to liquidate Modern dualism and discredit the notion that human civilization and nature are fundamentally different and separate.[14] Deep ecology, whose advocates demand that ecological systems be endowed with ethical value, is a case in point. Following in the footsteps of the Norwegian philosopher Arne Naess, deep ecologists seek to foster a radically new worldview and a new understanding of our relationship with nature, whereby we would stray away from Modern dualism so that we recognize that humans are part and parcel of Nature, not separate from it. According to this logic, once we come to terms with this ontological bond, we will come to the realization that we preserve ourselves by preserving nature.[15] Arguing from a different perspective, Bruno Latour has celebrated the Anthropocene as a configuration in which Modern dualism between the social and the natural is no longer tenable.[16] This inescapable truth, as he sees it, has propelled twenty-first-century humankind into a "postnatural period" in which hybridity, *in lieu of* dualism, is bound to be front and center.[17]

However, calling into question Modern dualism is far from being a consensus position. According to many scholars and public officials, formidable environmental challenges, such as climate change and biodiversity loss, can be addressed within a Modern framework, informed by scientific research and technological innovation. In his defense of "Enlightenment optimism," Steven Pinker claims that, in order to grapple with those challenges successfully, we actually need to pursue the Modern experiment further: "The key idea is that environmental problems, like other problems, are solvable, given the right knowledge."[18] What is called for, in other words, is an updated and better-informed version of Modern mastery of nature. That is because, were this balancing act to be achieved, we would be able to continue to reap the many benefits of Modernity (in the form of rising living standards, for example), while reducing our footprint on the natural world drastically. Such a rationale underpins the approach to decarbonization advocated by proponents of green growth, who are wont to assert that giving up on our constant pursuit of economic growth would be immoral, on the ground that the poorest and marginalized would be disproportionately hurt.[19] Consider, for example, Bill Gates's 2021 book, *How to Avoid a Climate Catastrophe*, in which climate change is described as an engineering problem that can be successfully contained thanks to human ingenuity and innovation.[20] In his view, nature remains a gigantic pool of resources to be manipulated at will to serve human needs.

There is a wealth of evidence to suggest that the recent public policy responses to climate change have been largely shaped by Steven Pinker's brand of Enlightenment optimism, rather than by a questioning of Modern dualism. The climate strategy adopted by the Biden administration no doubt testifies to this trend. In the summer of 2022, Congressional Democrats passed the Inflation Reduction Act, the main purpose of which is to bring about rapid and substantial decarbonization of the US economy by investing hundreds of billions of dollars into the development and deployment of a whole range of low-carbon technologies.[21] The basic idea behind this piece of legislation is that, thanks to industrial policy aimed at achieving green growth, the US consumer can continue to enjoy the material benefits of economic growth while lowering her carbon footprint. Far from pushing for a radically transformed understanding of our relationship with nature, President Biden has repeatedly insisted that streamlining and updating US energy systems and manufacturing processes would be the key to solving the climate crisis. John Kerry, his Special Presidential Envoy for Climate, even went so far as to call the transition to net zero "the greatest economic opportunity since the Industrial Revolution."[22]

In light of these conflicting perspectives on the path forward in the Age of the Anthropocene, it is hard to know whether Modern dualism is about to be abandoned altogether, or whether it could be rehabilitated by the proponents of ecological modernization. The emergence of solar geoengineering as an instrument to cope with a warming climate suggests that the latter scenario is a credible possibility.

What is solar geoengineering? Solar geoengineering comprises a set of emerging technologies that aim to reflect sunlight back into space to limit temperature increases temporarily. A relatively broad range of expressions – climate management,[23] climate intervention,[24] climate engineering,[25] albedo modification[26] – have been used by scholars to refer to these technologies. Put simply, Belter and Seidel define climate engineering as "deliberate, large-scale manipulation of the earth system with the intention of mitigating the effects of climate change."[27] It should be borne in mind that solar geoengineering does not help lower GHG concentrations in the atmosphere, and does not ameliorate ocean acidification. Several techniques fall into the solar geoengineering technology: urban whitewashing, marine cloud brightening, cirrus cloud thinning, stratospheric sulfate aerosols, and the use of orbital mirrors to reflect sunlight back into space.[28] Stratospheric sulfate aerosols, which is the most frequently discussed approach, would seek to mimic the cooling effect of volcanoes by pumping sulfur into the stratosphere, twenty kilometers above the surface of the Earth.[29] In effect, successfully deploying solar geoengineering would amount to devising a chemical thermostat for the Earth so that humankind could contain the rise in global temperatures and use the delay to actually decarbonize and build more climate-resilient infrastructures.[30]

Historically, solar geoengineering has not garnered nearly as much attention as mitigation and adaptation in the climate conversation. However, it began to be discussed more frequently after the Dutch geo-chemist and Nobel Prize winner Paul Crutzen wrote an article in support of research into solar radiation management in 2006. Because of persistent political inertia, which Crutzen deplored, research into solar geoengineering might prove necessary to provide future generations with a contingency plan to cope with the worst effects of climate change, at least temporarily.[31] Some environmentalists bemoaned Crutzen's move because, they feared, it could be instrumentalized by opponents of climate action to downplay the need for vigorous mitigation policies.[32] The fact remains that, since the late 2000s, an increasing number of influential institutions – such as the British Royal Society, the US National Academies of Sciences,[33] the Biden administration,[34] Harvard University,[35] the RAND Corporation,[36] and the Bill and Melinda Gates Foundation,[37] to name only the most prominent ones – have debated and published reports about the potential implications of solar geoengineering. Even the UN-sponsored IPCC has addressed solar geoengineering, although it remains deeply skeptical of its usefulness in the fight against climate change, because of the multiple risks and uncertainties that surround its potential deployment.[38]

One of the most formidable obstacles proponents of research on solar geoengineering have to surmount results from determined opposition to local experiments, which cannot be dispensed with to enhance our understanding of sulfur aerosol injection or marine cloud brightening for instance. In 2021, for example, the research team behind Harvard University's Stratospheric Controlled Perturbation Experiment, also known as SCoPEx, was unable to conduct an experiment in Sweden due to opposition from the Saami people.[39] In early 2023, a venture capital–backed startup called Make Sunsets unilaterally released aerosol-dispersing weather balloons in Mexico, but its experiment was quickly interrupted by the Mexican government.[40] Those examples are enough to realize how elusive the actual deployment of solar geoengineering on a global scale remains for now. Yet, interest at elite institutions, such as the Carnegie Climate Governance Initiative and the Global Overshoot Commission,[41] and the US Congress,[42] appears to be only growing. Meanwhile, the discussion about the legitimacy of solar geoengineering rages on.

Most advocates of research on solar geoengineering insist that emissions reduction ought to remain the centerpiece of our climate strategy and are quick to portray solar geoengineering as a hypothetical contingency plan to be activated if and when the global decarbonization effort falls short.[43] Hence, Ken Caldeira and David Keith have pushed for research efforts on the grounds that solar geoengineering could "be deployed rapidly" and prove to be "shockingly inexpensive."[44] Many reluctant supporters of research on solar geoengineering premise their stance on the belief that quick and substantial decarbonization will fail to materialize so that, sooner or later, solar geoengineering will emerge as a necessary evil in a world shaped by climate change.[45] Further, considering solar geoengineering is sometimes branded as an ethical gesture benefiting poor people in developing countries who, presumably, could use the reprieve offered by solar geoengineering to raise their living standards by burning fossil fuels as long as it remains economically advantageous to do so.[46]

Critics of solar geoengineering counter that the complexity of the Earthsystem makes such an enterprise deeply delusional and dangerous.[47] They also worry that touting geoengineering as a viable option creates a moral hazard, whereby opponents of decarbonization use the prospect of solar geoengineering as a pretext to postpone action.[48] Other frequently voiced concerns revolve around the risk of creating a path dependency, making it impossible for future generations to opt out of an established geoengineering program,[49] and multiple geopolitical risks, including "*predatory geoengineering*, where one country chooses a particular form of geoengineering mainly to disadvantage its geopolitical rivals,"[50] and the lack of an adequate international framework to regulate its orderly implementation.[51]

Solar geoengineering and the Modern vision of nature

The features of solar geoengineering as a set of emerging technologies and the arguments laid out by proponents of research suggest that the ideological assumptions that buttress the Modern vision of nature could end up being consolidated if solar geoengineering were to garner more attention in the years and decades to come. At least five characteristics of advocacy in favor of solar geoengineering dovetail with the Modern vision of nature:

1 *Techno-optimism and the ability of humankind to master nature.* Put simply, the case for solar geoengineering rests on a very optimistic faith in our ability to understand and manipulate highly complex systems – in this case, the so-called Earthsystem. In a quintessentially Modern fashion, this approach amounts to solving a problem by dominating nature – through science and its technological applications – instead of submitting and adapting to it. Central to the solar geoengineering proposition, according to Benjamin Hofbauer, is the notion that "to be more sustainable, humanity should increase its impact on nature, rather than step back from it."[52] Human control of nature, therefore, is not just necessary but also desirable. A recurrent corollary to this argument is a rejection of the precautionary principle, branded as an irrational refusal to strike a balance between the cost of action and the cost of inaction.[53]
2 *Faith in large-scale expert-driven solutions to environmental problems.* In his plea for research into solar geoengineering, David Keith takes issue with the

vision of technological change favored by many environmental and climate activists, like Naomi Klein, who seek to promote bottom-up, low-tech, and local approaches at the expense of large-scale and centralized ones. Keith counters that, in practice, large-scale technologies can sometimes better succeed in lowering our environmental footprint.[54] In any case, solar geoengineering, as "a first step towards 'earth systems' geoengineering,"[55] can only become a reality if we embrace what Frédéric Neyrat has called disapprovingly geoconstructivism, an approach to Earthsystem manipulation predicated on supercharging our techno-industrial capabilities.[56]

3 *Anthropocentrism*. Anthropocentrism, defined as a mode of ethical deliberation that ascribes intrinsic value to humans only, is arguably one of the mainstays of the Modern vision of nature. To the degree that they try to protect the Earth, solar geoengineering advocates do so with a view to advancing human interests:

> Environmental right or wrong depends on the consequences to humans. Although we have responsibilities regarding the natural world, we have no direct responsibility to the natural world. Environmental responsibility is, at bottom, a matter of prudence: we protect the environment for our own interests (of course, sometimes this can include such interests as aesthetic enjoyment or symbolic appreciation).[57]

Supporting solar geoengineering thus follows from an acceptance that resorting to large-scale and enduring manipulation of nature is morally justified as long as it is done to benefit humankind.

4 *Upholding a mechanistic vision of nature*. Geoconstructivism, as conceptualized by Frédéric Neyrat, is derived from a profoundly Modern mechanistic conception of nature.[58] Such a perspective informed former ExxonMobil CEO and US Secretary of State Rex Tillerson's remark to the effect that climate change is an engineering problem with engineering solutions.[59] In the geoengineer's scheme of things, the Earthsystem ought to be seen and dealt with as a soulless entity to be manipulated at will, thanks to human ingenuity. In his indictment of solar geoengineering, Mike Hulme points out that individuals and institutions – such as think tanks and defense agencies – who view the Earth as a machine are more likely than not to show interest in solar geoengineering.[60]

5 *Defending the* status quo *on behalf of social justice*. More often than not, calls for researching the potential of solar geoengineering are accompanied by pleas in favor of the world's poor, especially those living in developing countries. Were the deployment of solar geoengineering to be successful, David Keith contends, people in poor countries could begin or continue to use fossil fuels to spur economic development and improve their living standards, as people in rich countries did before them, without incurring the most debilitating effects of climate change.[61] Implicitly, this

> line of argument means that we cannot dispense with the Modern vision of nature lest we lose the material benefits it brought about from the industrial revolution onward.

In this chapter, it is my contention that, because the climate conversation is often couched in terms that are consistent with the Modern vision of nature, solar geoengineering will gain more and more traction as our climate predicament gets worse. This is not to suggest this Modern vision will go unchallenged. Yet, many participants in discussions over the best courses of action to handle climate change continue to reason within an ideological framework that could open the way for solar geoengineering as a much-needed contingency plan. The contentious and deeply fraught debate over solar geoengineering seems to lend credibility to such a scenario.

Is solar Geoengineering Inevitable?

To be sure, determined adversaries of solar geoengineering have been quick to denounce its dangerously hubristic nature, echoing Arne Naess's critique of the Modern pretension to guide and improve nature:

> The arrogance of stewardship consists in the idea of superiority which underlies the thought that we exist to watch over nature like a highly respected middleman between the Creator and the Creation. We know too little about what happens in nature to take up the task.[62]

This stance has informed the indictment of solar geoengineering made by Mike Hulme and Jack Stilgoe, among others.[63] Clive Hamilton also dismisses "the urge to mastery over nature [...] inscribed in the climate engineering project"[64] as "a Promethean urge"[65] and "monstrous hubris."[66] More broadly, most critics of solar geoengineering worry that even contemplating its deployment could prevent their contemporaries from thinking outside the box of Modern dualism and maintain the status quo, despite the advent of the Anthropocene.[67]

On the other side of the ideological spectrum are individuals and institutions of an ecomodernist persuasion. Although it originated in a US think tank called the Breakthrough Institute, founded in 2003, the word ecomodernism only came in common use from 2013 onward.[68] In their manifesto, ecomodernists describe the Earth as "a human planet" and tout the possibility of "creating a good Anthropocene."[69] From their perspective, the key is to embrace our role as stewards of nature through the aggressive pursuit of technological innovation. In practice, ecomodernists recommend that we do our utmost to decouple our economic activities from nature:

> modern technologies, by using natural ecosystem flows and services more efficiently, offer a real chance of reducing the totality of human impacts on the biosphere. To embrace these technologies is to find paths to a good

> Anthropocene. [...] Decoupling raises the possibility that societies might achieve peak human impact without intruding much further on relatively untouched areas. Nature unused is nature spared.[70]

Ecomodernists hope that such a vision could come to fruition through the deployment of technologies, such as nuclear plants and bioengineering, so that twenty-first-century humans will keep on enjoying the benefits of Modernity while lowering their footprint on the environment.

Although they do not necessarily describe themselves as such, many supporters of research into solar geoengineering do qualify as ecomodernists (and indeed, David Keith, one of the most prominent advocates of research is also one of the signatories of the Ecomodernist Manifesto). Some free-market advocates, such as the economists Stephen Dubner and Steven Levitt, have a tendency to be extremely sanguine about the prospects of solar geoengineering, which, they think, offers a "cheap and simple" remedy to climate change that could spare humankind a major economic destabilization brought about by a global decarbonization effort.[71] Other techno-optimists have made a more cautious and limited endorsement of solar geoengineering as one tool among others in the fight against climate change. Of paramount importance to the ecomodernist proposition is the notion that environmentalists should stop sacralizing nature, and should rather endorse the idea of coevolution: "Humans have been cocreators of the environment they inhabit. Any proposal to fix environmental problems by turning away from technology risks worsening them by attempting to deny the ongoing coevolution of humans and nature."[72] It bears emphasis that ecomodernists are not interested in solar geoengineering only, but as one among other technological options. Yet, the fact remains, their understanding of the proper connection between humans and nature makes solar geoengineering a distinct possibility.

The polarization between determined adversaries of solar geoengineering and ecomodernists is far from the whole story. Another group, which I call reluctant supporters of research, gives us a glimpse of what the fate of solar geoengineering may turn out to be in the age of the Anthropocene. Following in Paul Crutzen's footsteps, many participants in the solar geoengineering conversation subscribe to what Olaf Corry calls "Reluctant Geoengineering."[73] They bemoan a situation in which political apathy and lack of international cooperation are forcing them to contemplate solar geoengineering as a necessary plan B, which they are deeply uncomfortable with.[74] Typically, when referring to solar geoengineering, they will use metaphors like "chemo-therapy for the planet"[75] and "planetary methadone."[76] Those scholars systematically insist that mitigation must remain our overriding objective, only to concede that we also need to learn more about solar geoengineering in case we are left with no better option. Although they sound much less confident than ecomodernists about our ability to take on the task of manipulating the planet's temperatures efficiently, they leave their readers with the impression that they will eventually rally behind the ecomodernist position – albeit reluctantly. In part, this sense of inevitability stems from their adherence to an ideological framework still largely shaped and informed by Modern dualism.

The longer the techno-scientific *ethos* remains dominant and central, the more likely it is that solar geoengineering becomes a meeting point toward which ecomodernists and increasingly desperate climate activists converge, as the consequences of climate change further destabilize entire regions of the world. Were this scenario to materialize, it would create a paradox of the Anthropocene: instead of fostering a radical reorientation of our ways of inhabiting and interacting with the natural world, the advent of the Anthropocene may well lead a group of political, scientific, and corporate leaders to become even more radicalized in their quest to dominate and control nature. Such a scenario echoes Joel Wainwright's and Geoff Mann's reflections about the potential emergence of a "Climate Leviathan" in the twenty-first century – a national or supranational entity powerful enough to declare a state of climate emergency and impose unprecedented coercive measures worldwide so as to try to stabilize the Earth's climate. Climate Leviathan, Wainwright and Mann add, is likely to be "a beast of adaptation," if only because climate change is already underway and expected to get worse.[77] Unsurprisingly, they single out unilateral deployment of atmospheric geoengineering as one of the radical steps Climate Leviathan could end up taking.[78]

Conclusion

This chapter suggests that, because the Modern vision of nature is currently under siege, it does not necessarily mean that it will no longer wield any influence and shape the way we collectively handle nature. To many, mobilizing science and technology to cope with resource shortages, climate disruption, and biodiversity loss remains a very attractive proposition. Such a stance is likely to pave the way for the reaffirmation of Modern dualism, whereby the human species endeavors to master nature for the sake of its own preservation and flourishing. Mounting interest in solar geoengineering – including from scholars who remain deeply uncomfortable with it – appears to be a case in point.

Pondering the ways in which technological change sometimes ends up shaping and determining social and political outcomes, the political scientist Langdon Winner once asked: "As we 'make things work,' what kind of world are we making?"[79] One may wonder whether committing to solar geoengineering in order to contain climate change could create a path dependency, making the prospect of a departure from the human quest to conquer and control nature even more distant than it already is. The geophysicist James Lovelock has raised doubts regarding our true capabilities: "Before we start geoengineering, we have to raise the following question: are we sufficiently talented to take on what might become the onerous permanent task of keeping the Earth in homeostasis?" Were we to fail, "we could find ourselves in a Kafka-like world from which there is no escape."[80] If it comes to that, Modern dualism will have run its course, before humankind manages to avert its worst effects.

Notes

1 Jean-Baptiste Fressoz, *L'apocalypse joyeuse: une histoire du risque technologique* (Paris: Seuil, 2012), 294.
2 Climate Action Tracker, "Despite Glasgow Climate Pact, 2030 Climate Target Updates have stalled, Mid-Year Update," June 2022, https://climateactiontracker.org/documents/1051/CAT_2022-06-03_Briefing_MidYearUpdate_DespiteGlasgowTargetUpdatesStalled.pdf.
3 Gérald Hess, *Éthiques de la nature* (Paris: Presses Universitaires de France, 2013), 38.
4 Philippe Descola, *Par-delà nature et culture* (Paris: Gallimard, 2005), 37.
5 Pierre Charbonnier, *Abondance et liberté: une histoire environnementale des idées politiques* (Paris: La Découverte, 2020), 313–314.
6 Bill Devall and George Sessions, *Deep Ecology: Living as if Nature Mattered* (Salt Lake City: Gibbs Smith Publisher, 1985), 41.
7 Devall and Sessions, *Deep Ecology*, 52.
8 Bill McKibben, *The End of Nature*, 2nd ed. (New York: Anchor Books, 1999), 58.
9 J. Baird Callicott, "La Nature Est Morte, Vive la Nature!," *The Hastings Center Report*, no. 22 (1992): 73–90.
10 Paul Crutzen and Eugene Stoermer, "The Anthropocene," *Global Change Newsletter*, no. 41 (2000): 17–18, https://doi.org/10.12987/9780300188479-041.
11 Paul Wapner, "The Changing Nature of Nature: Environmental Politics in the Anthropocene," *Global Environmental Politics* 14, no. 4 (November 2014): 48–49, https://doi.org/10.1162/GLEP_a_00256.
12 Herman E. Daly and Joshua Farley, *Ecological Economics: Principles and Applications* (Washington DC: Island Press, 2004), 15.
13 Tim Jackson, *Prosperity without Growth: Foundations for the Economy of Tomorrow*, 2nd edition (London: Routledge, 2017), 62.
14 Catherine Larrère and Raphaël Larrère, *Du bon usage de la nature: pour une philosophie de l'environnement* (Paris: Alto Aubier, 1997), 162–163.
15 Eric Katz, Andrew Light and David Rothenberg, "Introduction: Deep Ecology as Philosophy," in *Beneath the Surface: Critical Essays in the Philosophy of Deep Ecology*, ed. Eric Katz, Andrew Light, and David Rothenberg (Cambridge: The MIT Press, 2000), ix.
16 Bruno Latour, *Face à Gaïa: Huit conférences sur le Nouveau Régime Climatique* (Paris: La Découverte, 2015), 159.
17 Latour, *Face à Gaïa*, 187–188.
18 Steven Pinker, *Enlightenment Now: The Case for Reason, Science, Humanism, and Progress* (New York: Viking, 2018), 121.
19 Robert Pollin, *Greening the Global Economy* (Cambridge: The MIT Press, 2015), 109.
20 Bill Gates, How to Avoid a Climate Catastrophe: The Solutions We Have and the Breakthroughs We Need (London: Allen Lane, 2021), 14.
21 For a detailed description of this piece of legislation, see Evergreen Action Policy Team, "Evergreen Explains: The Climate Impact of the Inflation Reduction Act," (August 23, 2022), https://www.evergreenaction.com/documents/The-Climate-Impact-of-the-IRA.pdf.
22 John Kerry, "John Kerry: Companies that quickly embrace green tech will clean up," *Financial Times*, November 17, 2021, https://www.ft.com/content/b044fe94-ae00-4363-9362-3906a9b92cf8.
23 Jay Michaelson, "Geoengineering and Climate Management from Marginality to Inevitability," *Tulsa Law Review* 46, no. 2 (Winter 2010): 222.
24 Committee on Geoengineering Climate, "Climate Intervention: Reflecting Sunlight to Cool Earth," The National Academies Press, Washington DC, 2015.
25 Christopher W. Belter and Dian J. Seidel, "A Biometric Analysis of Climate Engineering Research", WIREs Climate Change 4, no. 5 (2013): 418, https://doi.org/10.1002/wcc.229.

26 American Geophysical Union, "Position Statement on Climate Intervention," January 12, 2018, https://www.agu.org/-/media/Files/Share-and-Advocate-for-Science/Position-Statements/Climate-Intervention-Position-Statement-Final-2018-1.pdf.
27 Belter and Seidel, "A Biometric Analysis," 418.
28 Mike Hulme, Can *Science Fix Climate Science? A Case against Climate Engineering* (Cambridge: Polity, 2014), 7–8.
29 Hulme, *Can Science Fix*, 6.
30 Ken Caldeira and David Keith, "The Need for Climate Engineering Research," *Issues in Science and Technology* 27, no. 1 (Fall 2010): 57–62.
31 Paul Crutzen, "Albedo Enhancement by Stratospheric Sulfur Injections: A Contribution to Resolve a Policy Dilemma?," *Climatic Change* 77 (2006): 211–220, https://doi.org/10.1007/s10584-006-9101-y.
32 Oliver Morton, *The Planet Remade: How Geoengineering Could Change the World* (Princeton: Princeton University Press, 2015), 154.
33 National Academies of Sciences, "New Report Says U.S. Should Cautiously Pursue Solar Geoengineering Research to Better Understand Options for Responding to Climate Change Risks," March 25, 2021, https://www.nationalacademies.org/news/2021/03/new-report-says-u-s-should-cautiously-pursue-solar-geoengineering-research-to-better-understand-options-for-responding-to-climate-change-risks.
34 Scott Waldman and John Fialka, "White House preparing report on climate geoengineering," *E&E News*, February 13, 2023, https://subscriber.politicopro.com/article/eenews/2023/02/13/white-house-preparing-report-on-climate-geoengineering-00082329.
35 Harvard University Center for the Environment, "Harvard's Solar Geoengineering Research Program," https://geoengineering.environment.harvard.edu/.
36 Robert J. Lempert and Don Prosnitz, *Governing Geoengineering Research: A Political and Technical Vulnerability Analysis of Potential Near-Term Options* (Santa Monica: RAND Corporation, 2011), https://www.rand.org/pubs.
37 Eli Kintisch, *Hack the Planet: Science's Best Hope – or Worst Nightmare – for Averting Catastrophe* (Hoboken: John Wiley & Sons, 2010), 177.
38 IPCC, *Summary for Policymakers*, in *Global Warming of 1.5°C. An IPCC Special Report on the impacts of global warming of 1.5°C above pre-industrial levels and related global greenhouse gas emission pathways, in the context of strengthening the global response to the threat of climate change, sustainable development, and efforts to eradicate poverty*, ed. Masson-Delmotte, V., P. Zhai, H.-O. Pörtner, D. Roberts, J. Skea, P.R. Shukla, A. Pirani, W. Moufouma-Okia, C. Péan, R. Pidcock, S. Connors, J.B.R. Matthews, Y. Chen, X. Zhou, M.I. Gomis, E. Lonnoy, T. Maycock, M. Tignor, and T. Waterfield (Cambridge University Press: Cambridge, 2018) 14–15.
39 Alison Snyder and Andrew Freedman, "Warming up to solar geoengineering," *Axios*, March 25, 2021, https://www.axios.com/2021/03/25/geoengineering-climate-change-research-program.
40 Corbin Hiar, "Startup halts geoengineering operation in Mexico," *E&E News*, January 19, 2023, https://www.eenews.net/articles/startup-halts-geoengineering-operation-in-mexico/.
41 Andrew Freedman, "Hacking the planet may be closer than you think," *Axios*, December 10, 2022, https://www.axios.com/2022/12/10/geoengineering-research-hacking-planet.
42 Corbin Hiar, "Blocking sun rays finds support in the Senate," *E&E News*, March 2, 2023, https://www.axios.com/2022/12/10/geoengineering-research-hacking-planet.
43 Stephen H. Schneider, "Geoengineering: Could we or Should we Make it Work?" *Philosophical Transactions: Mathematical, Physical and Engineering Sciences* 366, no. 1882 (2008): 3857–3858; John G. Shepherd, "Geoengineering the Climate: An Overview and Update," *Philosophical Transactions: Mathematical, Physical and Engineering Sciences* 370, no. 1974 (2012): 4166–4175.
44 Caldeira and Keith, "Need for Climate," 58.

45 Michaelson, "Geoengineering and Climate," 221–260.
46 David Keith, *A Case for Climate Engineering* (Cambridge: The MIT Press, 2013), 137.
47 Hulme, Can *Science Fix*, 98–99.
48 Schneider, "Geoengineering," 3847.
49 Wil C.G. Burns, "Climate Geoengineering: Solar Radiation Management and its Implications for Intergenerational Equity," in *Climate Change Geoengineering: Philosophical Perspectives, Legal Issues and Governance Frameworks*, ed. Wil C.G. Burns and Andrew L. Strauss (New York: Cambridge University Press, 2013), 199–220.
50 Stephen M. Gardiner, "The Desperation Argument for Geoengineering," *Political Science and Politics* 46, no. 1 (2013): 31, https://doi.org/10.1017/S1049096512001424.
51 Schneider, "Geoengineering," 3857.
52 Benjamin Hofbauer, "Techno-Moral Change through Solar Geoengineering: How Geoengineering Challenges Sustainability," *Prometheus* 38, no. 1 (2022): 87.
53 Richard J. Zeckhauser and Gernot Wagner, "The Implications of Uncertainty and Ignorance," in *Governance of the Deployment of Solar Geoengineering*, ed. Robert N. Stavins and Robert C. Stowe (Cambridge (MA): Harvard Project on Climate Agreements, 2019), 107–111.
54 Keith, *A Case for*, 140–150.
55 Colin McInnes, "Time to Embrace Geoengineering: Beyond Planetary Boundaries," *Breakthrough Institute*, June 27, 2013, https://thebreakthrough.org/issues/energy/time-to-embrace-geoengineering.
56 Frédéric Neyrat, *La part indestructible de la Terre: critique du géoconstructivisme* (Paris: Seuil, 2016), 11–12.
57 Joseph R. DesJardins, *Environmental Ethics: An Introduction to Environmental Policy* (Belmont: Thomson, 2006), 96.
58 Neyrat, *La part indestructible*, 88.
59 Chris Mooney, "Rex Tillerson's View of Climate Change: It's an 'Engineering Problem'," *Washington Post*, December 14, 2016, https://www.washingtonpost.com/news/energy-environment/wp/2016/12/13/rex-tillersons-view-of-climate-change-its-just-an-engineering-problem/.
60 Hulme, *Can Science Fix*, 26.
61 Keith, *A Case for*, 137.
62 Arne Naess, *Ecology, Community and Lifestyle*, 2nd ed. (Cambridge: Cambridge University Press, 2003), 187.
63 Hulme, *Can Science Fix*, 111; Jack Stilgoe, *Experiment Earth: Responsible Innovation in Geoengineering* (London: Earthscan, 2015), 183.
64 Clive Hamilton, *Earthmasters: The Dawn of the Age of Climate Engineering* (New Haven: Yale University Press, 2013), 72.
65 Hamilton, *Earthmasters*, 18.
66 Hamilton, *Earthmasters*, 182.
67 Kevin Surprise, "Stratospheric Imperialism: Liberalism, (Eco)Modernization, and Ideologies of Solar Geoengineering Research," *EPE: Nature and Space* 3, no. 1 (2020): 143; Anne Fremaux and John Barry, "The 'Good Anthropocene' and Green Political Theory: Rethinking Environmentalism, Resisting Ecomodernism," in *Anthropocene Encounters: New Direction in Green Political Thinking*, ed. Frank Biermann and Eva Lövbrand (Cambridge: Cambridge University Press, 2019), 171–190.
68 Jonathan Symons, *Ecomodernism: Technology, Politics and the Climate Crisis* (Cambridge: Cambridge University Press, 2019), 8.
69 John Asafu-Adjaye et al., "An Ecomodernist Manifesto," Breakthrough Institute, April 2015, https://static1.squarespace.com/static/5515d9f9e4b04d5c3198b7bb/t/552d37bbe4b07a7dd69fcdbb/1429026747046/An+Ecomodernist+Manifesto.pdf.
70 Asafu-Adjaye et al., "An Ecomodernist Manifesto," 17–19.

71 Steven Levitt and Stephen J. Dubner, *SuperFreakonomics: Global Cooling, Patriotic Prostitutes and Why Suicide Bombers Should Buy Life Insurance*, 2nd ed. (New York: Harper, 2011), 194.
72 Michael Shellenberger and Ted Nordhaus, "Evolve: The Case for Modernization as the Road to Salvation," in *Love your Monsters: Postenvironmentalism and the Anthropocene*, ed. Michael Shellenberger (Oakland: Breakthrough Institute, 2011).
73 Olaf Corry, "The International Politics of Geoengineering: The Feasibility of Plan B for Tackling Climate Change," *Security Dialogue* 48, no. 4 (2017): 300.
74 Michaelson, "Geoengineering and Climate," 221–260; Peter Singer, *One World Now: The Ethics of Globalization*, 2nd ed. (New Haven: Yale University Press, 2016), 64–68; Mark G. Lawrence, "The Geoengineering Dilemma: To Speak or Not to Speak," *Climatic Change* 77, no. 3–4 (2006): 245–248.
75 Gernot Wagner and Martin L. Weitzman, *Climate Shock: The Economic Consequences of a Hotter Planet* (Princeton: Princeton University Press, 2015), 105.
76 Schneider, "Geoengineering," 3858.
77 Joel Wainwright and Geoff Mann, *Climate Leviathan: A Political Theory of our Planetary Future*, 2nd edition (London: Verso, 2020), 29.
78 Wainwright and Mann, *Climate Leviathan*, 148–149.
79 Langdon Winner, *The Whale and the Reactor: A Search for Limits in an Age of High Technology* (Chicago: The University of Chicago Press, 1986), 17.
80 James Lovelock, "A Geophysiologist's Thoughts on Geoengineering," *Philosophical Transactions: Mathematical, Physical and Engineering Sciences* 366, no. 1882 (2008): 3888.

Index

Pages in *italics* refer to figures, pages in **bold** refer to tables, and pages followed by n refer to notes.

For Product Safety Concerns and Information please contact our EU
representative GPSR@taylorandfrancis.com
Taylor & Francis Verlag GmbH, Kaufingerstraße 24, 80331 München, Germany

www.ingramcontent.com/pod-product-compliance
Ingram Content Group UK Ltd.
Pitfield, Milton Keynes, MK11 3LW, UK
UKHW020149200926
475702UK00016B/2630

* 9 7 8 1 0 3 2 4 7 2 4 1 6 *